THE

H

Seed Proteins

Annual Proceedings of the Phytochemical Society of Europe

Seed Proteins

Edited by

J. DAUSSANT
Laboratoire de Physiologie des Organes Végétaux,
CNRS 92190, Meudon, France

J. MOSSÉ
Laboratoire d'Etude des Protéines, INRA Versailles, France

J. VAUGHAN
Department of Biology, Queen Elizabeth College,
University of London, England

1983

ACADEMIC PRESS
A Subsidiary of Harcourt Brace Jovanovich, Publishers
London • New York
Paris • San Diego • San Francisco • São Paulo
Sydney • Tokyo • Toronto

ACADEMIC PRESS INC. (LONDON) LTD.
24/28 Oval Road
London NW1

United States Edition Published by
ACADEMIC PRESS INC.
111 Fifth Avenue
New York, New York 10003

British Library Cataloguing in Publication Data

Seed proteins.—(Annual proceedings of the
  Phytochemical Society of Europe, ISSN 0309–9393;
  no. 20)
  1. Seeds—Congresses
  I. Daussant, J.   II. Mossé, J.
  III. Vaughan, J.   IV. Series
  631.5'21   OK661

ISBN 0-12-204380-4

Text set in 10/12 pt Linotron 202 Times, printed and bound
in Great Britain at The Pitman Press, Bath

Contributors

D. Boulter, *Department of Botany, University of Durham, Durham, England*

R. R. D. Croy, *Department of Botany, University of Durham, Durham, England*

J. Daussant, *Laboratoire de Physiologie des Organes Végétaux C.N.R.S 4 ter, Route des Gardes, 92190 Meudon, France*

J. M. Field, *Biochemistry Department, Rothamsted Experimental Station, Harpenden, Herts, England*

G. Grant, *The Rowett Research Institute, Bucksburn, Aberdeen, Scotland*

J. Ingversen, *Department of Physiology, Carlsberg Laboratory, Gamie Carlsberg Vej 10, DK-2500 Copenhagen Valby, Denmark*

A. W. MacGregor, *Grain Research Laboratory, Canadian Grain Commission, 1404–303 Main Street, Winnipeg, Canada*

B. J. Miflin, *Biochemistry Department, Rothamsted Experimental Station, Harpenden, Herts, England*

J. Mikola, *Department of Biology, University of Jyvaskyla, Jyvaskyla 10, Finland*

J. Mossé, *Laboratoire d'Etude des Protéines, Physiologie et Biochimie Végétales, Centre INRA, 78000 Versailles, France*

R. L. Ory, *Southern Regional Research Center, USDA-SEA, AR P.O. Box 19687, New Orleans, USA*

P. I. Payne, *Plant Breeding Institute, Maris Lane, Trumpington, Cambridge, England*

J-C. Pernollet, *Laboratoire d'Etude des Protéines Physiologie, et Biochimie Végétales, Centre INRA, 78000, Versailles, France*

A. Pusztai, *The Rowett Research Institute, Bucksburn, Aberdeen, Scotland*

F. Salamini, *Istituto Biosintesi Vegetali C.N.R. Milano, Italy*

A. A. Sekul, *Southern Regional Research Center, USDA-SEA, AR P.O. Box 19687, New Orleans, USA*

P. R. Shewry, *Biochemistry Department, Rothamsted Experimental Station, Harpenden, Herts, England*

A. Skakoun, *Laboratoire de Physiologie des Organes Végétaux C.N.R.S. 4 ter, Route des Gardes, 92190 Meudon, France*

C. Soave, *Istituto Biosintesi Vegetali C.N.R. Milano, Italy*

J. C. Stewart, *The Rowett Research Institute, Bucksburn, Aberdeen, Scotland*

J. G. Vaughan, *Department of Biology, Queen Elizabeth College, London, England*

Preface

Storage proteins in seeds provide the amino acids for the growing seedlings in the early period of their life. Seed proteins represent the major source of proteins for both human and animal nutrition. The bulk of the proteins in seeds is formed by the storage proteins, therefore the nutritional quality of seed proteins largely depends on the amino acid composition of the storage proteins. In addition to their amino acid composition, the physicochemical properties of storage proteins are important because they play a part in food processing. Other proteins existing in more or less discrete amounts in the seeds are important too: enzymes are involved in food industries; lectins, which seem to play a role in recognition processes, are used in medical research; proteic enzyme inhibitors as well as proteins with allergenic properties affect the nutritional quality not only of seed proteins but also of other raw materials contained in seeds such as starch or lipids used in food products.

The characterization of seed proteins and the improvement of both quality and yield of seed proteins have been investigated for several decades. Outstanding advances have been achieved in the past few years, in particular as concerns the molecular biology and regulation of synthesis, the deposition, the subcellular localization, the degradation of seed proteins and the regulation of enzymatic activities. New developments in the methods of investigation of proteins also promoted progress in traditional research on seed proteins. It therefore seemed interesting to organize a symposium in order to confront different aspects of the research on seed proteins, traditional aspects as well as new trends.

This book is based on the twelve review lectures given at the symposium held in Versailles (France), September 1981, on the subject "Seed Proteins". The contributions concern the structure of seed proteins, their localization, their ontogenical evolution, their molecular biosynthesis and degradation, their improvement by breeding techniques and their technological properties. The first four chapters concern seed proteins with known biological activities: α-amylases, proteolytic enzymes and their inhibitors, lectins and allergens. The two following chapters deal with methodological aspects concerning seed protein analysis either for the study of proteins or as a tool in phylogeny and taxonomy. The last six chapters concern storage proteins in cereal and legumes. The symposium showed that the rapid evolution of all the topics treated is going to lead to new perspectives and applications. In particular, recent developments concerning the structure, the biosynthesis mechanisms and the localization of structural genes of seed proteins open new perspectives for the improvement of the quality of seed proteins in agronomic, nutritional and technological levels.

The symposium of the Phytochemical Society of Europe was organized in association with the Centre National de la Recherche Scientifique, the Institut National de la Recherche Agronomique and the EFRAC study group of the Parliamentary Assembly of the Council of Europe. We wish to express our thanks to the four organizations in the frame of which the meeting was prepared. The Editors thank Academic Press for their expert assistance in preparing this volume for publication.

January 1983

J. Daussant
J. Mossé
J. G. Vaughan

Contents

CHAPTER 1

Cereal α-amylases: Synthesis and Action Pattern

A. W. Macgregor

CHAPTER 2

Proteinases, Peptidases, and Inhibitors of Endogenous Proteinases in Germinating Seeds

J. Mikola

CHAPTER 3

Seed Lectins: Distribution, Location and Biological Role

A. Pusztai, R. R. D. Croy, G. Grant and J. C. Stewart

CHAPTER 4

Allergens in Oilseeds

R. L. Ory and A. A. Sekul

CHAPTER 5

Immunochemistry of Seed Proteins

J. Daussant and A. Skakoun

CHAPTER 6

The Use of Seed Proteins in Taxonomy and Phylogeny

J. G. Vaughan

CHAPTER 7

Structure and Location of Legume and Cereal Seed Storage Proteins

J-C. Pernollet and J. Mossé

CHAPTER 8

The Molecular Biology of Storage Protein Synthesis in Maize and Barley Endosperm

J. Ingversen

CHAPTER 9

Genetic Organization and Regulation of Maize Storage Proteins

C. Soave and F. Salamini

CHAPTER 10

Regulation of Storage Protein Synthesis and Deposition in Developing Legume Seeds

D. Boulter

CHAPTER 11

Breeding for Protein Quantity and Protein Quality in Seed Crops

P. I. Payne

CHAPTER 12

Cereal Storage Proteins and Their Effect on Technological Properties

B. J. Miflin, J. M. Field and P. R. Shewry

CHAPTER 1

Cereal α-Amylases: Synthesis and Action Pattern

A. W. MACGREGOR

Grain Research Laboratory, Canadian Grain Commission*
1404–303 Main Street, Winnipeg, Canada

I. Introduction

Starch, the chief source of carbohydrate in human diets, forms the major energy reserve in cereal grains. It is laid down in cereal endosperms in the form of discrete, insoluble particles called starch granules. When cereal grains require to mobilize their energy store they must first of all convert starch from an insoluble to a soluble form so that it may be metabolized. This is accomplished by the action of α-amylase, an enzyme that is able to hydrolyze intact starch granules with the production of soluble starch fragments. These may be further hydrolyzed to small saccharides by

---

* Paper no. 500 of the Canadian Grain Commission, Grain Research Laboratory, 1404–303 Main Street, Winnipeg, Canada R3C 3G8.

continued action of α-amylase and other carbohydrase enzymes. α-Amylase, therefore, is of prime importance in initial stages of starch degradation whether this occurs in naturally germinating cereal grains in the earth or under controlled industrial processing such as malting and brewing or to excess in the baking industry, causing severe technological problems. Obviously, the biosynthesis and action pattern of this enzyme are of wide importance and interest. This review attempts to gather together the latest information on these facets of cereal α-amylases.

II. α-Amylase Synthesis

A. BARLEY

The presence of α-amylase in barley kernels during early stages of development was first reported by Duffus (1969) and later confirmed by a number of workers (Bilderback, 1971; LaBerge et al., 1971; MacGregor et al., 1971a; Stoddart, 1971; Niku-Paavola et al., 1973; Allison et al., 1974). Enzyme activity increases shortly after anthesis, reaches a peak and then declines (Fig. 1). Variation in the date at which maximum enzyme activity occurs is probably caused by differences in growth rates of barley cultivars, length of growing season and the way in which stage of growth is determined by different workers.

Dissection studies showed the enzyme to be present mainly in the pericarp of developing kernels (MacGregor et al., 1972; Allison et al., 1974) and so it may be called pericarp α-amylase. Peaks of α-amylase activity that appear at a slightly later stage of growth in both aleurone and endosperm tissues (Allison et al., 1974) and in whole kernels (Meredith and Jenkins, 1973) could be caused by early stimulation of germination-type α-amylases as discussed for wheat (Section II.B.), rather than by pericarp α-amylase.

Although pericarp α-amylase appeared to be homogeneous on analysis by gel electrophoresis (Stoddart, 1971; MacGregor et al., 1974) it may be resolved into several bands of activity by isoelectric focusing (Fig. 2). These bands have low isoelectric points of pI 4·8–5·0. Whether or not this pattern is common to different barley cultivars has yet to be determined.

Application of chlorocholine chloride, an inhibitor of gibberellin synthesis (Briggs, 1968c), to barley kernels shortly after anthesis inhibits α-amylase synthesis (Duffus, 1969). Subsequent application of gibberellic acid (GA_3) removes this inhibition and indicates a direct effect of GA_3 on α-amylase synthesis. Immature kernels do contain appreciable amounts of gibberellin (Jones et al., 1963) and so this hormone may control not only α-amylase synthesis in the aleurone layer of germinating barley (Varner, 1964) but also pericarp α-amylase in immature kernels.

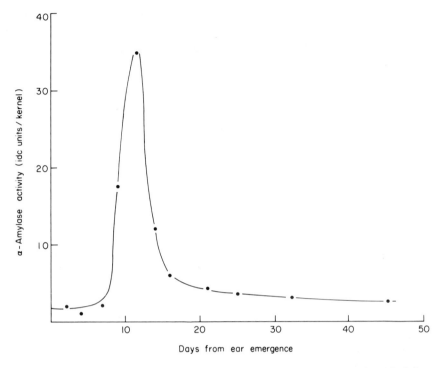

FIG. 1. α-Amylase activity in developing kernels of Conquest barley (modified from MacGregor *et al.*, 1971a).

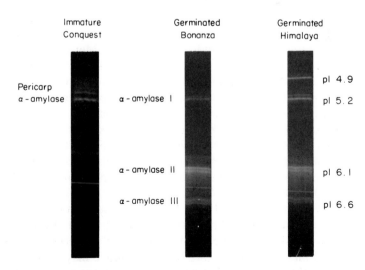

FIG. 2. Isoelectric focusing of α-amylases from developing Conquest barley, germinated Bonanza barley and germinated Himalaya barley.

α-Amylase activity declines rapidly in the pericarp at the same time that starch granules disappear from the same tissue (MacGregor et al., 1972). This suggests that when the temporary starch store is depleted there is no further requirement for α-amylase in the pericarp, the biosynthetic mechanism for the enzyme is shut off, perhaps by increasing levels of abscisic acid (Goldbach and Michael, 1976), and the enzyme is degraded. Only small amounts of α-amylase are present in mature barley kernels (Kneen, 1944; Greenwood and MacGregor, 1965) and it is unlikely that the enzyme exists in these kernels in an inactive form (Grabar and Daussant, 1964). Apparently, pericarp tissue is able to synthesize α-amylase for only a short period of time and for α-amylase induction experiments it would be important to use such tissues when they contain fully active biosynthetic systems.

The first detailed reports showing that barley kernels produce starch liquefying activity (or diastase) during germination were published almost one hundred years ago (Brown and Morris, 1890; Brown and Escombe, 1898; Haberlandt, 1914). At that time a controversy was started that has continued to the present day over the relative importance of aleurone and embryo tissues in the formation of diastase (or α-amylase) in germinating kernels. Little progress was made in understanding α-amylase synthesis in cereal grains until Yomo (1958) reported that a diffusable factor produced by barley embryos induced formation of hydrolytic enzymes in barley endosperms. Further work (Yomo, 1960; Paleg, 1960a,b) showed that embryoless barley treated with GA_3 produced α-amylase, indicating that the diffusable factor was probably GA_3. This led to the suggestion that GA_3, produced by barley embryos, migrates to the endosperm and there promotes formation of α-amylase (Paleg, 1960b, MacLeod and Millar, 1962). That α-amylase is formed in the aleurone and not in the whole endosperm was confirmed by observations that endosperms devoid of aleurone layers did not produce hydrolytic enzymes when treated with GA_3 (MacLeod and Millar, 1962) but aleurone layers did form α-amylase when similarly treated (Briggs, 1963, 1964; MacLeod et al., 1964; Varner, 1964; Yomo and Iinuma, 1964). The results of Briggs (1963, 1964) suggested strongly that α-amylase was formed in aleurone layers by de novo synthesis and not by activation of pre-existing enzyme precursors. This was proved correct by the elegant experiments of Varner (1964) and Chrispeels and Varner (1967). Since then, the idea of GA_3 migrating from the embryo to the aleurone and there activating α-amylase synthesis during germination of cereal grains has been accepted generally. This hypothesis is further supported by the finding of rapid GA_3 synthesis in germinating embryos (Cohen and Paleg, 1967; MacLeod and Palmer, 1967; Radley, 1967) and of a relationship between GA_3 synthesis in the embryo and α-amylase synthesis in the aleurone of germinating barley kernels (Groat and Briggs, 1969). This general description of α-amylase synthesis in cereal grains is certainly very enticing but it is not complete.

There is evidence that maximum synthesis of α-amylase in the aleurone requires a factor from the embryo other than GA_3 (see Section II.H.). However, this factor (or factors) has not yet been identified.

Results from early studies suggested that the embryo or, more precisely, the scutellum produced starch degrading enzymes (presumably α-amylase) during initial stages of germination (Brown and Morris, 1890). However, in later studies, first with embryoless half-seeds (Brown and Escombe, 1898; Haberlandt, 1914) and then with kernel slices treated with GA_3 (MacLeod and Millar, 1962; Briggs, 1963) starch degradation occurred immediately under the aleurone layers. These findings along with the discovery of gibberellins and their remarkable effect on aleurone layers led to the idea that starch breakdown in the endosperm of germinating cereal grains is carried out primarily by α-amylase synthesized in the aleurone (Palmer, 1974; Palmer and Bathgate, 1976). In addition, because barley aleurone cells are homogeneous, do not grow, divide or differentiate when incubated, and synthesize readily identifiable proteins such as α-amylase when stimulated with the hormone GA_3, they have been used widely to investigate GA_3-stimulated synthesis of α-amylase as an example of hormonal control over protein synthesis. Unfortunately, these studies have focused attention on aleurone cells as a source of α-amylase and the potential for α-amylase synthesis in other tissues of cereal grains has not been thoroughly investigated.

Despite some results to the contrary (MacLeod and Palmer, 1966), there is abundant evidence of α-amylase synthesis in the embryo (Briggs, 1964, 1968a,b; Groat and Briggs, 1969; Briggs and Clutterbuck, 1973). This enzyme may account for at least 15% of the total α-amylase present in germinated barley kernels. In addition, recent results strongly suggest that the embryo is a major source of α-amylase during initial stages of germination (Gibbons, 1979, 1980; Okamoto et al., 1980). Furthermore, starch breakdown starts adjacent to the scutellar surface and moves through the endosperm parallel to this surface (Briggs, 1972; MacGregor, 1980). Only during later stages of germination is starch degraded under the aleurone layer by α-amylase synthesized and excreted by this tissue (Gibbons, 1980; MacGregor, 1980). The idea of the whole pattern of endosperm modification, including starch degradation, being under the complete control of hydrolytic enzymes from the aleurone (Palmer and Bathgate, 1976; Palmer, 1980) may have to be revised.

Germinated barley α-amylase is not a single entity as suggested in one report (Mitchell, 1972), but a complex mixture of isoenzymes (Frydenberg and Nielsen, 1965; van Onckelen and Verbeek, 1969; MacGregor, 1977). These can be separated conveniently into three main groups by isoelectric focusing as shown in Fig. 2 for Bonanza and Himalaya barleys. The enzyme pattern is not changed substantially by the addition of exogenous GA_3 (van Onckelen and Verbeek, 1969; MacGregor, 1978a) but the rate of α-amylase synthesis and the total amount synthesized may be increased (Verbeek et al., 1969; MacGregor, 1978a). Embryoless half-seeds and aleurone layers treated with GA_3 also produce a number of different α-amylases (Tanaka and Akazawa, 1970; Bilderback, 1974) but fewer than do whole kernels (Momotani and Kato, 1967; van Onckelen and Verbeek, 1969; Jacobsen et al., 1970; McMasters, 1974). This suggests that either the embryo synthesizes unique α-amylase isoenzymes or that a factor from the embryo other than GA_3 is necessary to induce synthesis of all α-amylase isoenzymes in aleurone cells. If such a factor exists it is unlikely to be a gibberellin since all common, naturally occurring gibberellins activate barley aleurone layers to synthesize the same α-amylase isoenzymes (McMasters, 1974).

Using isoelectric focusing Momotani and Kato (1972) detected eighteen bands of α-amylase activity in embryoless half-seeds treated with GA_3. The major components of this complex system could be segregated into three main groups of activity similar to those obtained from germinated kernels (Fig. 2). At low concentrations of GA_3 the major α-amylase groups (II and III in Fig. 2) were not synthesized in aleurone layers but small amounts of α-amylase I were detected. In the absence of GA_3, embryoless barley half-seeds produce small amounts of α-amylase (Paleg, 1970; van Onckelen et al., 1974) that, subsequently, have been shown to be, predominantly α-amylase I (MacGregor, 1976). Thus α-amylase I would appear to be under different genetic control to the other isoenzyme groups. Quite obviously, more detailed quantitative information is required on the relative amounts of these enzymes (certainly of the major groups if not of the individual isoenzymes) synthesized by whole kernels, embryos and aleurone layers with and without added GA_3 to more fully understand their site and mechanism of synthesis.

The presence in barley and malt of small amounts of an amylase differing in physical, hydrolytic and immunochemical properties from normal cereal α- and β-amylases has been reported (Niku-Paavola and Nummi, 1971; Niku-Paavola, 1977). However, the enzyme was later identified as an α-amylase on the basis of its action pattern on amylose (MacGregor et al., 1979). Although it does not appear to be of fungal origin (Niku-Paavola and Heikkinen, 1975) it is not found consistently in barley or malt kernels suggesting that it either originates from some contaminant on the kernels or is induced within these kernels by a contaminant. No information is available on the site or mechanism of synthesis of this enzyme in barley or malt.

B. WHEAT

Sandstedt and Beckord (1946) reported the presence of α-amylase in the pericarp of developing kernels of wheat. Later workers confirmed that immature wheat does contain an α-amylase (Guilbot and Drapron, 1963; Olered, 1967). Olered and Jönsson (1968, 1970) discussed the significance of the finding that non-sprouted wheat could contain α-amylase and the possible technological importance of this led to a flurry of reports during the ensuing ten years on the formation and properties of the enzyme. Unfortunately, no consistent approach to the problem was adopted leading to different methods being used for separating, detecting and naming the enzyme as well as for preferentially inhibiting interfering β-amylase. This has led to a very confused situation. In addition, at least three groups of α-amylase appear to be present in wheat kernels during growth and maturation (Marchylo *et al.*, 1980a). This finding is important to an understanding of fluctuations in α-amylase levels during growth and maturation of wheat and so it will be discussed in some detail and related to previous reports on the enzyme.

Using isoelectric focusing and specific, sensitive methods for detecting α-amylase, Marchylo *et al.* (1980a) showed that the α-amylase present in wheat up to 15–20 days after anthesis is associated largely with the pericarp (Fig. 3). Activity was detected in other tissues, especially the seed coat, but

FIG. 3. Isoelectric focusing of α-amylases from developing pericarp tissue and whole kernels of Glenlea wheat (modified from Marchylo *et al.*, 1980a).

this may have been caused by contamination from the pericarp during dissection of the previously frozen kernels. This enzyme appears in wheat kernels shortly after anthesis, reaches a peak 10–20 days later and then declines to a low level (Fig. 4). A similar profile of α-amylase activity in

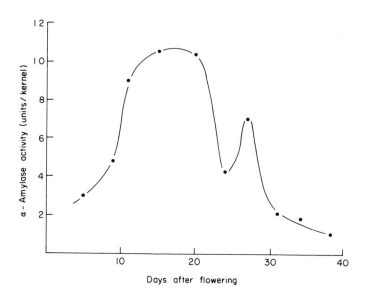

FIG. 4. α-Amylase activity in developing kernels of Glenlea wheat (modified from Marchylo *et al.*, 1980a).

immature kernels has been found by other workers (Olered and Jönsson, 1970; Lacroix *et al.*, 1976; Meredith and Jenkins, 1973). This is the true pericarp α-amylase detected by Sandstedt and Beckord (1946), Banks *et al.* (1972), Meredith and Jenkins (1973), Kruger (1972a) and Iliev (1974). The enzyme has a low pI and is, presumably, the Group II α-amylase of Sargeant (1979), the D1 α-amylase of Daussant and Renard (1976), the α-I, II and III of Kruger (1972a) and the "green" α-amylase of Olered and Jönsson (1970). In addition, using electrophoretic techniques, a highly cathodic α-amylase component (D2 or α-amylase III) has been detected in developing kernels of wheat (Daussant and Renard, 1976; Daussant, 1978) and triticale (Daussant and Hill, 1979). Furthermore, these authors showed that this enzyme is immunochemically distinct from other α-amylases found in both developing and germinating kernels of wheat and triticale. Further investigation is required to characterize this α-amylase component and to determine its location within developing kernels.

Initially, pericarp α-amylase was shown to contain only one or two components (Olered and Jönsson, 1970; Alexandrescu and Mihailescu, 1970). More recently, as many as six have been detected (Marchylo *et al.*, 1980a) but qualitative analysis indicates the presence of only two to three major components, as found by other workers (Kruger, 1972a; Iliev, 1974;

Daussant, 1978; Sargeant, 1979), with the remaining isoenzymes present in only trace amounts (Fig. 3). This, however, should be verified by quantitative experiments. The pattern and number of α-amylase components found may depend on the wheat cultivar studied (Marchylo *et al.*, 1980a).

The mechanism of synthesis of pericarp α-amylase in wheat has not been investigated. Gibberellic acid does not induce α-amylase synthesis in pericarps isolated 25 days after anthesis (Marchylo *et al.*, 1980b) but this does not eliminate the possibility that the hormone is implicated in pericarp α-amylase synthesis. Perhaps by this stage of development the mechanism for α-amylase synthesis in pericarp tissue has been closed down.

As pericarp α-amylase declines in activity (Fig. 3) two new groups of α-amylase appear in immature kernels (Marchylo *et al.*, 1980a; Sargeant, 1979). They must be responsible for the small peak of activity observed in Fig. 4 at 24 days after anthesis. The stage of growth at which these enzymes are first detected varies with wheat cultivar (Marchylo *et al.*, 1980a). Neither of these two enzyme groups is present in the pericarp (Fig. 3) and so should be clearly distinguished from pericarp α-amylase. One group has the same pI as pericarp α-amylase and a similar pattern of components, while the other has a pI range of 6·0–6·5 (Sargeant, 1979). These new enzymes are similar in electrophoretic properties to two major groups of α-amylase found in sprouted and germinated wheat (Sargeant, 1979; Marchylo *et al.*, 1980a). Therefore, it is tempting to conclude that the two groups of α-amylase found in immature kernels occur because of pre-germination and are the same enzymes found in germinated wheat. This working hypothesis will be assumed until more definitive evidence on the identities of the enzymes is obtained!

More work is required, using a number of wheat cultivars grown under different conditions, to establish whether or not the appearance in immature wheat of the two germination type α-amylases is a general phenomenon. That immature aleurone and embryo tissues are capable of synthesizing germination-type α-amylases in the presence and absence of exogenous GA_3 has been established (Marchylo *et al.*, 1980b) but tissue age is critical. α-Amylase synthesis increases with tissue age. This increased ability of older tissues to synthesize α-amylase could be due to decreased levels of abscisic acid (ABA) in such tissues. Abscisic acid, an inhibitor of α-amylase synthesis (Chrispeels and Varner, 1966) increases in amount during early stages of kernel development and then decreases as kernels mature (King, 1976; Radley, 1976). High levels of ABA during early stages of growth could prevent tissues from synthesizing germination-type α-amylase but, as the hormone level drops, synthesis could occur if other conditions such as kernel moisture content were appropriate. This could explain the inability of Nicholls (1979) to induce α-amylase synthesis in immature, de-embryonated kernels. If these were too immature, the amount of exogenous GA_3 used may not have been sufficient to overcome high levels of ABA in the tissues. However, when the kernels were slowly desiccated or kept moist for prolonged periods (8 days) then ABA levels may have been reduced to a sufficiently low level to allow α-amylase synthesis to take place. Gibberellin levels in growing kernels show one or two peaks, depending on wheat cultivar (Rejowski, 1964; Radley, 1976), but these do not appear to coincide with the development of pericarp α-amylase or of germination-type α-amylases. However, it would be premature at this stage to rule out the possibility of GA_3 involvement with α-amylase synthesis in immature wheat kernels.

α-Amylase activity, then, found in wheat kernels during later stages of growth is almost certainly caused by germination-type α-amylases (Daussant et al., 1979). The so-called reactivation of "green" α-amylase during this period (Olered and Jönsson, 1970; Olered, 1976) is probably caused by synthesis of these enzymes and not reactivation of pericarp α-amylase. These germination-type enzymes are probably responsible also for the apparent periodicity of α-amylase synthesis (Jenkins et al., 1974) and for the peaks of heat labile and stable α-amylases reported by Meredith and Jenkins (1973). These enzymes were differentiated on the basis of preferential heat stability. However, heat inactivation of amylases is complex and the results are difficult to interpret because the extent of inactivation depends not only on the temperature and time of heating but also on the pH and protein content of the extract (Greenwood and MacGregor, 1965). Therefore, results based on differential heat treatments should be treated with caution until they are confirmed by more precise methodology.

The marked increase in α-amylase content of wheat during germination (Kneen, 1944) is due to de novo synthesis of the enzyme (Daussant and Abbott, 1969; Daussant and Corvazier, 1970). Addition of GA_3 to germinating kernels (Fleming and Johnson, 1961; Jeffers and Rubenthaler, 1974) and to embryoless kernels (Moro et al., 1963; Khan et al., 1973) increases α-amylase synthesis in much the same way as in barley (Section II.A.). However, in some wheat cultivars, α-amylase synthesis is very low even in the presence of GA_3 (McMaster, 1976; Gale and Marshall, 1973). This property is being used by plant breeders to produce wheat cultivars having high sprouting resistance and, therefore, low α-amylase content even under poor harvesting conditions thus giving such wheats a technological advantage. No information is available on how such wheats mobilize their endosperm reserves during germination!

α-Amylase from germinated wheat, like that from germinated barley, is heterogeneous, as mentioned previously. Two main components were detected in initial electrophoretic studies (Olered and Jönsson, 1970; Daussant and Corvazier, 1970) but later work showed each component to be heterogeneous (Kruger, 1972b; Khan *et al.*, 1973; Warchalewski and Tkachuk, 1978) with a total of eight (Kruger, 1972b; Daussant, 1978) to twelve (Khan *et al.*, 1973) different bands of α-amylase. Using isoelectric focusing techniques, however, even more bands have been detected recently (Marchylo *et al.*, 1980b; Sargeant, 1980). These can be split conveniently into two main groups, a minor component with pI values of 4·5–4·8 and a major component having pI values of 6·0–6·5 (Sargeant, 1980). Obviously, the precise number of α-amylase bands found depends on the resolving power of the separation method used, the sensitivity of the method used for detecting α-amylase, the extent of germination (Kruger, 1972b; Khan *et al.*, 1973; Sargeant, 1980) and, possibly, the cultivar of wheat analyzed.

Because each research group tends to use its own separation technique and nomenclature system it is difficult to critically compare results from different groups (e.g. Khan *et al.*, 1973; Sargeant, 1979; Marchylo *et al.*, 1981). A unified approach to this problem of the nomenclature of α-amylase isoenzymes from cereal grains should be attempted by those active in this area. Because of the simplicity, wide-spread use and high resolving power of isoelectric focusing perhaps this technique could be used to determine the pI values of, at least, the major isoenzymes. These values should then be used in published reports to identify particular isoenzymes being discussed. However, extreme care must be taken to ensure that each isoenzyme thus characterized is real and not an artifact formed by either the extraction method or isoelectric focusing technique used.

C. TRITICALE

The potentially high yields of triticale are often not realized because of extensive kernel shrivelling as the crop approaches maturity. Because an association has been found between kernel shrivelling and α-amylase content (Muntzing, 1963; Klassen *et al.*, 1971), the nature of α-amylases in developing triticale has been investigated by a number of research groups.

α-Amylase activity increases shortly after anthesis, reaches a peak and then declines to low levels in most lines examined (Dedio *et al.*, 1975; Jenkins and Meredith, 1975; Lorenz and Welsh, 1976). This initial α-amylase is associated primarily with the pericarp (Dedio *et al.*, 1975) and so to avoid unnecessary confusion may be referred to as pericarp α-amylase. It is similar in electrophoretic and immunochemical properties to the pericarp α-amylase of wheat (Daussant and Hill, 1979).

In some triticale lines, especially those very susceptible to kernel shrivelling, α-amylase activity redevelops during later stages of maturation (Klassen *et al.*, 1971; Dedio *et al.*, 1975; Jenkins and Meredith, 1975; Lorenz and Welsh, 1976). The exact stage of growth at which this increase in enzyme activity occurs varies from one cultivar to another but it tends to occur when kernel growth ceases and rapid kernel desiccation takes place (King *et al.*, 1979). This α-amylase is not associated with the pericarp but is found predominantly in aleurone and endosperm tissues (Dedio *et al.*, 1975). It is heterogeneous and essentially similar to α-amylases found in germinated triticale (Silvanovich and Hill, 1977; Daussant and Hill, 1979) consisting of several isoenzymes that can be segregated into two main groups on the basis of their isoelectric points. The low pI group is similar to the original pericarp α-amylase but is synthesized in the embryo or aleurone along with the higher pI group. This sequence of α-amylase synthesis is similar to that discussed for wheat (Section II.B.). Increased α-amylase activity in triticale during later stages of maturation, therefore, is caused by early stimulation of the α-amylase biosynthetic mechanism that normally is not activated until mature seeds are subjected to germination conditions. Kernels harvested at this stage of growth are able to synthesize significant amounts of α-amylase in the absence of both embryo and exogenous GA_3 (King *et al.*, 1979). These kernels must already contain sufficient GA_3 to initiate and sustain α-amylase synthesis. Also, ABA levels fall rapidly in triticale kernels as desiccation takes place and there is a direct relationship between falling ABA levels and increased ability of triticale half-seeds to synthesize α-amylase (King *et al.*, 1979). High levels of ABA present in kernels during early stages of growth may well block α-amylase synthesis but when this block is removed synthesis can occur if sufficient amounts of GA_3 are available. Normally, this is not the case until germination takes place, but mature kernels showing susceptibility to premature α-amylase synthesis may already contain adequate levels of GA_3. Therefore, kernels highly susceptible to sprouting may contain excessive amounts of GA_3 rather than abnormally low levels of ABA.

Desiccation plays an important and, apparently, complex role in kernel susceptibility to sprouting (Evans *et al.*, 1975; Nicholls, 1979, 1980). In general, once seeds of triticale and of other cereal grains (King, 1976; Bilderback, 1971) attain maximum grain weight and are then desiccated either artificially or naturally, they become more susceptible to sprouting and α-amylase synthesis is readily stimulated. This whole area of interaction of plant growth hormones, grain desiccation and α-amylase biosynthesis in triticale requires in-depth studies of all of these factors in the same sample of grain.

D. RYE

Immature rye kernels appear to contain small amounts of pericarp α-amylase (Jenkins and Meredith, 1975; Lorenz and Welsh, 1976; Rao *et al.*, 1976) that is heterogeneous (Wilp and Buschbeck, 1978). Although the enzyme level falls as the kernels mature it does not decrease to the low levels found in wheat and barley and may even tend to increase during later stages of maturation (Hagberg and Olered, 1975; Jenkins and Meredith, 1975; Lorenz and Welsh, 1976). α-Amylase isoenzymes present during final stages of kernel maturation appear to be similar to those found in germinated rye (Wilp and Buschbeck, 1978). Therefore, α-amylase found in mature rye may arise through premature stimulation of germination-type α-amylase as in the case of triticale (Section II.C.).

Embryoless half-seeds of rye form α-amylase in response to GA_3 but significant amounts of enzyme are rapidly formed in the absence of GA_3 (Palmer, 1970; Mierzwinska, 1977). Mature kernels, obviously, do not have to wait for hormones synthesized in the embryo before commencing α-amylase synthesis under conditions conducive to germination. Sufficient GA_3 or other necessary hormones must already be present in the non-embryo part of mature kernels. Therefore, α-amylase synthesis in rye, as in triticale, can be stimulated very rapidly.

α-Amylase activity develops rapidly in germinating rye (Ballance and Manners, 1975) with the formation of a number of isoenzymes. Two main components were detected by continuous electrophoresis (Manners and Marshall, 1972), at least five or six were found by gel electrophoresis (Wagenaar and Lugtenborg, 1973; Wilp and Buschbeck, 1978) but, surprisingly, only three were found after isoelectric focusing analysis (Möttönen, 1975). Obviously, the enzyme is heterogeneous and the number of components found may depend not only on the methodology used but also on the cultivar of rye studied (Wilp and Buschbeck, 1978).

Embryonic as well as aleurone tissues of germinating rye kernels contain α-amylase (Wilp and Buschbeck, 1978) showing that rye embryos are capable of synthesizing the enzyme. There appears to be no difference in the pattern of α-amylase components formed in the two tissues.

E. OATS

There is evidence of pericarp α-amylase in developing oat kernels but the quantity is very small (Meredith and Jenkins, 1973). Mature samples of some oat cultivars appear to contain significant amounts of α-amylase (Kneen, 1944) probably caused by premature synthesis of germination-type enzyme rather than by a carry-over of pericarp α-amylase (Smith and Bennett, 1974).

During germination α-amylase activity increases markedly (Kneen, 1944; Palmer, 1970; Manners and Yellowlees, 1973). It is fairly safe to assume that the enzyme is synthesized *de novo* but no direct confirmation of this has been reported. Although exogenous GA₃ does stimulate α-amylase synthesis in both whole kernels and endosperm slices of oats the response is much less than in other cereal grains (Palmer, 1970; Pomeranz and Shands, 1974). Embryos of oats, rye, wheat and barley are more effective than GA₃ in stimulating α-amylase synthesis in oat endosperms suggesting that these embryos may excrete, in addition to GA₃, other hormones that activate oat aleurone (Palmer, 1970).

α-Amylase of germinated oats contains several components and the isoenzyme pattern is, to some extent, variety dependent (Smith and Bennett, 1974).

F. RICE

Shortly after anthesis, α-amylase activity increases in rice kernels and then decreases to low levels as the kernels mature. Peak enzymatic activity occurs at different times for different cultivars (Baun *et al.*, 1970; Shinke *et al.*, 1978). The enzyme appears to be heterogeneous (Baun *et al.*, 1970).

During germination there is a significant increase in the α-amylase content of rice kernels (Murata *et al.*, 1968; Palmiano and Juliano, 1972; Williams and Peterson, 1973). This is probably caused by *de novo* synthesis of the enzyme as is found in barley and wheat. Although there is little direct experimental evidence to confirm such a hypothesis, findings such as GA₃-stimulated α-amylase synthesis in embryoless half-seeds (Tanaka and Akazawa, 1970; Palmiano and Juliano, 1972) and inhibition of α-amylase synthesis by cycloheximide (Palmiano and Juliano, 1972) do support the idea. Enzyme formation is considerably slower than in wheat, barley or oats (Chen *et al.*, 1973) but it can be increased significantly by adding GA₃ to germinating kernels (Roy *et al.*, 1973; Chen *et al.*, 1973). This suggests that GA₃ synthesis in rice embryos may be a limiting factor in α-amylase synthesis during germination.

The enzyme is heterogeneous and appears to contain two main and several minor components but there is some uncertainty in differentiating completely between α-amylase and β-amylase bands (Tanaka *et al.*, 1970; Okamoto and Akazawa, 1978, 1979). Germinated whole kernels and embryoless half-seeds treated with GA₃ give similar, but not identical, α-amylase patterns indicating that, at least for α-amylase synthesis, exogenous GA₃ is a reasonable replacement for the rice embryo (Tanaka *et al.*, 1970). Not all α-amylase components appear to be synthesized at the same rate nor at the same time during germination (Tanaka *et al.*, 1970) but quantitative determinations of each component must be made to confirm this.

During initial stages of germination (12–48 h) α-amylase migrates into the endosperm from the scutellar epithelium and although α-amylase is detected in the aleurone after 48 h this enzyme does not appear to move into the endosperm until after 4 days of germination (Okamoto and Akazawa, 1979). Therefore, the scutellar epithelium appears to be the initial site of α-amylase synthesis in germinating rice. There appears to be no significant difference in isoenzyme composition between α-amylases formed in the embryo or in aleurone layers (Okamoto and Akazawa, 1979). Presumably, synthesis of α-amylase is under the control of GA_3 in both tissues. Unfortunately, no information is available on the relative amount of α-amylase formed by the two tissues.

G. MAIZE

Immature maize kernels contain α-amylase (Chao and Scandalios, 1969) but changes in enzyme activity with stage of growth have not been determined. The enzyme is heterogeneous with the major component having a pI of 4·8 and other minor components having similar values (Chao and Scandalios, 1969; Scandalios *et al.*, 1978; Goldstein and Jennings, 1978). This enzyme appears to be similar to the pericarp α-amylase of wheat and barley but the tissue in which it is synthesized within maize kernels is not known. However, unlike pericarp α-amylase from other cereals the maize enzyme persists in kernels throughout growth and maturation.

The increased α-amylase activity found in germinated kernels appears to arise through further synthesis of the low pI enzyme first detected in immature kernels (Chao and Scandalios, 1971, 1972). Although enzyme heterogeneity increases during germination only trace amounts of α-amylases having higher pIs, similar to those found in kernels of other germinated cereals, have been detected (Chao and Scandalios, 1971; Scandalios, 1974).

Embryoless maize half-seeds synthesize almost equal amounts of α-amylase in the presence and absence of exogenous GA_3 (Harvey and Oaks, 1974; Goldstein and Jennings, 1975). This finding may not apply to all cultivars of maize (Ingle and Hageman, 1965) but it is significant and does warrant further investigation. Abscisic acid inhibits α-amylase synthesis but this inhibition can be overcome by the addition of GA_3 (Harvey and Oaks, 1974), suggesting that GA_3 is involved in the synthesis of α-amylase in germinating maize kernels. Presumably, endosperms of maize kernels contain sufficient amounts of GA_3 or other necessary hormones to stimulate completely α-amylase synthesis in aleurone cells and so do not require embryo participation. Interestingly, embryoless barley seeds, when incubated in water, produce small amounts of α-amylase that is predominantly the low pI component of germinated barley α-amylase. In this case, however, addition of GA_3 does stimulate synthesis of all α-amylase isoenzymes (MacGregor, 1976).

Significant amounts of α-amylase are formed in embryos of germinating kernels (Dure, 1960; Goldstein and Jennings, 1975) but quantitative data on the isoenzyme pattern of this enzyme compared to that formed in aleurone cells are not available.

H. MOLECULAR BASIS FOR α-AMYLASE SYNTHESIS

Recent reviews (Jacobsen, 1977; Jacobsen et al., 1979; Ho, 1979) have covered this topic adequately so only a brief outline will be given here.

Most studies on α-amylase synthesis have been carried out using the relatively simple system of isolated aleurone layers stimulated with GA_3. Although barley aleurones are commonly used, aleurone layers from other cereal grains should yield comparable results (MacLeod et al., 1964; Phillips and Paleg, 1972; Palmer, 1970). There are some inherent disadvantages in using this simplified system rather than studying α-amylase synthesis in intact germinating kernels but these will be discussed at the end of the section.

Exogenous GA_3 has no effect on total RNA synthesis in aleurone layers (Chandra and Varner, 1965; Jacobsen and Zwar, 1974) but it does enhance the synthesis of total mRNA (Ho and Varner, 1974). Furthermore, the activity of mRNA specific for α-amylase, and α-amylase activity both increase shortly after the addition of GA_3, reach a peak and then decline (Higgins et al., 1976; Higgins et al., 1977). These results suggest that GA_3 is required to transcribe mRNA for α-amylase synthesis. Studies have shown that sufficient α-amylase mRNA is formed in aleurone layers during the first 12 h after GA_3 addition to maintain α-amylase synthesis in the presence of transcription inhibitors (Chrispeels and Varner, 1967; Goodwin and Carr, 1972; Ho and Varner, 1974). Therefore, inhibition of α-amylase synthesis after this time must involve translation and not transcription processes. Although GA_3 may not be required for the translation of α-amylase mRNA the hormone may increase the efficiency of translation (Higgins et al., 1976).

There is evidence that although GA_3 enhances the synthesis of α-amylase it may reduce synthesis of other proteins and their mRNA species as well (Varner *et al.*, 1976). Therefore, GA_3 may promote synthesis of α-amylase and other hydrolytic enzymes by suppressing synthesis of other proteins and so making available for α-amylase synthesis a higher proportion of the pool of amino acids and nucleic acids and of the protein synthetic machinery present in aleurone cells.

Abscisic acid completely inhibits α-amylase synthesis in GA_3-stimulated aleurone layers when it is added at the same time as GA_3 (Chrispeels and Varner, 1966) as well as 12 h later (Ho and Varner, 1976). This suggests that ABA does not act on transcription but on some post transcription event. Later studies (Jacobsen *et al.*, 1979) indicate that ABA is probably involved in both transcription and translation. Formation of non-α-amylase mRNA and protein is enhanced by ABA (Ho and Varner, 1976; Jacobsen *et al.*, 1979) so this hormone may act by preferentially suppressing the synthesis of α-amylase mRNA species. Obviously, GA_3 and ABA acting together could exert very tight control over α-amylase synthesis in aleurone layers. Secretion of α-amylase from aleurone cells appears to be dependent on GA_3 also (Chrispeels and Varner, 1967) but the exact mechanism of this process is poorly understood (Chrispeels, 1976; Tomos and Laidman, 1979). Some evidence suggests that the enzyme diffuses in a soluble form from its site of synthesis through the cytoplasm to the plasma membrane (Jones, 1972; Chen and Jones, 1974a, 1974b; Jones and Chen, 1976). However, isolation of vesicles containing α-amylase from both barley (Firn, 1975) and wheat (Gibson and Paleg, 1972, 1975, 1976) aleurone tissues strongly suggests that α-amylase is transported within aleurone cells in membrane bound vesicles. These vesicles appear to accumulate in those areas of aleurone cells that are close to the endosperm (Vigil and Ruddat, 1973), prompting the conclusion that α-amylase is transported in a controlled direction from its site of synthesis, the endoplasmic reticulum, to the plasma membrane (Chrispeels, 1976). Because GA_3 appears to be involved in the formation of α-amylase-containing vesicles (Vigil and Ruddat, 1973) transportation of α-amylase within aleurone cells may be under GA_3 control.

Finally, the enzyme is moved through the plasma membrane (mechanism unknown) and secreted across the cell wall. This latter process probably occurs after dissolution of the cell wall by cell wall degrading enzymes synthesized under the control of GA_3 (Jones 1969; Briggs, 1973). A summary of possible stages in the synthesis and secretion of α-amylase in aleurone cells is shown in Fig. 5.

Although new insight into the synthesis of α-amylase in cereal grains has been obtained from studies on isolated aleurone layers caution must be exercised when using this information to explain α-amylase synthesis in whole kernels. It is unlikely that GA_3 is a complete replacement for the embryo, and other hormones that affect α-amylase synthesis may well be

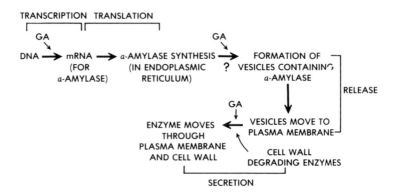

FIG. 5. Possible steps in the synthesis and secretion of α-amylase in aleurone cells.

excreted from the embryo (Petridis *et al.*, 1965; MacLeod *et al.*, 1966; Groat and Briggs, 1969; Palmer, 1970; Clutterbuck and Briggs, 1973). Embryos from barley and, presumably, other cereal grains do synthesize α-amylase (Briggs, 1964; Briggs and Clutterbuck, 1973) but more studies are required to determine the mechanism of synthesis and the contribution made by embryo α-amylase to the total enzyme synthesized during germination. For practical reasons, aleurone layers are usually isolated from barley half-seeds after 2–3 days of steeping but the effect of this steeping on subsequent metabolic events within aleurone cells has not been thoroughly evaluated. There is evidence that α-amylase synthesis is accelerated in aleurone layers that have been hydrated before being treated with GA_3 (MacLeod *et al.*, 1964; Yung and Mann, 1967). Normally, aleurone layers with attached seed coat (inner pericarp, testa, hyaline layer) have been used for studies on α-amylase synthesis and all metabolic events have been ascribed to the action of GA_3 on the aleurone. However, recent results with wheat showing that α-amylase synthesis is significantly decreased in aleurone layers devoid of testa (Marchylo *et al.*, 1981) indicate that the testa may play an active part in α-amylase synthesis in cereal grains. Aleurone layers for studies on α-amylase synthesis are usually isolated from the barley cultivar, Himalaya, but germinated kernels of this cultivar have a unique α-amylase pattern (Fig. 2). Component 1 contains a strong, low pI band of activity that is rarely found in germinated samples of other barley cultivars. Therefore, α-amylase synthesis in Himalaya may be atypical. Finally, studies using only aleurone layers do not take into account the rate of water uptake or of synthesis and movement of GA_3 to the aleurone layer within intact kernels before stimulation of aleurone cells can commence. Therefore, the rate of α-amylase synthesis in GA_3-stimulated aleurone layers cannot be related directly to the rate of synthesis in intact kernels.

No explanation has yet been given for the large number of α-amylase isoenzymes found in germinated cereal grains (Momotani and Kato, 1967; Marchylo *et al.*, 1981). Some isoenzymes may be artifacts formed by extraction, purification or separation techniques but others may well have a genetic basis. Little progress has, as yet, been made in this area (Scandalios, 1974).

III. α-AMYLASE ACTION PATTERN

A. DEGRADATION OF STARCH COMPONENTS IN SOLUTION

Only the action pattern of cereal α-amylases will be discussed in this review. It must be remembered that α-amylases from different sources such as fungi, bacteria and mammals do not have the same action pattern as cereal α-amylases (Greenwood and Milne, 1968b) and comments made in this review may not apply to those enzymes. Although detailed studies have not been made on all cereal α-amylases, results obtained thus far suggest a similar action pattern for most enzymes in this group (Greenwood and Milne, 1968a). Therefore, the enzymes will be discussed together and not individually.

Amylose, rather than starch, has been the preferred substrate for most studies on amylase action pattern because the products of hydrolysis are linear and so are more easily analyzed. α-Amylase attacks the α-(1→4) bonds in long amylose molecules in a random fashion (Fig. 6) and quickly decreases their size. This rapid phase of hydrolysis, or dextrinization, is characterized by a fast decrease in amylose viscosity, loss of ability to stain with iodine, production of large maltodextrins and slow formation of small sugars. This is followed by the slow phase of hydrolysis, or saccharification, in which there is a steady formation of small sugars (Myrbäck and Neumüller, 1950).

When α-amylase hydrolyzes a particular α-(1→4) linkage in amylose the reaction may proceed in one of two ways. The two amylose fragments produced may diffuse away from the enzyme and be re-attacked subsequently in a random fashion (multichain attack). Alternatively, one fragment may be retained by the enzyme and re-attacked close to a chain-end one or more times before being released (multiple attack). The latter mechanism leads to the formation of small saccharides during early stages of amylose hydrolysis whereas the former mechanism does not. These two possibilities have been thoroughly discussed by Thoma (1976a) and Banks and Greenwood (1977). It seems likely that cereal α-amylases degrade amylose or starch by the multichain attack mechanism (Banks *et al.*, 1970; Banks and Greenwood, 1977).

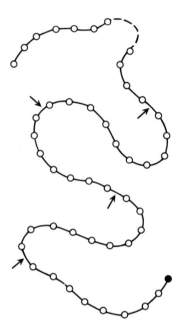

Fɪɢ. 6. Initial random hydrolysis of amylose by cereal α-amylases. –○–, α-(1 → 4) linked D-glucose residue; –●–, reducing end-group; ↑ , linkage hydrolyzed by α-amylase.

Formation of maltodextrins during the hydrolysis of amylose by an α-amylase from malted barley (MacGregor, 1978b) is shown in Fig. 7. During the initial rapid phase of hydrolysis amylose is degraded to a mixture of maltodextrins in which the larger members (G_7 and higher) predominate. As hydrolysis proceeds these larger dextrins disappear until, first G_8 and then G_7 are the largest dextrins present. Finally, after prolonged hydrolysis, G_7 is hydrolyzed completely leaving a mixture of G_6 and smaller dextrins, indicating that these are highly resistant to further hydrolysis, as suggested by Okada *et al.* (1969). Similar results have been obtained with the main α-amylase component of malted barley (MacGregor *et al.*, 1971b), barley pericarp α-amylase (MacGregor *et al.*, 1974) a "new" α-amylase from barley (MacGregor *et al.*, 1979) and α-amylase from germinated rice (Murata *et al.*, 1968). In addition, using quantitative paper chromatography, Greenwood and Milne (1968b, 1968c) showed that G_6 accumulates in digests of amylose hydrolyzed by α-amylases from oats, rye, wheat, malted wheat and malted barley. Therefore, the final distribution of maltodextrins shown in Fig. 7 is fairly typical of the end-products formed when amylose is hydrolyzed by cereal α-amylases. Enormous amounts of pure α-amylase would be required to reduce such a digest to glucose and maltose so, for practical purposes, the

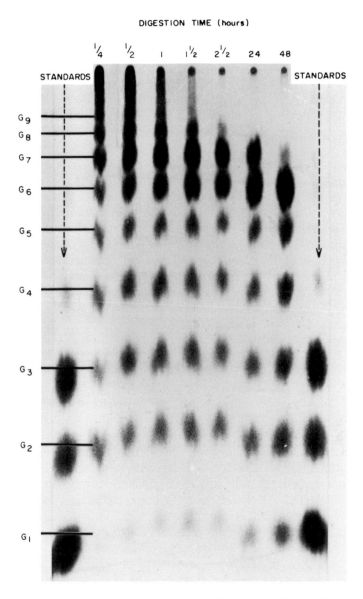

FIG. 7. Products of amylose hydrolysis by α-amylase I from malted barley. G_1, G_2 . . . . G_9 represent glucose, maltose . . . . maltononaose. (MacGregor, 1978b).

end-products are a mixture of glucose through G_6 and not glucose and maltose as generally accepted. Results showing the end-products to be glucose and maltose may have been achieved with impure enzyme since small traces of β-amylase would rapidly reduce these maltodextrins to a mixture of glucose, maltose and maltotriose and, finally, to glucose and maltose. Highly specific methods are now available for α-amylase preparation (Silvanovich and Hill, 1976) and sensitive immunochemical methods can be used to detect β-amylase impurities in α-amylase preparations (MacGregor et al., 1979). These should be utilized when preparing α-amylase for critical kinetic studies.

If all glucosidic bonds in amylose were equally readily hydrolyzed by α-amylase then glucose would be the sole hydrolytic product. This is obviously not so (Fig. 7). Bird and Hopkins (1954) suggested that the distribution of end-products could be explained if the second to the fifth bonds from the non-reducing end and the first bond from the reducing end of amylose molecules were more resistant, than all other bonds, to hydrolysis by α-amylase. This idea was refined by Greenwood and Milne (1968c) who suggested that: (a) the five bonds nearest the non-reducing end of a saccharide molecule are resistant to α-amylase; (b) the bond at the reducing end is half as likely to be attacked as other bonds; (c) the penultimate bond at the reducing end is twice as readily hydrolyzed as other bonds; (d) all other bonds are equally readily hydrolyzed. A theoretical scheme for the hydrolysis of maltodecaose (G_{10}) incorporating these assumptions is shown in Fig. 8.

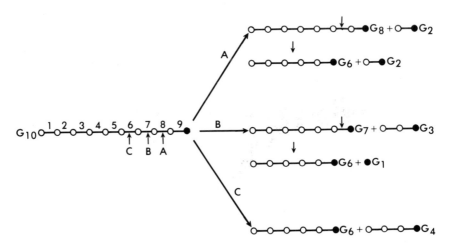

FIG. 8. Hydrolysis of G_{10} (maltodecaose) by cereal α-amylase. The three most probable hydrolytic routes are (A) initial hydrolysis at linkage 8; (B) initial hydrolysis at linkage 7; (C) initial hydrolysis at linkage 6.

Theory predicts that G_{10} should be hydrolyzed preferentially via route A to give initially G_8 and G_2 but G_8 would be further hydrolyzed to give G_6 and G_2. Final products would then be G_6 and G_2. However, routes B and C are possible also. Route B would produce G_6, G_3 and G_1 whereas route C would produce G_6 and G_4.

Theoretical results obtained from such a scheme agree reasonably well with experimental data (Greenwood and Milne, 1968b). These show that G_2 and G_6 are the most abundant hydrolytic products. Much more complex mathematical models have now been presented to explain the pattern of end-products formed by α-amylase hydrolysis of amylose (Thoma, 1976b; Thoma and Allen, 1976; Allen and Thoma 1976a, 1976b; Hiromi, 1970; Nitta *et al.*, 1971; Iwasa *et al.*, 1974; Aoshima *et al.*, 1974). However, these have not yet been applied to cereal α-amylases.

On the basis of Fig. 8, hydrolysis of amylose by cereal α-amylases may not appear to be random. However, in a large amylose molecule containing several hundred glucose residues only about seven bonds are attacked in a non-random manner. Therefore, initial hydrolysis is essentially random. As hydrolysis proceeds the number of saccharide molecules increases, their size decreases and a much higher proportion of the total bonds is attacked in a non-random manner. At this stage a high proportion of the bonds are resistant to α-amylase attack and so the rate of hydrolysis falls. These resistant bonds are not inherently different to the other bonds in amylose or maltosaccharide molecules. They are resistant because they are at, or near, the ends of amylose or maltodextrin molecules and the enzyme appears to find difficulty in forming a complex with chain-ends such that scission of these particular bonds can take place.

Cereal α-amylases hydrolyze amylopectin and amylopectin β-limit dextrin but at a reduced rate compared to the hydrolysis of amylose (Greenwood and MacGregor, 1965). Amylopectin is a highly branched molecule containing relatively short unit chains (average chain length is 20–26 glucose residues for cereal amylopectins) joined by α-$(1 \rightarrow 6)$ bonds (Greenwood and Thomson, 1962; Manners and Bathgate, 1969). These bonds are not only resistant to attack by α-amylase but they also confer resistance to adjacent α-$(1 \rightarrow 4)$ bonds (Manners, 1962). Therefore, many α-$(1 \rightarrow 4)$ bonds in amylopectin are resistant to α-amylase hydrolysis. A similar argument applies to amylopectin β-limit dextrin.

Initial hydrolytic products from amylopectin are large dextrins (Bird and Hopkins, 1954) but on prolonged hydrolysis by α-amylases from malted barley (Hall and Manners, 1978) and malted rye (Manners and Marshall, 1971) a mixture of small saccharides and larger branched dextrins is obtained. A small branched saccharide, 6^3-α-D-glucosylmaltotriose, was found as a reaction product in both digests. Because of the apparent resistance to α-amylase attack of α-$(1 \rightarrow 4)$ bonds close to chain ends and α-$(1 \rightarrow 6)$ branch points it is not surprising that only a very small amount of such a product is formed.

B. DEGRADATION OF STARCH GRANULES

Many studies on the hydrolysis of starch granules by α-amylases have been carried out to obtain information on the structure of granules. Very often α-amylases from fungi and bacteria rather than from cereal grains have been used because such enzymes are readily available and digest cereal starches more rapidly than do cereal amylases (Sandstedt and Gates, 1954). Therefore, only limited information is available on the action of cereal amylases on their respective starches. α-Amylases from germinated wheat and barley have been studied the most intensively because of the importance of malted barley α-amylase in the malting and brewing industries and the deleterious effect of α-amylase from sprouted wheat in the baking industry.

In general, a cereal α-amylase will hydrolyze intact starch granules from any cereal grain but only very slowly compared to the hydrolysis of the same starch in solution (Sandstedt and Gates, 1954). The rate of attack varies with the source of both the enzyme and the starch (Sandstedt and Gates, 1954; Sandstedt and Ueda, 1969). The susceptibility of starch granules from barley to malt α-amylase increases as the extent of starch gelatinization increases (Coulter and Potter, 1972; Slack and Wainwright, 1980) and this is probably a general phenomenon. α-Amylase components from the same source may hydrolyze starch granules at different rates (Fig. 9). α-Amylase I, the minor α-amylase component from malted barley, is significantly more efficient than the major component, α-amylase II, in hydrolyzing large granules from barley starch (MacGregor and Ballance,

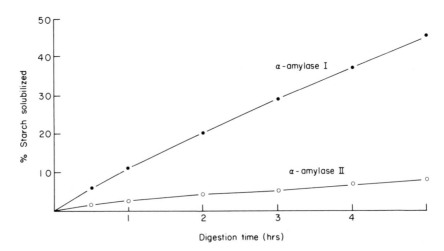

FIG. 9. Hydrolysis of large starch granules from barley by α-amylases I and II from germinated barley (modified from MacGregor and Ballance, 1980).

1980). This difference is obviously a function of the enzymes and not of the starch. Surprisingly, the minor component of wheat α-amylase does not appear to attack wheat starch granules but it does hydrolyze, almost completely to maltose, high molecular weight saccharides produced by the major α-amylase component of wheat on starch granules (Sargeant and Walker, 1978). These results suggest that this particular α-amylase has a unique action pattern.

α-Amylase hydrolysis of starch granules may be increased by the addition of β-amylase (Maeda *et al.*, 1978a) and further increased by adding R-enzyme, a starch debranching enzyme, to the α- and β-amylase mixture (Maeda *et al.*, 1978b). β-Amylase does not attack intact starch granules (Sandstedt *et al.*, 1937) nor is there evidence to suggest that R-enzyme does. Presumably, these enzymes hydrolyze the soluble products formed by α-amylase action on starch granules to small saccharides for which α-amylase has no affinity. Thus, a higher proportion of the α-amylase present is available to attack the starch granules.

At 35°C, α-amylases from malted barley hydrolyze small barley starch granules faster than large granules (MacGregor and Ballance, 1980). The reason for this is not obvious but must be due to a difference in the physical properties of the two types of granule. At 65°C, large granules are hydrolyzed faster than small granules by malt α-amylase (Bathgate and Palmer, 1973; Slack *et al.*, 1979). However, at this temperature a higher proportion of large granules will have gelatinized and so will be in solution and, therefore, more susceptible to α-amylase (MacGregor and Ballance, 1980).

It is difficult to analyze initial products of α-amylase hydrolysis of starch granules because such products may be hydrolyzed again before they can be separated from α-amylase. However, small saccharides and large molecular weight dextrins have been detected as hydrolytic products from wheat starch hydrolyzed by wheat α-amylase (Sargeant and Walker, 1978) and in digests of barley starch and malt α-amylase (Maeda *et al.*, 1978a). More thorough analyses of these products should give us a better understanding of the mechanism by which α-amylases attack starch granules.

Another way to gain insight into the mechanism of α-amylase hydrolysis of starch granules is to compare the properties of amylose and amylopectin from intact and amylase degraded starch granules. This has been done for oats and malted oats (Manners and Bathgate, 1969) and for barley and malted barley (Greenwood and Thomson, 1959; Kano, 1977; Nakamura, 1978) but the differences found were small. In kernels of malt a small area of the endosperm adjacent to the embryo contains highly degraded granules, the area under the aleurone contains some degraded granules but most of the granules in the endosperm interior remain relatively intact (MacGregor, 1980). Therefore, malt starch is a heterogeneous mixture and any difference in the properties of degraded granules will be diluted by the effect of intact granules. It is not surprising, then, that only small differences have been detected in the properties of starches isolated from cereals and their respective malts. Results obtained thus far do not clearly show that α-amylase hydrolyzes preferentially either the amylose or amylopectin components in intact granules. Detailed studies on uniformly degraded starch granules are required to answer this question.

Using light microscopy, Sandstedt and his group made a careful and detailed study of the action of malt α-amylase on starches from wheat and rye (Sandstedt, 1955). These workers documented many features of this interaction, such as discrete enzyme attack on the granule surface, formation of radial holes into the granule centre, preferential hydrolysis of some of the concentric layers within the granule and preferential digestion of the granule interior. These have now become very familiar through studies using scanning electron microscopy (SEM). Such studies show that α-amylases attack, initially, the equatorial groove of large starch granules from wheat and barley then form discrete tunnels into the granule and digest the central portions (Evers and McDermott, 1970; Dronzek et al., 1972; Jones and Bean, 1972; Kiribuchi and Nakamura, 1973; Lineback and Ponpipom, 1977). Small granules are digested in a different manner. In wheat, the enzyme enters small granules via one or two localized sites and then completely digests the granule interior (Dronzek et al., 1972). In barley, however, small granules are digested by surface erosion and only rarely are holes seen on granule surfaces (Kiribuchi and Nakamura, 1973; MacGregor and Ballance, 1980). This significant difference in the mechanism of hydrolysis of large and small granules from barley starch is not caused by different enzymes since the same enzyme can produce both effects (Fig. 10).

Other cereal starches are degraded in different ways. In corn and rice, for example, the granule surface becomes pitted and roughened but α-amylase does not appear to penetrate into the granule interior (Kiribuchi and Nakamura, 1974, 1975). No pitting or channelling is apparent in degraded oat starch but the granule surface becomes roughened (Lineback and Ponpipom, 1977). Granules of sorghum starch become highly pitted, roughened and sponge-like (Rasper et al., 1974) and resemble degraded granules from waxy barley starch (MacGregor and Ballance, 1980). There is no evidence that differences in the mechanism of degradation of cereal starches are caused by inherent differences in the respective amylases. They result from variations in the physical and perhaps chemical composition of the various starches.

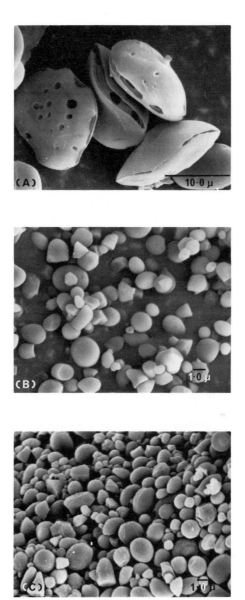

FIG. 10. Scanning electron photomicrograph of barley starch granules. (A) Large granules degraded by α-amylase. (B) Intact small granules. (C) Small granules degraded by α-amylase.

IV. Summary

Immature kernels of most cereals contain α-amylase activity that normally falls to low levels about half-way through kernel development. Because this activity is associated mainly with pericarp tissue it has been called pericarp α-amylase. α-Amylase levels may increase again during later stages of kernel development but this is probably caused by premature synthesis of germination-type α-amylase. This enzyme contains two main components, one of which is similar in electrophoretic properties to pericarp α-amylase, but it originates in kernel tissues other than the pericarp. Kernels of rye and triticale are particularly susceptible to premature development of germination-type α-amylase but a convincing explanation of this must await further studies on possible interactions between seed desiccation and hormones such as GA_3 and ABA that control α-amylase synthesis.

α-Amylase is formed by hormone controlled *de novo* synthesis in germinating cereal grains but the precise roles played by aleurone, seed coat and embryo tissues have yet to be clarified. The enzyme is heterogeneous and some components may be more efficient than others in hydrolyzing starch granules.

The end-products of amylose hydrolysis by cereal α-amylases are a mixture of saccharides from glucose to maltohexaose, with maltose and maltohexaose predominating. This distribution of end-products may be explained by assuming that all $\alpha(1 \rightarrow 4)$ bonds in amylose are equally readily hydrolyzed except for the 5 bonds nearest the non-reducing end of the molecule (more resistant to hydrolysis), the bond at the reducing end (more resistant to hydrolysis) and the penultimate bond at the reducing end (more easily hydrolyzed).

α-Amylases degrade intact granules of cereal starches to form a complex mixture of saccharides. Degraded granules exhibit a variety of physical changes that depend more on the source of starch than the source of enzyme.

References

Alexandrescu, V. and Mihailescu, F. (1970). *Rev. Roum. Biochim.* **7**, 3–8.
Allen, J. D. and Thoma, J. A. (1976a). *Biochem. J.* **159**, 105–120.
Allen, J. D. and Thoma, J. A. (1976b). *Biochem. J.* **159**, 121–132.
Allison, M. J., Ellis, R. P. and Swanston, J. S. (1974). *J. Inst. Brew., London* **80**, 488–491.
Aoshima, H., Manabe, T., Hiromi, K. and Hatano, H. (1974). *Biochim. Biophys. Acta* **341**, 497–504.
Ballance, G. M. and Manners, D. J. (1975). *Biochem. Soc. Trans.* **3**, 989–991.

Banks, W. and Greenwood, C. T. (1977). *Carbohydr. Res.* **57**, 301–315.
Banks, W., Greenwood, C. T. and Khan, K. M. (1970). *Carbohydr. Res.* **12**, 79–87.
Banks, W., Evers, A. D. and Muir, D. D. (1972). *Chem. Ind.* 573–574.
Bathgate, G. N. and Palmer, G. H. (1973). *J. Inst. Brew., London* **79**, 402–406.
Baun, L. C., Palmiano, E. P., Perez, C. M. and Juliano, B. O. (1970). *Plant Physiol.* **46**, 429–434.
Bilderback, D. E. (1971). *Plant Physiol.* **48**, 331–334.
Bilderback, D. E. (1974). *Plant Physiol.* **53**, 480–484.
Bird, R. and Hopkins, R. H. (1954). *Biochem. J.* **56**, 86–99.
Briggs, D. E. (1963). *J. Inst. Brew., London* **69**, 13–19.
Briggs, D. E. (1964). *J. Inst. Brew., London* **70**, 14–24.
Briggs, D. E. (1968a). *Phytochemistry* **7**, 513–529.
Briggs, D. E. (1968b). *Phytochemistry* **7**, 531–538.
Briggs, D. E. (1968c). *Phytochemistry* **7**, 539–554.
Briggs, D. E. (1972). *Planta* **108**, 351–358.
Briggs, D. E. (1973). *In* "Biosynthesis and its Control in Plants" (B. V. Milborrow, ed.), pp. 219–277. Academic Press, New York.
Briggs, D. E. and Clutterbuck, V. J. (1973). *Phytochemistry* **12**, 1047–1050.
Brown, H. T. and Escombe, F. (1898). *Proc. R. Soc. London* **63**, 3–25.
Brown, H. T. and Morris, G. H. (1890). *J. Chem. Soc.* **57**, 458–528.
Chandra, G. R. and Varner, J. E. (1965). *Biochim. Biophys. Acta* **108**, 583–592.
Chao, S. E. and Scandalios, J. G. (1969). *Biochem. Genet.* **3**, 537–547.
Chao, S. E. and Scandalios, J. G. (1971). *Genetics* **69**, 47–61.
Chao, S. E. and Scandalios, J. G. (1972). *Mol. Gen. Genet.* **115**, 1–9.
Chen, M.-C., Kao, C.-H. and Tang, W.-T. (1973). *Agron. Dept. Nat. Taiwan Un. Taipei.* **14**, 124–130.
Chen, R. and Jones, R. L. (1974a). *Planta* **119**, 193–206.
Chen, R. and Jones, R. L. (1974b). *Planta* **119**, 207–220.
Chrispeels, M. J. (1976). *Annu. Rev. Plant. Physiol.* **27**, 19–38.
Chrispeels, M. J. and Varner, J. E. (1966). *Nature* **212**, 1066–1067.
Chrispeels, M. J. and Varner, J. E. (1967). *Plant Physiol.* **42**, 1008–1016.
Clutterbuck, V. J. and Briggs, D. E. (1973). *Phytochemistry* **12**, 537–546.
Cohen, D. and Paleg, L. G. (1967). *Plant Physiol.* **42**, 1288–1296.
Coulter, P. R. and Potter, O. E. (1972). *J. Inst. Brew., London* **78**, 444–449.
Daussant, J. (1978). *Ann. Immunol. (Paris)* **129C**, 215–232.
Daussant, J. and Abbott, D. C. (1969). *J. Sci. Fd. Agric.* **20**, 633–637.
Daussant, J. and Corvazier, P. (1970). *FEBS Lett.* **7**, 191–194.
Daussant, J. and Hill, R. D. (1979). *Physiol. Plant* **45**, 255–259.
Daussant, J. and Renard, M. (1976). *Cereal Res. Commun.* **4**, 201–212.
Daussant, J., Renard, M. and Hill, R. D. (1979). *In* "Recent Advances in the Biochemistry of Cereals" (D. L. Laidman and R. G. Wyn Jones, eds.), pp. 345–348. Academic Press, London.
Dedio, W., Simmonds, D. H., Hill, R. D. and Shealy, H. (1975). *Can. J. Plant Sci.* **55**, 29–36.
Dronzek, B. L., Hwang, P. and Bushuk, W. (1972). *Cereal Chem.* **49**, 232–239.
Duffus, C. M. (1969). *Phytochemistry* **8**, 1205–1209.
Dure, L. S. (1960). *Plant Physiol.* **35**, 925–934.
Evans, M., Black, M. and Chapman, J. (1975). *Nature* **258**, 144–145.
Evers, A. D. and McDermott, E. E. (1970). *Die Stärke* **22**, 23–26.
Firn, R. D. (1975). *Planta* **125**, 227–233.

Fleming, J. R. and Johnson, J. A. (1961). *J. Agric. Food Chem.* **9**, 152–155.
Frydenberg, O. and Nielsen, G. (1965). *Hereditas* **54**, 123–139.
Gale, M. D. and Marshall, G. A. (1973). *Ann. Bot.* (London) **37**, 729–735.
Gibbons, G. C. (1979). *Carlsberg Res. Commun.* **44**, 353–366.
Gibbons, G. C. (1980). *Carlsberg Res. Commun.* **45**, 177–184.
Gibson, R. A. and Paleg, L. G. (1972). *Biochem. J.* **128**, 367–375.
Gibson, R. A. and Paleg, L. G. (1975). *Aust. J. Pl. Physiol.* **2**, 41–49.
Gibson, R. A. and Paleg, L. G. (1976). *J. Cell Sci.* **22**, 413–425.
Goldbach, H. and Michael, G. (1976). *Crop Sci.* **16**, 797–799.
Goldstein, L. D. and Jennings, P. H. (1975). *Plant Physiol.* **55**, 893–898.
Goldstein, L. D. and Jennings, P. H. (1978). *New Phytol.* **81**, 233–242.
Goodwin, P. B. and Carr, D. J. (1972). *In* "Plant Growth Substances, 1970" (D. J. Carr, ed.), pp. 371–377. Springer Verlag, Berlin.
Grabar, P. and Daussant, J. (1964). *Cereal Chem.* **41**, 523–532.
Greenwood, C. T. and MacGregor, A. W. (1965). *J. Inst. Brew., London* **71**, 405–417.
Greenwood, C. T. and Milne, E. A. (1968a). *Die Stärke* **20**, 101–107.
Greenwood, C. T. and Milne, E. A. (1968b). *Adv. Carbohydr. Chem.* **23**, 281–366.
Greenwood, C. T. and Milne, E. A. (1968c). *Die Stärke* **20**, 139–150.
Greenwood, C. T. and Thomson, J. (1959). *J. Inst. Brew., London* **65**, 346–353.
Greenwood, C. T. and Thomson, J. (1962). *J. Chem. Soc.* 222–229.
Groat, J. I. and Briggs, D. E. (1969). *Phytochemistry* **8**, 1615–1627.
Guilbot, A. and Drapron, R. (1963). *Annls. Physiol. Vég. Paris.* **5**, 5–18.
Haberlandt, G. (1914). *In* "Physiological Plant Anatomy" pp. 505–507. Translated from 4th German Edition by M. Drummond. MacMillan and Co., Ltd., London.
Hagberg, A. and Olered, R. (1975). *Hodowla Rosl. Aklim. Nasienn.* **19**, 581–592.
Hall, R. S. and Manners, D. J. (1978). *Carbohyd. Res.* **66**, 295–297.
Harvey, B. M. R. and Oaks, A. (1974). *Planta* **121**, 67–74.
Higgins, T. J. V., Zwar, J. A. and Jacobsen, J. V. (1976). *Nature* **260**, 166–169.
Higgins, T. J. V., Zwar, J. A. and Jacobsen, J. V. (1977). *In* "Acides nucléiques et synthèse des protéines chez les végétaux", Colloques Internationaux C.N.R.S. (J. H. Weil and L. Bogorad, eds.), pp. 481–486. Paris.
Hiromi, K. (1970). *Biochem. Biophys. Res. Commun.* **40**, 1–6.
Ho, D. T-H. (1979). *In* "Molecular Biology of Plants" (I. Rubenstein ed.), pp. 217–240. Academic Press, New York.
Ho, D. T-H. and Varner, J. E. (1974). *Proc. Natl. Acad. Sci. U.S.A.* **71**, 4783–4786.
Ho, D. T-H. and Varner, J. E. (1976). *Plant Physiol.* **57**, 175–178.
Iliev, S. V. (1974). *R. Acad. Bulg. Sci.* **27**, 113–116.
Ingle, J. and Hageman, R. H. (1965). *Plant Physiol.* **40**, 672–675.
Iwasa, H., Aoshima, K., Hiromi, K. and Hatano, H. (1974). *J. Biochem. (Tokyo)* **75**, 969–978.
Jacobsen, J. V. (1977). *Annu. Rev. Plant Physiol.* **28**, 537–564.
Jacobsen, J. V. and Zwar, J. A. (1974). *Aust. J. Plant Physiol.* **1**, 343–356.
Jacobsen, J. V., Scandalios, J. G. and Varner, J. E. (1970). *Plant Physiol.* **45**, 367–371.
Jacobsen, J. V., Higgins, T. J. V. and Zwar, J. A. (1979). *In* "The Plant Seed: Development, Preservation, and Germination" (I. Rubenstein, R. L. Phillips, C. E. Green and B. G. Gengerbach, eds.), pp. 241–262. Academic Press, New York.

Jeffers, H. C. and Rubenthaler, G. L. (1974). *Cereal Chem.* **51**, 772–779.
Jenkins, L. D. and Meredith, P. (1975). *N.Z. J. Sci.* **18**, 189–194.
Jenkins, L. D., Loney, D. P. and Meredith, P. (1974). *Cereal Chem.* **51**, 718–733.
Jones, R. L. (1969). *Planta* **88**, 73–86.
Jones, R. L. (1972). *Planta* **103**, 95–109.
Jones, F. T. and Bean, M. M. (1972). *Microscope* **20**, 333–340.
Jones, R. L. and Chen, R. (1976). *J. Cell Sci.* **20**, 183–198.
Jones, D. F., MacMillan, J. and Radley, M. (1963). *Phytochemistry* **2**, 307–314.
Kano, Y. (1977). *Bull. Brew. Sci.* **23**, 9–14.
Khan, A. A., Verbeek, R., Waters, E. C. and van Onckelen, H. A. (1973). *Plant Physiol.* **51**, 641–645.
King, R. W. (1976). *Planta* **132**, 43–51.
King, R. W., Salminen, S. O., Hill, R. D. and Higgins, T. J. V. (1979). *Planta* **146**, 249–255.
Kiribuchi, S. and Nakamura, M. (1973). *Denpun Kagaku* **20**, 193–200.
Kiribuchi, S. and Nakamura, M. (1974). *Denpun Kagaku* **21**, 299–306.
Kiribuchi, S. and Nakamura, M. (1975). *Protein, Nucleic Acid Enzyme* **20**, 46–59.
Klassen, A. J., Hill, R. D. and Latter, E. N. (1971). *Crop. Sci.* **11**, 265–267.
Kneen, E. (1944). *Cereal Chem.* **21**, 304–314.
Kruger, J. E. (1972a). *Cereal Chem.* **49**, 379–390.
Kruger, J. E. (1972b). *Cereal Chem.* **49**, 391–398.
LaBerge, D. E., MacGregor, A. W. and Meredith, W. O. S. (1971). *Can. J. Plant. Sci.* **51**, 469–477.
LaCroix, L. J. Waikakul, P. and Young, G. M. (1976). *Cereal. Res. Commun.* **4**, 139–146.
Lineback, D. R. and Ponpipom, S. (1977). *Die Stärke* **29**, 52–60.
Lorenz, K. and Welsh, J. R. (1976). *Lebensm.-Wiss. u.-Technol.* **9**, 7–10.
MacGregor, A. W. (1976). *Cereal Chem.* **53**, 792–796.
MacGregor, A. W. (1977). *J. Inst. Brew., London* **83**, 100–103.
MacGregor, A. W. (1978a). *J. Am. Soc. Brew. Chem.* **36**, 1–5.
MacGregor, A. W. (1978b). *Cereal Chem.* **55**, 754–765.
MacGregor, A. W. (1980). *MBAA Tech. Quart.* **17**, 215–221.
MacGregor, A. W. and Ballance, D. L. (1980). *Cereal Chem.* **57**, 397–402.
MacGregor, A. W., LaBerge, D. E. and Meredith, W. O. S. (1971a). *Cereal Chem.* **48**, 255–269.
MacGregor, A. W., LaBerge, D. E. and Meredith, W. O. S. (1971b). *Cereal Chem.* **48**, 490–498.
MacGregor, A. W., Gordon, A. G., Meredith, W. O. S. and LaCroix, L. (1972). *J. Inst. Brew., London* **78**, 174–179.
MacGregor, A. W., Thompson, R. G. and Meredith, W. O. S. (1974). *J. Inst. Brew., London* **80**, 181–187.
MacGregor, A. W., Daussant, J. and Niku-Paavola, M.-L. (1979). *J. Sci. Fd. Agric.* **30**, 1071–1076.
MacLeod, A. M. and Millar, A. S. (1962). *J. Inst. Brew., London* **68**, 322–332.
MacLeod, A. M. and Palmer, G. H. (1966). *J. Inst. Brew., London* **72**, 580–589.
MacLeod, A. M. and Palmer, G. H. (1967). *Nature* **216**, 1342–1343.
MacLeod, A. M., Duffus, J. H. and Johnston, C. S. (1964) *J. Inst. Brew., London* **70**, 521–528.
MacLeod, A. M., Duffus, J. H. and Horsfall, D. J. L. (1966). *J. Inst. Brew., London* **72**, 36–37.
McMaster, G. J. (1976). *Cereal Res. Commun.* **4**, 227–230.

McMasters, D. (1974). Ph.D. Thesis, University of Arkansas.
Maeda, I., Kiribuchi, S. and Nakamura, M. (1978a). *Agric. Biol. Chem.* **42**, 259–267.
Maeda, I., Nikuni, Z., Taniguchi, H. and Nakamura, M. (1978b). *Carbohydr. Res.* **61**, 309–320.
Manners, D. J. (1962). *Adv. Carbohydr. Chem.* **17**, 371–430.
Manners, D. J. and Bathgate, G. N. (1969). *J. Inst. Brew., London* **75**, 169–175.
Manners, D. J. and Marshall, J. J. (1971). *Carbohydr. Res.* **18**, 203–209.
Manners, D. J. and Marshall, J. J. (1972). *Die Stärke* **24**, 3–8.
Manners, D. J. and Yellowlees, D. (1973). *J. Inst. Brew., London* **79**, 377–385.
Marchylo, B. A., LaCroix, L. J. and Kruger, J. E. (1980a). *Can. J. Plant Sci.* **60**, 433–443.
Marchylo, B. A., LaCroix, L. J. and Kruger, J. E. (1980b). *Cer. Res. Commun.* **8**, 61–68.
Marchylo, B. A., LaCroix, L. J. and Kruger, J. E. (1981). *Plant Physiol.* **67**, 89–91.
Meredith, P. and Jenkins, L. D. (1973). Cereal Chem. **50**, 243–254.
Mierzwinska, T. (1977). *Acta Soc. Bot. Pol.* **46**, 69–78.
Mitchell, E. D. (1972). *Phytochemistry* **11**, 1673–1676.
Momotani, Y. and Kato, J. (1967). *Plant and Cell Physiol.* **8**, 439–445.
Momotani, Y. and Kato, J. (1972). *In* "Plant Growth Substances 1970" (D. J. Carr, ed.), pp. 352–355. Springer Verlag, Berlin.
Moro, M. S., Pomeranz, Y. and Shellenberger, J. A. (1963). *Phyton* **20**, 59–64.
Möttönen, K. (1975). *Die Stärke* **27**, 346–352.
Muntzing, A. (1963). *In* "Recent plant breeding research. Svalof, 1946–1961." pp. 167–178.
Murata, T., Akazawa, T. and Fukuchi, S. (1968). *Plant Physiol.* **43**, 1899–1905.
Myrbäck, K. and Neumüller, G. (1950). *In* "The Enzymes. Chemistry and Mechanism of Action". (J. B. Sumner, K. Myrbäck, eds.), pp. 653–724. Academic Press, New York.
Nakamura, M. (1978). *Kagaku to Seibutsu* **16**, 626–640.
Nicholls, P. B. (1979). *Aust. J. Plant Physiol.* **6**, 229–240.
Nicholls, P. B. (1980). *Aust. J. Plant Physiol.* **7**, 645–653.
Niku-Paavola, M.-L. (1977). *J. Sci. Fd. Agric.* **28**, 728–738.
Niku-Paavola, M.-L. and Heikkinen, M. (1975). *J. Sci. Fd. Agric.* **26**, 239–242.
Niku-Paavola, M.-L. and Nummi, M. (1971). *Acta Chem. Scand.* **25**, 1492–1493.
Niku-Paavola, M.-L., Nummi, M. and Skakoun, A. (1973). *Bios.* **4**, 354–356.
Nitta, Y., Mizushima, M., Hiromi, K. and Ono, S. (1971). *J. Biochem. (Tokyo)* **69**, 567–576.
Okada, S., Kitahata, S., Higashihara, M. and Fukumoto, J. (1969). *Agric. Biol. Chem.* **33**, 900–906.
Okamoto, K. and Akazawa, T. (1978). *Agric. Biol. Chem.* **42**, 1379–1384.
Okamoto, K. and Akazawa, T. (1979). *Plant Physiol.* **63**, 336–340.
Okamoto, K., Kitano, H. and Akazawa, T. (1980). *Plant and Cell Physiol.* **21**, 201–204.
Olered, R. (1967). *Vaextodling* **23**, 106.
Olered, R. (1976) *Cereal Res. Commun.* **4**, 195–199.
Olered, R. and Jönsson, G. (1968). *Getreide U. Mehl* **18**, 95–98.
Olered, R. and Jönsson, G. (1970). *J. Sci. Fd. Agric.* **21**, 385–392.
Paleg, L. G. (1960a). *Plant Physiol.* **35**, 293–299.
Paleg, L. G. (1960b). *Plant Physiol.* **35**, 902–906.

Paleg, L. G. (1970). *Proc. Inst. Brew. (Australia and New Zealand Section)* 131–139.

Palmer, G. H. (1970). *J. Inst. Brew., London* **76**, 378–380.

Palmer, G. H. (1974). *J. Inst. Brew., London* **80**, 13–30.

Palmer, G. H. (1980). *In* "Cereals for Food and Beverages. Recent Progress in Cereal Chemistry and Technology." (G. E. Inglett and L. Munck, eds.), pp. 301–338. Academic Press, New York.

Palmer, G. H. and Bathgate, G. N. (1976). *In* "Advances in Cereal Science and Technology". (Y. Pomeranz, ed.), Vol. 1. pp. 237–324. A.A.C.C. Inc., St. Paul, Minnesota, USA.

Palmiano, E. P. and Juliano, B. O. (1972). *Plant Physiol.* **49**, 751–756.

Petridis, C., Verbeek, R. and Massart, L. (1965). *J. Inst. Brew., London* **71**, 469.

Phillips, M. and Paleg, L. G. (1972). *In* "Plant Growth Substances 1970" (D. J. Carr, ed.) pp. 396–406. Springer Verlag, Berlin.

Pomeranz, Y. and Shands, H. L. (1974). *J. Food Sci.* **39**, 950–952.

Radley, M. (1967). *Planta* **75**, 164–171.

Radley, M. (1976). *J. Exp. Bot.* **27**, 1009–1021.

Rao, V. R., Mehta, S. L. and Joshi, M. G. (1976). *Phytochemistry* **15**, 893–895.

Rasper, V., Perry, G. and Duitschaever, C. L. (1974). *Can. Inst. Fd. Sci. Technol. J.* **7**, 166–174.

Rejowski, A. (1964). *Bull. Acad. Pol. Sci.* **12**, 233–236.

Roy, T., Ghose, B. and Sircar, S. M. (1973). *J. Exp. Bot.* **24**, 1064–1068.

Sandstedt. R. M. (1955). *Cereal Chem. (Supplement)* **32**, 17–47.

Sandstedt. R. M. and Beckord, O. C. (1946). *Cereal Chem.* **23**, 548–559.

Sandstedt, R. M., Blish, M. J., Mecham, D. K. and Bode, C. E. (1937). *Cereal Chem.* **14**, 17–34.

Sandstedt, R. M. and Gates, R. L. (1954). *Food Res.* **19**, 190–199.

Sandstedt, R. M. and Ueda, S. (1969). *J. Jpn. Soc. Starch Sci.* **17**, 215–228.

Sargeant, J. G. (1979). *In* "Recent Advances in the Biochemistry of Cereals" (D. L. Laidman and R. G. Wyn Jones, eds.), pp. 339–343. Academic Press, London.

Sargeant, J. G. (1980). *Cereal Res. Commun.* **8**, 77–86.

Sargeant, J. G. and Walker, T. S. (1978). *Die Stärke* **30**, 160–163.

Scandalios, J. G. (1974). *Ann. Rev. Plant Physiol.* **25**, 225–258.

Scandalios, J. G., Chao, S. E. and Melville, J. C. (1978). *J. Hered.* **69**, 149–154.

Shinke, R., Yamaguchi, T. and Nishira, H. (1978). *Sci. Rep. Fac. Agric. Kobe Univ.* **13**, 141–146.

Silvanovich, M. P. and Hill, R. D. (1976) *Anal. Biochem.* **73**, 430–433.

Silvanovich, M. P. and Hill, R. D. (1977). *Cereal Chem.* **54**, 1270–1281.

Slack, P. T., Baxter, E. D. and Wainwright, T. (1979) *J. Inst. Brew.*, **85**, 112–114.

Slack, P. T. and Wainwright, T. (1980). *J. Inst. Brew.*, **86**, 74–77.

Smith, J. B. and Bennett, M. D. (1974). *J. Sci. Fd. Agric.* **25**, 67–71.

Stoddart, J. L. (1971). *Planta* **97**, 70–82.

Tanaka, Y. and Akazawa, T. (1970). *Plant Physiol.* **46**, 586–591.

Tanaka, Y., Ito, T. and Akazawa, T. (1970). *Plant Physiol.* **46**, 650–654.

Thoma, J. A. (1976a). *Biopolymers* **15**, 729–746.

Thoma, J. A. (1976b). *Carbohydr. Res.* **48**, 85–103.

Thoma, J. A. and Allen, J. D. (1976). *Carbohyd. Res.* **48**, 105–124.

Tomos, A. D. and Laidman, D. L. (1979). *In* "Recent Advances in the Biochemistry of Cereals" (D. L. Laidman and R. G. Wyn Jones, eds.), pp. 119–146. Academic Press, London.

van Onckelen, H. A. and Verbeek, R. (1969). *Planta* **88,** 255–260.
van Onckelen, H. A., Verbeek, R. and Khan, A. A. (1974). *Plant. Physiol.* **53,** 562–568.
Varner, J. E. (1964). *Plant Physiol.* **39,** 413–415.
Varner, J. E., Flint, D. and Mitra, R. (1976). *In* "Genetic Improvement of Seed Proteins", pp. 309–328. Natl. Acad. Sci., Washington.
Verbeek, R., van Onckelen, H. A. and Gaspar, T. (1969). *Physiol. Plant.* **22,** 1192–1199.
Vigil, E. L. and Ruddat, M. (1973). *Plant. Physiol.* **51,** 549–558.
Wagenaar, S. and Lugtenborg, T. F. (1973). *Phytochemistry* **12,** 1243–1247.
Warchalewski, J. R. and Tkachuk, R. (1978). *Cereal Chem.* **55,** 146–156.
Williams, J. F. and Peterson, M. L. (1973). *Crop. Sci.* **13,** 612–615.
Wilp, R. and Buschbeck, R. (1978). *Biochem. Physiol. Pflanz.* **173,** 270–278.
Yomo, H. (1958). *Hakko Kyokaishi* **16,** 444–448.
Yomo, H. (1960). *Hakko Kyokaishi* **18,** 603–607.
Yomo, H. and Iinuma, H. (1964). *Proc. Am. Soc. Brew. Chem.* 97–102.
Yung, K-H. and Mann, J. D. (1967). *Plant Physiol.* **42,** 195–200.

CHAPTER 2

Proteinases, Peptidases, and Inhibitors of Endogenous Proteinases in Germinating Seeds

J. MIKOLA

Department of Biology, University of Jyväskylä, Jyväskylä 10, Finland

I. Introduction

In germinating seeds we can, at least conceptually, separate three stages of proteolysis. First there is an initial hydrolysis to provide free amino acids for the synthesis of the enzymatic machinery required for the conversion of the various, mainly insoluble reserve substances to forms suitable for transport. Next there is the bulk hydrolysis of the main reserve proteins to provide amino acids to the growing seedling. Finally, during the senescence of the reserve-depleted storage tissue, the cellular proteins, including the various parts of the aforementioned mobilization machinery, are broken down to provide the last ration of amino acids to the seedling before the onset of autotrophic growth. It seems that in most seeds a different proteolytic machinery, at least in the quantitative sense, functions at each of those stages.

The proteolytic enzymes and inhibitors detected in germinating seeds can be separated into seven groups. There seem to be two groups of proteinases, one already present in resting seeds and another appearing during germination. Three groups of peptidases (acid carboxypeptidases, naphthylamidases, and alkaline peptidases) are apparently present in all seeds but their roles are quantitatively different in different seeds and at different stages of germination. Then there are inhibitors of both exogenous and endogenous proteinases. According to current views the former act only as substrates in the proteolysis during germination while the latter group is an essential part of the proteolytic machinery, at least in some seeds.

The literature on the occurrence and properties of these enzymes and inhibitors is quite extensive (Ryan, 1973; Ashton, 1976; Richardson, 1977). However, for most seeds data is available only on one enzyme or one or a few groups of enzymes or inhibitors. Therefore, in this short review I will deal mainly with three species, each representing one of the main groups of higher plants: barley, a monocot; mung bean, a dicot; and Scots pine, a gymnosperm. For barley and Scots pine extensive data is available on all the seven groups of substances. For mung bean data is available on all the groups of enzymes and inhibitors except the alkaline peptidases. This lack, however, is more than compensated, by the wealth of information on the properties and functions of the main proteinase and its inhibitors provided by the group of Dr. Chrispeels during the last six years. There is, of course, a number of excellent pieces of work on other species but these will be mentioned only when they make essential additions to the general picture given by the three species selected. There is no need to describe the structures of barley grain and mung bean, but the seed of Scots pine is probably less familiar. It is a small oilseed, the decoated seed weighing only about 3 mg. There is a small embryo surrounded by a haploid, living endosperm. The protein content is about 40%, and at least the bulk of this is located in typical protein bodies both in the endosperm and the embryo (Simola, 1974). In the following review, I will first deal with the properties of each of the seven groups of enzymes and inhibitors and thereafter try to describe their obvious, probable or possible functions in the mobilization of seed proteins in the three species for which most information is available.

II. Proteinases

Before starting to describe the proteolytic enzymes present in germinating seeds, I want to make two general comments. The first concerns the terms used. In the literature the three terms, proteinase, protease, and endopeptidase, are used more or less as synonyms. Logically the term endopeptidase is the most precise, but unfortunately nature is not quite as precise, as most endopeptidases also have smaller (e.g. trypsin) or greater (e.g. subtilisin) exopeptidase activity. Therefore, in this paper I will use the term *proteinase* for all enzymes which have endopeptidase activity (ability to hydrolyse internal peptide bonds in proteins with the production of peptides) and the term *peptidase* for all enzymes which do not have endopeptidase activity but liberate C– or N–terminal amino acid residues from peptides and/or from proteins. (Dipeptidylaminopeptidases which liberate N-terminal dipeptide units from proteins have thus far not been detected in plants).

Secondly, I want to point out that , despite a great number of studies, no proteinases resembling trypsin or chymotrypsin have been found in any plant. On the other hand, most of the N-substituted amino acid esters which are frequently used as convenient, synthetic substrates for trypsin and chymotrypsin are rapidly hydrolysed by several acid plant carboxypeptidases (Mikola and Pietilä, 1972; Matoba and Doi, 1975). This coincidence has produced some confusion in the literature on the occurrence of proteinase activity in plant extracts.

A. PROTEINASES PRESENT IN RESTING SEEDS

Little proteinase activity has been detected in most resting seeds. The pH optima are acid, and the activities tend to be higher in the embryonic axis than in storage tissues. However, the activities are inconveniently low, and little information has been available on the enzymes responsible for the activities. Only the enzyme of hempseeds has been extensively purified (St. Angelo et al., 1969a,b; St. Angelo and Ory, 1970); unfortunately, studies on this most interesting proteinase have not been pursued further (apparently because of a good intention to protect mankind from hemp). Quite recently Dr. Doi working on rice in Kyoto and Dr. Salmia working on Scots pine at Helsinki made a series of highly interesting observations which may mean a small breakthrough in this field. Resting seeds of both plants contain acid proteinase activity which is largely inhibited by pepstatin A, a highly specific inhibitor of acid proteinases with a pepsin-like mechanism of catalytic action (Doi et al., 1980a,b; Salmia, 1981a). During germination the pepstatin sensitive activity remains unchanged despite great increases in total acid proteinase activity (Doi, private communications; Salmia, 1981b). In Scots pine the pepstatin sensitive enzyme is not affected by the inhibitors of endogenous proteinases which act on the enzymes which appear during germination.

So, there is an obvious candidate for the role of the initiator proteinase. No direct information on the intracellular localization of the enzyme is available, but the acid pH optimum suggests localization in some kind of protein bodies or vacuoles. Anyway, these results are fairly new, and they immediately raise two further questions: Do corresponding enzymes occur in resting dicot seeds, particularly in the axial tissues, where some reserve albumins are broken down very early during germination (Manickam and Carlier, 1980). Secondly, is the enzyme active already during dormancy, and if so, what mechanism does prevent its action before germination?

B. PROTEINASES APPEARING DURING GERMINATION

The pepstatin-sensitive proteinases do not increase in activity during germination. So which proteinases account for the high increases of proteinase activity during the period of rapid reserve protein degradation? Here the activities are really high, the enzymes from several seeds resemble each other, and the main enzyme in mung beans has been isolated and extensively characterized (Baumgartner and Chrispeels, 1977). The main properties of these enzymes are listed in Table 1.

TABLE I

Some general properties of proteinases appearing in storage tissues of seeds during germination[a]

pH optima at 3·5 to 5·5 depending on enzyme, substrate and buffer
–SH enzymes inhibited by pHMB and NEMI but not affected by (a) DFP or PMSF, (b) EDTA or o-phenanthroline, or (c) pepstatin A
Molecular weights in the range 20 000 to 40 000
Broad specificity: haemoglobin, casein, gelatin, endogenous reserve proteins
Inhibited by the inhibitors of endogenous proteinases present in resting and germinating seeds (barley, mung bean, pine)
Synthetized de novo during germination (barley, maize, mung bean; indirect evidence for several other species)
Localized in protein bodies/vacuoles
Activities (units per mg fresh or dry weight) much higher than corresponding activities in vegetative tissues

[a] Selected references: Sundblom and Mikola, 1972; Harvey and Oaks, 1974a, b; Chrispeels et al., 1976; Abe et al., 1977; Baumgartner and Chrispeels, 1977; Tully and Beevers, 1978; Salmia, 1981a,b.

The enzymes are acid –SH proteinases with relatively small molecular weights. Although direct evidence is available only for a few species, a wealth of indirect evidence suggests that these enzymes are synthesized de novo during germination in all seeds. In barley these enzymes are located mainly in the starchy endosperm where the internal pH is close to pH 5. In the cotyledons of germinating mung beans the enzyme occurs in protein bodies where the internal pH probably is also near 5. In some vegetative tissues, as in the mesophyll cells of wheat leaves corresponding acid –SH proteinases occur in the central vacuoles (Waters *et al.*, 1982).

In Scots pine (Salmia, 1981b) and mung bean these enzymes cannot be properly assayed at early stages of germination because they are strongly inhibited by the endogenous proteinase inhibitors. As will be discussed in detail later, *in vivo* the enzymes and inhibitors apparently occur in different compartments. On homogenization this compartmentation is broken down and the probable high *in vivo* activities are masked by the inhibitors *in vitro*. This may be more generally the reason why in several germinating seeds the rapid degradation of reserve proteins begins before proteinase activities measured from extracts begin to increase rapidly.

Knowledge on the specificity of these enzymes is scanty and indirect. They certainly have endopeptidase activity as for instance when mung bean vicilin is incubated at pH 5·1 with the pure mung bean proteinase, the first reaction products are large peptides which are progressively broken down to smaller peptides (Baumgartner and Chrispeels, 1977). Secondly, their specificity seems quite broad, as they are able to digest, in addition to the reserve proteins, several exogenous proteins.

Recently the group of Harry Beevers has reported a few remarkable observations concerning the acid –SH proteinase present in germinating castor beans.

First, when the central vacuoles, which are formed through the stepwise fusion of autolysing protein bodies, were isolated at the last stage of reserve mobilization the acid proteinase was located in the vacuoles (Nishimura and Beevers, 1978). Second, the enzyme was found to inactivate several cytoplasmic enzymes during extraction and storage near pH 7 (Alpi and Beevers, 1981). This finding has three implications. It shows that the enzyme has a broad specificity as it is able to digest several native proteins. As a consequence it suggests that the enzyme is ideally suited to the senescence related degradation of cellular proteins. Finally, it shows that the enzyme is really dangerous to the cytoplasm. Therefore, the cytoplasm has to be protected from the enzyme either by strict compartmentation or by inhibitors present in cytosol.

Third, the enzyme was found to be inhibited by another highly specific proteinase inhibitor, leupeptin (Alpi and Beevers, 1981). In rice (Doi *et al.*, 1980a) and barley (Sopanen, unpublished) the small proteinase activity present in resting seeds is not affected by leupeptin, but leupeptin-sensitive, acid, –SH proteinases appear during germination. So there is some obvious promise that leupeptin will prove a useful marker inhibitor of the corresponding proteinases also in other seeds.

III. INHIBITORS OF ENDOGENOUS PROTEINASES

Most seeds contain high inhibitor activity against trypsin, chymotrypsin, and microbial alkaline proteinases (Richardson, 1977). Several dozens of these inhibitors have been obtained in pure form and extensively characterized. The inhibitors are usually relatively stable, small-molecular proteins, and they form stoichiometric, inactive complexes with the respective proteinases. During the early years it was frequently suggested that the physiological function of these inhibitors was to prevent the action of pre-formed seed proteinases during the resting state. However, in every case a test was made with a pure inhibitor, it had no effect on the endogenous proteinase activity appearing during germination. By now, the best hypothesis seems to be that these inhibitors instead of a physiological role have an ecological role protecting the seeds against some predators: animals, insects or microbes.

However, when carefully tested several resting seeds have been found to contain other proteins which inhibit the acid proteinases present in extracts of germinating seeds. For instance the barley grain contains relatively high amounts of two trypsin inhibitors (Mikola and Suolinna, 1969; Kirsi and Mikola, 1971; Mikola and Kirsi, 1972), two groups of inhibitors affecting chymotrypsin and microbial alkaline proteinases (Mikola and Suolinna, 1971; Boisen et al., 1981), and a more specific subtilisin inhibitor (Yoshikawa et al., 1976). None of these is active on endogenous proteinases. However, in the resting grains there is also small inhibitor activity against the proteinase activity present in germinated grains (Kirsi and Mikola, 1971). The behaviour of the inhibitors of endogenous proteinases during germination is shown in Fig. 1. There is little proteinase activity in the resting seed, mainly due to the pepstatin-sensitive enzyme(s) (Sopanen, unpublished), and an about 20-fold increase due to –SH proteinases, occurs during germination. The resting seeds contain small inhibitor activity against the –SH enzymes, and this activity disappears before the rapid increase of proteinase activity. However, the inhibitor activity in resting seeds corresponds only to about 5 to 10% of the proteinase activity in germinating seeds. Consequently, the amount of inhibitors is far too small to account for the high increase of proteinase activity during germination. Moreover, it has been known for some time (Jacobsen and Varner, 1967) that the barley proteinases are synthesized de novo during germination. On the other hand, the inhibitors are abundant enough to protect the tissues from the small proteinase activity present during the resting stage. The inhibitors acting on trypsin and microbial serine proteinases (and chymotrypsin) show a different behaviour (Fig. 2). The activities are very high in resting seeds, definite decreases coincide with the increase of proteinase activity, but high inhibitor activities are still present

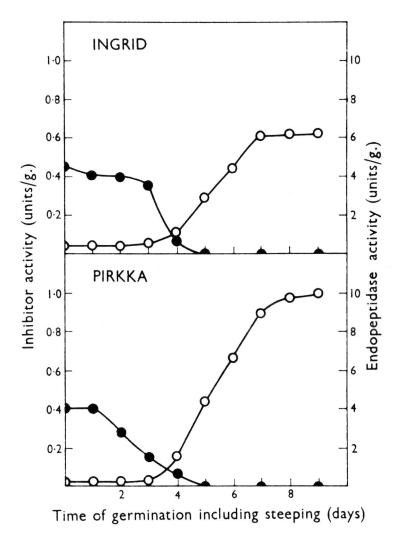

Fig. 1. Changes in the activities of proteinases (○) and inhibitors of endogenous proteinases (●) during germination in two cultivars of barley (Mikola and Enari, 1970). Note the 10-fold difference between the two activity scales.

after the proteinase activity has reached the maximal level. This seems a further demonstration that these inhibitors do not control or affect the proteinase activity.

For mung bean really nice data has been obtained by Chrispeels and Baumgartner. The resting seeds contain inhibitor activity against both trypsin and the major acid–SH proteinase appearing during germination. The two inhibitors have been separated; one inhibits trypsin (Chrispeels

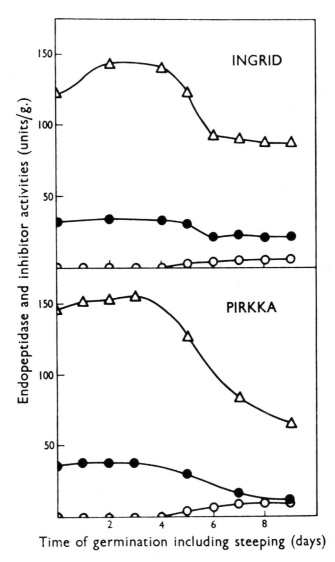

FIG. 2. Changes in the activities of proteinases (○), trypsin inhibitors (●), and inhibitors of the alkaline proteinase of *Aspergillus oryzae* (△) during germination in two cultivars of barley (Mikola and Enari, 1970).

and Baumgartner, 1978) and the other the mung bean proteinase (Baumgartner and Chrispeels, 1976). The latter inhibitor appears to be localized in the cytosol whereas the main proteinase is in the protein bodies safe from the inhibitor (Baumgartner and Chrispeels, 1977). Here the function of the inhibitor seems obvious. It will protect the cytoplasm from the very high and unspecific proteinase activity present in protein bodies turning later to vacuoles.

The situation in pineseeds is even simpler. These primitive seeds do not contain inhibitor activity against exogenous proteinases but the resting seeds have high inhibitor activity against the two acid –SH proteinases appearing during germination (Salmia and Mikola, 1980; Salmia, 1981a). However, the inhibitors do not affect the pepstatin-sensitive acid proteinase which is present already in resting seeds. The inhibitor activities are due to at least four inhibitors (Salmia, 1980). During germination the inhibitors disappear not before but rather in parallel with the transfer of total nitrogen from the endosperm to the seedling (Salmia, 1980). In extracts of germinating seeds the inhibitor activities are rapidly destroyed by some enzymatic reactions (Fig. 3). These decreases are much more rapid in the extracts than *in vivo*; this suggests that in the germinating seed the inhibitors and the inactivating enzyme(s) occur in different compartments. At the same time the proteinase activities increase spontaneously; the increases tend to be higher than the corresponding decreases of inhibitor activity expecially in the assays made at pH 7. However, the inactivation of inhibitors seems to be the major (and probably only) cause of the increases of the proteinase activities. Data on the intracellular localization of the enzymes and inhibitors is not available, but all the indirect evidence suggest a similar compartmentation and function as in mung beans.

The operation of a corresponding system in yeast cells has been convincingly demonstrated by the group of Dr. Holzer in Freiburg (Matern *et al.* 1974; Matern *et al.*, 1979; Müller and Holzer, 1980). The vacuoles of the cells contain two acid proteinases called A and B, and the cytosol contains specific inhibitors of the both proteinases. In extracts of whole cells the proteinases are present in excess, and proteinase A inactivates the inhibitors of proteinase B while proteinase B inactivates the inhibitor of proteinase A. As a consequence, the proteinase activities of fresh extracts increase during storage. The systems present in mung bean and Scots pine accordingly resemble the yeast system, which has been much more extensively characterized. However, a possibility of homology seems excluded as yeast proteinase A is inhibited by pepstatin and proteinase B is a serine proteinase while all the plant proteinases inhibited by the endogenous inhibitors are –SH proteinases.

IV. Peptidases

The main properties of the three groups of plant peptidases mentioned before have been compiled on Table II and will be discussed below.

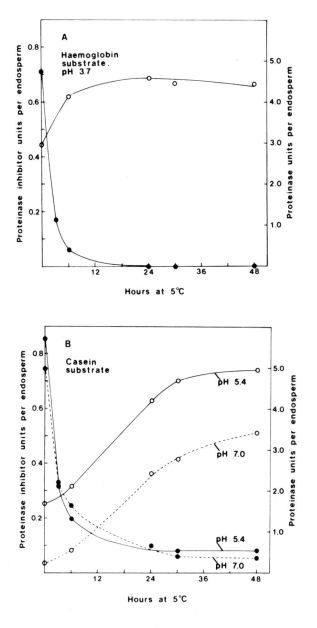

FIG. 3. Changes in the activities of pineseed proteinases (○) and their inhibitors (●) during dialysis of an extract of "germinating" endosperm at 5°C (Salmia and Mikola, 1980). Note the difference between the activity scales.

TABLE II
Main groups of peptidases detected in plant tissues

	Acid carboxy-peptidases[a]	Naphthyl-amidases[b]	Alkaline peptidases[c]
Number of enzymes in different tissues	1–5	2–4	2–3
Marker substrates	Z-Phe-Ala	Leu-β-NA	Leu-Tyr
	Z-Phe-Phe	Arg-β-NA	Leu-Gly-Gly
	Z-Ala-Arg	Phe-β-NA	Leu-NH$_2$
	Z-Gly-Pro-Ala	Pro-β-NA	Ala-Gly
		Leu-p-nitroanilide	Ala-Pro
pH optima	4–6	6–8	8–10
Inhibitors	DFP (PMSF)	pHMB	pHMB
		phenanthroline	phenanthroline
Natural substrates	Oligopeptides, proteins	Dipeptides, oligopeptides	Dipeptides, oligopeptides
Intracellular localization	Vacuoles, protein bodies	Cytosol	Cytosol?

[a] Selected references: Matoba and Doi, 1975; Salmia and Mikola, 1976a; Preston and Kruger, 1976, 1977; Mikola and Mikola, 1980; Mikola and Mikola, 1981.
[b] Selected references: Kolehmainen and Mikola, 1971; Elleman, 1974; Hejgaard and Bøg-Hansen, 1974.
[c] Selected references: Sopanen and Mikola, 1975; Salmia and Mikola, 1975; Sopanen, 1976.

A. ACID CARBOXYPEPTIDASES

The acid carboxypeptidases are characterized by "lysosomal" pH optima at 4 to 6 and complete inactivation by diisopropylfluorophosphate (DFP), a marker inhibitor not affecting any of the other well known plant peptidases or proteinases. In two vegetative tissues, the giant cells of the alga *Nitella* (Doi *et al.*, 1975) and the mesophyll cells of wheat leaves (Waters *et al.*, 1982) acid carboxypeptidase activity is localized exclusively in the central vacuoles. Moreover, in the endosperm of germinating castor bean acid carboxypeptidase activity is again found in the central vacuoles formed by fusion of protein bodies (Nishimura and Beevers, 1978). In yeast cells a corresponding acid carboxypeptidase is localized in vacuoles (Matern *et al.*, 1974), while in mammalian tissues a closely similar enzyme, called cathepsin A (Kawamura *et al.*, 1975), occurs in lysosomes (McDonald and Schwabe, 1977). So, the evidence seems overwhelming that these enzymes are lysosomal carboxypeptidases in nature.

B. NAPHTHYLAMIDASES

The distinguishing feature of the naphthylamidases (also called arylami-dases and neutral peptidases) is their ability to hydrolyse peptide bonds between amino acids and aromatic amines (β-naphthylamine, p-nitroaniline, 4-(phenylazo)-phenylamine). In the few cases when the enzymes have been sufficiently purified they have been shown to act also on natural di- and oligopeptides (Kolehmainen and Mikola, 1971; Elleman, 1974). In mesophyll cells of wheat leaves these enzymes are localized partly in the cytosol and partly in chloroplasts (Waters *et al.*, 1982). These enzymes are present in all resting seeds and they are very easy both to assay and to stain on gels. Therefore, they are by far the most studied seed peptidases. However, the activities in most seeds are relatively small, they do not increase during germination, and corresponding activities are generally higher in actively growing tissues.

C. ALKALINE PEPTIDASES

We have used the term alkaline peptidases for a group of three amino- or dipeptidases which (a) are abundant in germinating barley, (b) have no or very low activity on the marker substrates of the naphthylamidases, and (c) have pH optima in the range 7·5 to 10. At least two of these enzymes are also abundant in resting and germinating seeds of Scots pine (Salmia and Mikola, 1975), peanut (Mikola, 1976), and kidney bean (Möttönen and Mikola, unpublished). These enzymes show high activities also in all vegetative plant tissues. Unfortunately, their assays are not so convenient as those of the other peptidases, and very few people have included them in studies on seed peptidases.

Two of these enzymes have been purified and characterized from germinating barley (Sopanen and Mikola, 1975; Sopanen, 1976). The dipeptide, Leu-Tyr, is a good marker substrate for the first enzyme as it is hydrolysed very slowly or not at all by the other peptidases at high pH. The enzyme is strikingly similar to the well known mammalian leucine aminopeptidases isolated for instance from pig kidney and bovine ocular lens in respect to substrate specificity, effects of inhibitors, molecular size and unusual stability. Therefore we have called it barley leucine aminopeptidase although there is a danger of confusion, as several authors use this name for naphthylamidases acting on Leu-β-naphthylamine or Leu-p-nitroanilide. The intracellular localization of this enzyme has been studied in wheat leaves (Waters *et al.*, 1982); the enzyme seems to be localized mainly in the cytosol but partly in chloroplasts.

Another dipeptide, Ala-Gly, seems an equally good marker substrate for the other enzyme (Sopanen, 1976). This is a specific dipeptidase which again is closely similar to mammalian dipeptidases.

The third enzyme, also abundant in barley, can be assayed using Ala-Pro as a marker substrate. The enzyme is different from all the peptidases enumerated above (L. Mikola, unpublished) but it has not yet been purified.

One factor which has apparently contributed to the lack of interest on these enzymes in plants is their high pH optima; where could such enzymes function? However, the enzymes are abundant in all living plant tissues studied, at least the pine enzymes have considerable activity also at pH 7, and the pH dependence *in vivo* may be different from that found in extracts.

V. MOBILIZATION OF SEED PROTEINS DURING GERMINATION

A. BARLEY

The sequence of events in the mobilization of seed proteins is in some respects easier to study in cereal grains than in most other seeds because the various processes normally taking place in different organelles of single cells are partly divided between three different tissues: enzyme synthesis and secretion in the aleurone layer (and partly in the scutellum), storage and hydrolysis in the starchy endosperm, and uptake and metabolism of the hydrolysis products in the scutellum (for a detailed review see Mikola, 1981).

The starchy endosperm is a non-living tissue and apparently there is no compartmentation left during germination. It shows high activities of various hydrolytic enzymes with pH optima at 4 to 6. These include at least two acid –SH proteinases (Sundblom and Mikola, 1972) and five acid carboxypeptidases (Mikola and Mikola, 1980). The alkaline peptidases and naphthylamidases, on the contrary, are lacking completely (Mikola and Kolehmainen, 1972). The internal pH stays close to pH 5 apparently due to secretion of malic acid by the aleurone layer (Mikola and Virtanen, 1980). Accordingly, the whole tissue corresponds to a huge secondary lysosome: it contains a large battery of acid hydrolases, the pH is about 5, and its only function is complete hydrolysis of its contents to provide building block molecules for use in other tissues.

The insoluble reserve proteins are hydrolysed to soluble peptides by the proteinases, and free amino acids are formed by the action of the carboxypeptidases on these peptides. However, the proteinases and carboxypeptidases are not able to hydrolyse dipeptides and probably some types of tripeptides. Therefore, the product of the hydrolysis is apparently a mixture of free amino acids and small peptides. The scutellum, however, possesses active uptake systems both for amino acids (Sopanen *et al.*, 1980) and for small peptides (Sopanen, 1979, and references cited therein). Moreover, it contains high activities of the alkaline peptidases and moderate activities of the naphthylamidases (Mikola and Kolehmainen, 1972). The peptides taken up by the scutellum are immediately hydrolysed probably by the alkaline and neutral peptidases acting in the cytosol. Accordingly, the hydrolysis of the reserve proteins present in the starchy endosperm occurs definitely in two stages. First there is an acid ("lysosomal") initial hydrolysis where the insoluble reserve proteins are degraded *in situ* to a mixture of amino acids and small peptides by acid proteinases and carboxypeptidases. Then the peptides pass to the scutellum where they are hydrolysed to amino acids by neutral and/or alkaline peptidases probably in the cytosol.

The aleurone layer of the barley grain is a highly interesting tissue as there the abundant reserve proteins are packed in typical protein bodies (Jones, 1969) and relatively high activities of all the three groups of peptidases are present (Mikola and Kolehmainen, 1972). It is tempting to speculate that a similar two-stage hydrolysis occurs there within single cells; the acid proteinases and carboxypeptidases acting inside the protein bodies turning into vacuoles could produce a mixture of amino acids and peptides, and the peptides could pass to the cytosol where the hydrolysis could be completed by the neutral and alkaline peptidases.

B. MUNG BEAN

The mung bean is a typical, relatively small starchy legume seed. The bulk of the reserve proteins occurs in protein bodies within the cotyledonary cells but there are protein bodies also in the axis (Manickam and Carlier, 1980). The resting seeds contain small acid proteinase activity (Chrispeels and Boulter, 1975) but it has not been characterized. During germination the acid proteinase activity increases dramatically due to *de novo* synthesis of an acid –SH proteinase which is not present in resting seeds (Baumgartner and Chrispeels, 1977). The enzyme is apparently synthetized on rough endoplasmic reticulum and transported within small vesicles to protein bodies. The enzyme seems to act only within the protein bodies turning into vacuoles (Van der Wilden *et al.*, 1980, and references cited therein) while its specific inhibitor is in the cytosol (Baumgartner and Chrispeels, 1976). The protein bodies contain acid carboxypeptidase activity already during the resting stage and the activity increases especially during the senescence of the cotyledons (Chrispeels and Boulter, 1975). The protein bodies do not contain naphthylamidase activity although it is relatively abundant in the cytosol (Chrispeels and Boulter, 1975; Van der Wilden *et al.*, 1980). Data on the alkaline peptidases is not available. However, they probably are there as the corresponding enzymes are abundant in the cotyledons of peanuts (Mikola, 1976) and kidney beans

(Möttönen and Mikola, unpublished). As a whole there is obvious similarity between the starchy endosperm of barley grain and the protein body in mung bean cotyledon. Accordingly, a two-stage proteolysis is again implicated: an acid initial hydrolysis in the protein bodies and final hydrolysis of peptides in the cytosol.

C. SCOTS PINE

The seed of the Scots pine is, thanks to the comprehensive Ph.D. Thesis of Mrs. Salmia (1981c), the only seed for which detailed data on the occurrence of all groups of proteolytic enzymes and inhibitors has been gathered in same conditions of germination. The results are summarized in Fig. 4. The endosperm in the resting seed contains small acid proteinase activity mainly due to the pepstatin-sensitive enzyme. During germination the activity increases, and this increase is entirely due to two acid –SH proteinases (Salmia 1981a,b). The "resting" endosperm also contains high activities of both the alkaline peptidases (Salmia and Mikola, 1975) and naphthylamidases (Salmia and Mikola, 1976b). The alkaline peptidases increase during the period of rapid reserve protein mobilization, and then both activities decrease again during the senescence of the endosperm. Carboxypeptidase activity, on the contrary, is very small in resting seeds, increases slowly during rapid mobilization, but shows a great increase during senescence (Salmia and Mikola, 1976a).

No data is available on the intracellular localization of the various enzymes, but the widely different pH optima again suggest at least two different compartments. Then following the previous line of interpretation, during the initial stage some reserve proteins would be hydrolysed to soluble peptides within protein bodies or some other acid compartment by the pepstatin-sensitive proteinase. Then the soluble peptides would pass to the cytosol where rapid hydrolysis to amino acids would be effected by the alkaline and neutral peptidases. During the main period of reserve mobilization *de novo* synthesis of the acid –SH proteinases and their transport to the protein bodies would accelerate the hydrolysis. During the last stage the vacuoles would retain the proteinases, gain carboxypeptidases, and the bulk of the amino acids would be produced in the vacuoles. At the same time the reserve protein would be depleted, and the main substrate of the proteinases would be the cellular proteins including the aminopeptidases and proteinase inhibitors which disappear at this stage.

Addendum

In discussions after the talk Dr. Jørn Hejgaard wondered whether the inhibitors of endogenous proteinases might be identical or related to the

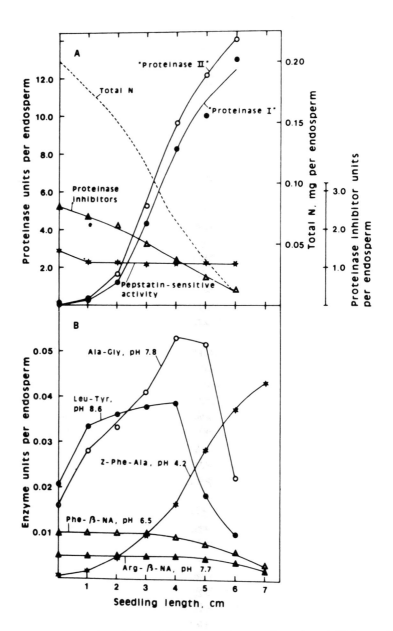

FIG. 4. (A) Changes in the activities of different proteinases and proteinase inhibitors and (B) Activities on the marker substrates of different peptidases in the endosperm of Scots pine during germination (Salmia, 1981c).

papain inhibitors detected in several seeds (e.g. Abe *et al.*, (1980) *Agric. Biol. Chem.* **44**, 685–686; Vartak *et al.*, (1980) *Arch. Biochem. Biophys.* **204**, 134–140); this seems possible and is certainly worthy of investigation. Secondly, Dr. Willem Van der Wilden raised the possibility that the alkaline peptidases with their "strange" pH optima could be located in cell walls; this is another point which should be clarified.

REFERENCES

Abe, M., Arai, S. and Fujimaki, M. (1977). *Agric. Biol. Chem.* **41**, 839–899.
Alpi, A, and Beevers, H. (1981). *Plant Physiol.* **67**, 499–502.
Ashton, F. M. (1976). *Annu. Rev. Plant Physiol.* **27**, 95–117.
Baumgartner, B. and Chrispeels, M. J. (1976). *Plant Physiol.* **58**, 1–6.
Baumgartner, B. and Chrispeels, M. J. (1977). *Eur. J. Biochem.* **77**, 223–233.
Boisen, S., Andersen, C. Y. and Hejgaard, J. (1981). *Physiol. Plant.* **52**, 167–176.
Chrispeels, M. J. and Baumgartner, B. (1978). *Plant Physiol.* **61**, 617–623.
Chrispeels, M. J. and Boulter, D. (1975). *Plant Physiol.* **55**, 1031–1037.
Chrispeels, M. J., Baumgartner, B. and Harris, N. (1976). *Proc. Natl. Acad. Sci. USA.* **73**, 3168–3172.
Doi, E., Ohtsuru, C. and Matoba, T. (1975). *Plant Sci. Lett.* **4**, 243–247.
Doi, E., Shibata, D., Matoba, T. and Yonezava, D. (1980a). *Agric. Biol. Chem.* **44**, 435–436.
Doi, E., Shibata, D., Matoba, T. and Yonezava, D. (1980b). *Agric. Biol. Chem.* **44**, 741–747.
Elleman, T. C. (1974). *Biochem. J.* **141**, 113–118.
Harvey, B. R. M and Oaks, A. (1974a). *Plant Physiol.* **54**, 449–452.
Harvey, B. R. M. and Oaks, A. (1974b). *Plant Physiol.* **54**, 453–457.
Hejgaard, J. and Bøg-Hansen, T. C. (1974). *J. Inst. Brew, London* **80**, 436–442.
Jacobsen, J. V. and Varner, J. E. (1967). *Plant Physiol.* **42**, 1596–1600.
Jones, R. L. (1969). *Planta* **85**, 359–375.
Kawamura, Y., Matoba, T., Hata, T., and Doi, E. (1975). *J. Biochem.* **77**, 729–737.
Kirsi, M. and Mikola, J. (1971). *Planta (Berl.)* **96**, 281–291.
Kolehmainen, L. and Mikola, J. (1971). *Arch. Biochem. Biophys.* **145**, 633–642.
Manickam, A. and Carlier, A. R. (1980). *Planta* **149**, 234–240.
Matern, H., Betz, H. and Holzer, H. (1974). *Biochem. Biophys. Res. Commun.* **60**, 1051–1057.
Matern, H., Weiser, U. and Holzer, H. (1979), *Eur. J. Biochem.* **101**, 325–332.
Matoba, T. and Doi, E. (1975). *J. Biochem.* **77**, 1297–1303.
McDonald, J. K. and Schwabe, C. (1977). *In* "Proteinases in Mammalian Cells and Tissues" (A. J. Barrett, ed.), pp. 329–335. North-Holland Publishing Co., Amsterdam.
Mikola, J. (1976). *Physiol. Plant.* **36**, 255–258.
Mikola, J. (1981). *Abhdlg. Akad. Wiss. DDR, Abt. Math., Naturwiss., Techn.* **5N**, 153, 161.
Mikola, J., Enari, T-M. (1970). *J. Inst. Brew.* **76**, 182–188.
Mikola, J. and Kirsi, M. (1972). *Acta Chem. Scand.* **26**, 787–795.
Mikola, J. and Kolehmainen, L. (1972). *Planta (Berlin)* **104**, 167–177.

Mikola, L. and Mikola, J. (1980). *Planta* **149**, 149–154.
Mikola, L. and Mikola, J. (1981). *Plant Physiol.* **67**, Supplement, abstract 84.
Mikola, J. and Pietilä, K. (1972). *Phytochemistry* **11**, 2977–2980.
Mikola, J. and Suolinna, E-M. (1969). *Eur. J. Biochem.* 555–560.
Mikola, J. and Suolinna, E-M. (1971). *Arch. Biochem. Biophys.* **144**, 567–575.
Mikola, J. and Virtanen, M. (1980). *Plant Physiol.* 65, supplement, abstract 783.
Müller, M. and Holzer, H. (1980). *In* "Enzyme Regulation and Mechanism of Action" (P. Mildner and B. Ries, eds), pp. 339–349. Pergamon Press, Oxford and New York.
Nishimura, M. and Beevers, H. (1978). *Plant Physiol.* **62**, 44–48.
Preston, K. R. and Kruger, J. E. (1976). *Plant Physiol.* **58**, 516–520.
Preston, K. and Kruger, J. (1977). *Phytochemistry* **16**, 525–528.
Richardson, M. (1977). *Phytochemistry* **16**, 159–169.
Ryan, C. A. (1973). *Annu. Rev. Plant. Physiol.* **24**, 173–196.
Salmia, M. A. (1980). *Physiol. Plant.* **48**, 266–270.
Salmia, M. A. (1981a). *Physiol. Plant.* **51**, 253–258.
Salmia, M. A. (1981b). *Physiol. Plant.* **53**, 39–47.
Salmia, M. A. (1981c). *Publ. Dept. Bot. Univ. Helsinki* **7**, 1–33.
Salmia, M. A. and Mikola, J. (1975). *Physiol. Plant.* **33**, 261–265.
Salmia, M. A. and Mikola, J. (1976a). *Physiol. Plant.* **36**, 388–392.
Salmia, M. A. and Mikola, J. (1976b). *Physiol. Plant.* **38**, 73–77.
Salmia, M. A. and Mikola, J. (1980).*Physiol. Plant.* **48**, 126–130.
Simola, L. K. (1974). *Acta Bot. Fenn.* **103**, 1–31.
Sopanen, T. (1976). *Plant Physiol.* **57**, 867–871.
Sopanen, T. (1979). *Plant Physiol.* **64**, 570–574.
Sopanen, T. and Mikola, J. (1975). *Plant Physiol.* **55**, 809–814.
Sopanen, T., Uuskallio, M., Nyman, S. and Mikola, J. (1980). *Plant Physiol.* **65**, 249–253.
St. Angelo, A. J. and Ory, R. L. (1970). *Phytochemistry* **9**, 1933–1938.
St. Angelo, A. J., Ory, R. L. and Hansen, H. J. (1969a). *Phytochemistry* **8**, 1135–1138.
St. Angelo, A. J., Ory, R. L. and Hansen, H. J. (1969b). *Phytochemistry* **8**, 1873–1877.
Sundblom, N-O. and Mikola, J. (1972). *Physiol. Plant.* **27**, 281–284.
Tully, R. E. and Beevers, H. (1978). *Plant Physiol.* **62**, 746–750.
Van der Wilden, W., Herman, E. M. and Chrispeels, M. J. (1980). *Proc. Natl. Acad. Sci. USA.* **77**, 428–432.
Waters, S. P., Noble, E. R. and Dalling, M. J. (1982). *Plant Physiol.* **69**, 575–579.
Yoshikawa, M., Iwasaki, T., Fujii, M, and Oogaki, M. (1976). *J. Biochem.* **79**, 765–773.

CHAPTER 3

Seed Lectins: Distribution, Location and Biological Role

A. PUSZTAI, R. R. D. CROY[*], G. GRANT AND J. C. STEWART

THE ROWETT RESEARCH INSTITUTE, BUCKSBURN, ABERDEEN, SCOTLAND

I. INTRODUCTION

The general occurrence in plants of proteins which agglutinate human or animal red cells has been well established since Stillmark's original observation of the haemagglutination caused by extracts of the toxic *Ricinus* seeds (Stillmark, 1888). The results obtained by Stillmark already indicated some selectivity in the agglutination of red cells from different animals and this observation was corroborated and further extended by Landsteiner and Raubitschek (1908). A number of these phytohaemagglutinins were later also shown to be specific in their reactions with red cells of human blood groups ABO and MN (Renkonen, 1948; Boyd and Reguera, 1949). Thus, to call attention to their specificity these blood antigen specific agglutinins and precipitins were called lectins (Boyd and Shapleigh, 1954).

It was noticed early (Sumner and Howell, 1936) that agglutination by concanavalin A was inhibited by cane sugar. However, it was Watkins and Morgan (1952) who first observed that simple sugars were capable of neutralizing normal agglutinins in a specific manner. These observations

---

[*] Botany Department, University of Durham, Durham, England.

are now well established and this sugar specificity serves as the basis for the most general definition of lectins (Kocourek and Hořejši, 1981) as sugar-binding proteins or glycoproteins of non-immune origin which are devoid of enzymatic activity towards the sugars to which they bind. However, the cell-wall-bound and generally occurring so-called β-lectins (Jermyn and Yeow, 1975), which have been reviewed recently, will not be discussed here.

Since these original discoveries and over the years, lectins have been shown to exhibit a great number of unusual chemical and biological properties. Thus some lectins are mitogenic for lymphocytes (Nowell, 1960). Lectins also precipitate polysaccharides and glycoproteins in a specific manner, a property which is being increasingly used for lectin purification by affinity chromatography. Lectins also appear to have some selectivity for transformed cells (Aub *et al.*, 1963) and thus are useful reagents for cell fractionation and for studies of membrane topography (Sharon and Lis, 1972; Lis and Sharon, 1977). Some lectins are also toxic for animals. The poor nutritive value of some legume seed proteins is frequently attributed to their high lectin content (Jaffè, 1980; Pusztai, 1980). The great volume of work done in the past ten years has been directed mainly towards studies of these and other interesting properties of lectins (Goldstein and Hayes, 1978). Progress in these topics, particularly those related to cells and membranes from animals has been very substantial and sometimes spectacular. Progress however has been much slower in elucidating the biological role and the synthesis by plant tissues of most of the lectins described up to date. Therefore in the following chapter a short survey of the distribution and biological properties of lectins in the plants, where they naturally occur, will be given. The starting point for this work is the excellent review by Callow (1975).

II. Occurrence in Flowering Plants

Since the discovery of non-toxic and agglutinating substances in seeds of bean, pea, lentil and vetch by Landsteiner and Raubitschek (1908) extracts from the seeds of a great number of higher plant species have been surveyed for the presence of agglutinins. This interest was greatly and further stimulated by the first findings of blood group specific agglutinins in lima beans (Boyd and Reguera, 1949) and in other species of Leguminosae (Renkonen, 1948). Full details of these and later investigations are described in the many excellent reviews published since (Krüpe, 1956; Mäkela, 1957; Saint Paul, 1961; Boyd, 1963; Tobiška, 1964; Boyd, 1970; Toms and Western, 1971; Sharon and Lis, 1972; Gold and Balding, 1975; Goldstein and Hayes, 1978; Lis and Sharon, 1981).

Of all the plants examined, generally those within the Leguminosae were the richest source of agglutinins (Renkonen, 1948; Mäkela, 1957). According to Toms and Western (1971) about 95% of the lectins reported since 1948 were from the seeds of legumes. The available data on agglutinin distribution in the Leguminosae was consistent with some aspects of botanical classification (Hutchinson, 1964) while inconsistent with others (Toms and Western, 1971). However in such studies the correctness of seed identification, the age of the seed sample, the selection of the right extraction procedures and a wide enough choice of test erythrocytes all present methodological problems which have serious effects on the final results. Some of these problems can be avoided by rigorous standardization of the procedures. The precision and reliability of the data however would be improved by the measurement of exact physicochemical and immunochemical properties of the lectins in addition to the usual serological tests for the three types of reactivity distribution: blood group specific, non-specific and non-haemagglutinating.

To ascertain the value of such an approach, seeds from a small collection within the genus *Phaseolus* were first tested by measurement of agglutination with a wide choice of cells, some of which had been treated with proteolytic enzymes (Table I). The results were generally in agreement with those from previous studies (Mäkela, 1957; Tobiška, 1959; Boyd et al., 1961; Toms and Western, 1971) with the possible exception of some samples of *Ph. ricciardianus* and *Ph. mungo* for which variable results had been recorded previously. The results also showed that, with a few important exceptions, seed samples which had activity at all were active towards all cells and that the highest activity was usually obtained with pronase-treated rat cells. The important exceptions were firstly *Ph. lunatus* and *Ph. leucanthius* which showed blood group A specificity but otherwise had little activity towards normal cells unless treated with pronase, secondly *Ph. vulgaris cv.* "Pinto III" which reacted strongly only with pronase-treated rat cells but poorly with all other cells tested. Of the 28 samples tested 9 were poorly active. However, without a much more rigorous control on the origin and the identification of these seed samples it would be very difficult to establish whether the small residual activity was genuine and how much variability within the species could be expected.

Further information about lectin content and the chemical nature of the lectins in the 28 samples was obtained from the results of SDS-PAGE experiments by comparing the polypeptide subunit composition of their respective seed proteins with each other and with those of the three basic types of *Ph. vulgaris* (Fig. 1). It is now generally accepted that, based on the results of SDS-PAGE experiments and as a first approximation, most *Ph. vulgaris* cultivars (type-1, "Processor" in Fig. 1) contain two basic types of lectin subunit of about 30 000 subunit weight, designated E (erythro) and L (leuco) (Yachnin and Svenson, 1972; Pusztai and Watt, 1974; Pusztai

TABLE I

Nitrogen content and agglutination titre of *Phaseolus* bean seeds. Extracts (aqueous, pH 8) of seed samples (nominal concentration of 50 mg/ml) were serially diluted with 0·9% NaCl solution, incubated with various blood cells at room temperature for 16 h after which the settled cells were examined for clumping under microscope. The results are expressed as μg/ml seed sample in the last tube to give a strong 2+ agglutination. (N.A. = not agglutinated up to 12·5 mg/ml concentration). Cells for the tests were prepared from heparinized blood samples, some of which were treated with pronase (1 mg per 10 ml of 4% erythrocyte; 25° for 1 h) or with trypsin (0·1 mg per 10 ml undiluted erythrocyte, 25° for 1 h) and washed three times with 0·9% NaCl solution in a centrifuge.

Species/cultivar	N%	Rabbit	Pronase-treated Rat	Pig	Trypsin-treated Cow	Sheep	Human O+	Human AB+	Human lymphocytes O+	Human lymphocytes AB+
1. Ph. vulgaris "Processor"	3·5	25	1·5	50	25	100	400	400	6250	1600
2. Ph. coccineus	4·0	50	3	50	50	400	200	200	3200	1600
3. Ph. multiflorus	6·1	25	0·8	25	25	200	12·5	25	1600	800
4. Ph. multiflorus	3·4	100	1·5	50	12·5	400	400	200	3200	3200
5. Ph. lunatus	3·7	N.A.	12·5	12500	6250	N.A.	N.A.	3200	N.A.	12500
6. Ph. hysterinus	4·1	25	1·5	50	50	200	12·5	25	3200	1600
7. Ph. darwinius	3·4	25	6·2	25	100	100	50	100	1600	800
8. Ph. tuberosus	4·3	12·5	3	25	50	100	200	200	1600	1600
9. Ph. rebra	4·4	12·5	1·5	50	50	200	200	200	1600	800
10. Ph. acutifolius latifolius	4·6	50	50	25	400	1600	200	100	1600	3200
11. Ph. caffer	4·7	6	3	25	25	200	200	200	200	400
12. Ph. cerasiformis	4·0	12·5	3	50	50	200	400	200	400	200
13. Ph. calcaratus	3·3	12500	3200	12500	12500	N.A.	N.A.	N.A.	N.A.	N.A.
14. Ph. aborigineus	4·4	6	1·5	100	50	200	400	200	400	800
15. Ph. angularis	3·6	12500	1600	N.A.	6200	12500	400	N.A.	12500	N.A.
16. Ph. aconitifolius	5·7	N.A.	1600	N.A.	12500	N.A.	N.A.	N.A.	N.A.	12500
17. Ph. formosus	4·8	12·5	1·5	50	100	200	200	25	200	400
18. Ph. leucanthius	4·0	12500	6	N.A.	6250	12500	N.A.	50	1600	1600
19. Ph. lathyroides	3·2	12500	800	N.A.	12500	12500	N.A.	N.A.	N.A	12500
20. Ph. ricciardianus	4·5	N.A.	3	100	100	200	400	100	400	400
21. Ph. aureus	3·8	12500	3200	N.A.	12500	N.A.	N.A.	N.A.	N.A.	N.A.
22. Ph. acutifolius	3·8	6	1·5	25	200	400	100	25	800	100
23. Ph. filiformis	4·1	3200	800	12500	6250	N.A.	N.A.	N.A.	N.A.	N.A.
24. Ph. mungo	4·5	25	1·5	50	100	200	100	800	800	100
25. Ph. latifolius	4·8	50	25	50	800	1600	100	800	3200	1600
26. Ph. trilobus	3·2	12500	800	N.A.	12500	N.A.	N.A.	N.A.	N.A.	N.A.
27. Ph. ritensis	4·0	12500	400	12500	N.A.	N.A.	N.A.	N.A.	N.A.	12500
28. Ph. semierectus	4·3	12500	3200	N.A.	N.A.	N.A.	N.A.	N.A.	N.A.	N.A.
29. Ph. vulgaris "Pinto III"	4·2	6250	50	12500	N.A.	N.A.	N.A.	N.A.	3200	6250

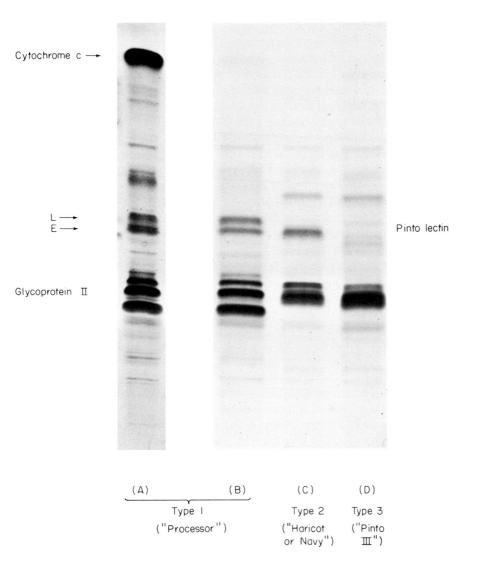

FIG. 1. SDS-PAGE patterns of the three basic types of *Phaseolus vulgaris* bean seeds. The ground seeds were extracted with 2% SDS-0·1% 2-mercaptoethanol (nominal concentration 15 mg/ml); heated at 100° for 15 min and 10 μl of the samples applied to the gels: (A) type-1, ("Processor"); (B) type-1, ("Processor"); (C) type-2 ("Haricot" or "Navy" bean); (D) type-3 ("Pinto III"). Running conditions were: discontinuous system based that on Laemmli (1970), stacking gel, 4%, separating gel: 17·6%; 66 mA for 4·5 h. Gels were run (A) in the vertical apparatus by Hoefer Scientific Instruments, Model SE 500 (U.S.A.) and (B), (C), and (D) in the horizontal set-up, "Multiphor" by LKB (Gt, Britain Ltd).

and Stewart, 1978; Felsted *et al.*, 1981). However, as shown by the results of isoelectric focusing experiments these subunits do not represent homogeneous species (Pusztai and Watt, 1974; Pusztai and Stewart, 1978; Felsted *et al.*, 1981). Indeed, with improvements in SDS-PAGE techniques both subunits were also resolved into at least two closely spaced bands (Fig. 1). Recent results (Felsted *et al.*, 1981) have also indicated size heterogeneity of the E subunit. Type-2 ("Haricot" or "Navy" bean in Fig. 1) beans give a distinct lectin pattern (Pusztai and Watt, 1974). Here the L subunit is missing from the SDS-PAGE patterns and these beans are not mitogenic though the isolated lectins agglutinated lymphocytes (Pusztai and Watt, 1974). The albumin lectins in these beans were made up of two subunits in a ratio of 3 (30 000 subunit weight) to 1 (35 000 subunit weight), while globulin lectins were made up of four subunits of 30 000 subunit weight. Finally, type-3 beans ("Pinto III" in Fig. 1) contained neither E nor L subunits but contained small amounts of a number of isolectins different from those found in type-1 or in type-2 beans. These lectins had a dimeric structure and were composed of two subunits of about 29 000 subunit weight (Pusztai *et al.*, 1982a,b) instead of the usual four subunits per molecule as in the other types of lectins.

A comparison of the protein subunit patterns of the 28 samples (Fig. 2) fully confirmed and extended the results obtained with the agglutination technique. The nine species within the *Phaseolus* genus which had low agglutination titres, *Ph. calcaratus*, *Ph. angularis*, *Ph. aconitifolius*, *Ph. lathyroides*, *Ph. semierectus*, *Ph. aureus*, *Ph. filiformis*, *Ph. trilobus*, *Ph. ritensis* also had subunit patterns quite different from those of the three *Ph. vulgaris* types. In addition, some of these nine samples showed similarities to one another, such as for example *Ph. lathyroides* to *Ph. semierectus* or some similarities, such as these two to *Ph. calcaratus*. Good agreement was also found between the agglutination reactivity and the protein subunit patterns of *Ph. lunatus* and *Ph. leucanthius*. These patterns were also quite different from the rest of the *Phaseolus* species examined. The *Ph. coccineus*, the two *Ph. multiflorus* samples and the *Ph. darwinius* beans were also very similar to each other. Although these beans showed as good agglutinating activities (Table I) as those of type-1 or type-2 beans, their lectin subunit patterns were clearly, though slightly, different from those of all other beans in Fig. 2.

Finally, 13 samples in Fig. 2 were related to one of the three basic types of *Ph. vulgaris*. There were two type-2 beans: *Ph. acutifolius latifolius* and *Ph. latifolius* and only one type-3 bean, "Pinto III". The other ten samples all belonged to type-1 on the basis of their lectin subunit patterns. These results confirm and extend the results of Derbyshire *et al.* (1976) obtained by similar methods.

Further information on the nature of the lectins in the seed samples was obtained by immunochemical tests. Extracts of the 28 seed samples were reacted in double diffusion experiments with immune sera raised against E-type, L-type and type-3 lectins respectively (Fig. 3). The nine poorly agglutinating species reacted with none of the test sera. Thus these seeds contained no (or very little) lectins structurally related to any of the three basic *Phaseolus* lectin types. *Ph. lunatus* and *Ph. leucanthius* gave very slight cross-reactivity with the anti-type-3 lectin serum or anti-L lectin serum and practically none with anti-E lectin serum. Again the agreement with the results of the other tests was good. The results also fully confirmed previous findings based on immunochemical methods indicating substantial differences between the proteins of *Ph. vulgaris* and *Ph. lunatus* seeds (Kloz and Klozova, 1974). The *Ph. coccineus*-type beans gave a close cross-reaction with all three types of antibodies. These lectins are closely related to, though not identical with the *Ph. vulgaris*-type lectins. The two samples of type-2 beans gave no reactions characteristic for the L subunits, thus confirming the results of SDS-PAGE experiments. The rest of the beans reacted with all these sera as expected.

In conclusion, as the molecular structure of the lectins purified up to date (e.g. Goldstein and Hayes, 1978) can be very diverse and thus occasionally the only link found between them is their reactivity with red cells, the determination of molecular and immunochemical properties could yield more precise information about possible phylogenetic relationships of the lectins than one would obtain from the results of agglutination tests alone. Thus the combination of such data would also increase the reliability of potential classification systems of plant species.

III. Location and Biological Role

A. LOCATION IN THE PLANT

The biological roles postulated for lectins in plants are not at present supported by direct experimental evidence. Since any role proposed for the lectins would have to be compatible with their *in vivo* location, studies of their occurrence, distribution, localization and their ontogeny in the plant are of great importance.

It was established very early that the best source of agglutinins was from the seeds of higher plants (for early references see Gold and Balding, 1975; Callow, 1975). Thus it was found by immunochemical studies that both in *Lens culinaris* (Howard *et al.*, 1972) and in *Phaseolus vulgaris* (Mialonier *et al.*, 1973) lectins were formed only in the ripening and mature seeds. Most of these and other observations tend generally to confirm the early findings of von Eisler and von Portheim (1911) that agglutinins are present mainly

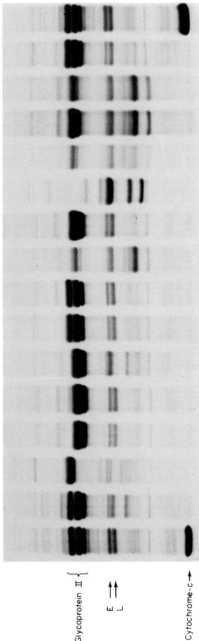

1 *Ph. vulgaris* cv. "Processor"

Ph. vulgaris cv. "Creamy coffee"

2 *Ph. coccineus*

3 *Ph. multiflorus*

4 *Ph. multiflorus*

5 *Ph. lunatus*

6 *Ph. hysterinus*

7 *Ph. darwinius*

8 *Ph. tuberosus*

9 *Ph. rebra*

10 *Ph. acutifolius latifolius*

11 *Ph. caffer*

12 *Ph. cerasiformis*

13 *Ph. calcaratus*

14 *Ph. aborigineus*

1 *Ph. vulgaris* cv. "Processor"

Glycoprotein II

E
L

Cytochrome-c

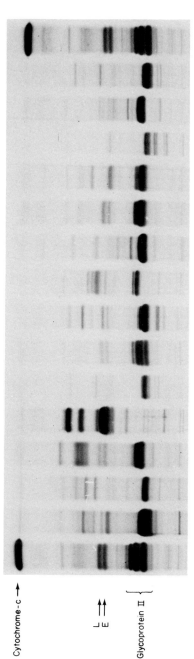

Cytochrome-c →

L
E ↑↑

Glycoprotein II

1   *Ph. vulgaris* cv. "Processor"

28   *Ph. semierectus*

27   *Ph. ritensis*

26   *Ph. trilobus*

25   *Ph. latifolius*

24   *Ph. mungo*

23   *Ph. filiformis*

22   *Ph. acutifolius*

21   *Ph. aureus*

20   *Ph. ricciardianus*

19   *Ph. lathyroides*

18   *Ph. leucanthius*

17   *Ph. formosus*

16   *Ph. aconitifolius*

15   *Ph. angularis*

1   *Ph. vulgaris* cv "Processor"

FIG. 2. SDS-PAGE patterns obtained with the samples from the collection of bean seeds within the *Phaseolus* genus. The gels were run in the horizontal system (Multiphor) and the conditions were the same as described in Fig. 1. The name of the samples is given on the patterns.

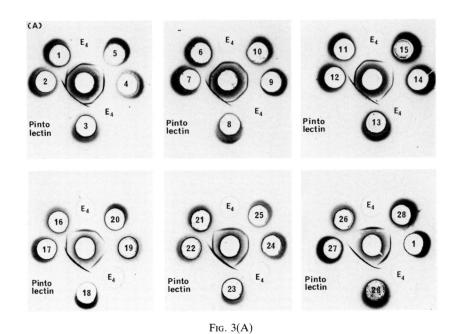

FIG. 3(A)

FIG. 3(B)

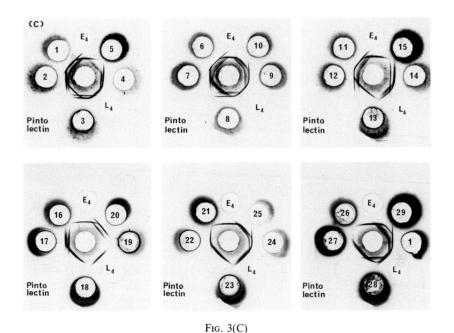

FIG. 3(C)

FIG. 3. Double diffusion patterns obtained with seed samples from the *Phaseolus* collection. Centre wells contained between 50–70 μl of purified (A) anti-"Pinto" seed (type-3) lectin serum; (B) anti-E_4 lectin serum and (C) anti-L_4 lectin serum. The peripheral wells contained standards (pure "Pinto III" seed lectin, 50 μg; E_4 lectin, 25–50 μg and L_4-lectin, 25–50 μg) or aqueous (pH 8) extracts of the various seeds, (30–60 μl) as indicated in the figure. The numbers of the samples in the wells are the same as in Table 1 and in Fig. 2. The gels were incubated at room temperature for 24 h, then dried, washed with 0·9% NaCl solution and stained with Coomassie-Blue.

in those parts of the plants which contain reserve nutrients and that they disappear from these tissues at the same time as the stored reserve proteins. However there are a number of notable exceptions known. For example, in wheat grains the lectin was restricted to the germ (none in the endosperm) and after germination the lectin level remained high for at least 34 days. The lectin was associated almost excusively with the rapidly growing tissues of roots and shoots (Mishkind *et al.*, 1980). In the young seedlings of peanut, though over 90% of the lectin was found in the cotyledons, the remainder was in hypocotyls, stems and leaves, while the young roots contained practically none (Pueppke, 1979). In the family of Solanaceae, for example in potatoes, most of the lectin activity was associated with the tubers (Marinkovich, 1964) and not with the seeds. Moreover, in *Phytolacca americana* most of the lectin activity was reported to be in the leaves and stems of the plant (Farnes *et al.*, 1964). More examples of lectins in the vegetable parts of plants have recently been referred to by Rougè and Pére (1982). The recent finding of an immunochemically cross-reacting material, a potential prolectin, in the leaves and stems of *Dolichos biflorus* (Talbot and Etzler, 1978) is of special importance. Although this protein does not agglutinate red cells it is structurally related to the seed lectin and has certain similarities to it in its sugar-binding specificity (Etzler and Borrebaeck, 1980).

B. ULTRASTRUCTURAL LOCATION

Seed lectins were usually located in the cytoplasm. Thus Mialonier *et al.* (1973) found by using immunoperoxidase methods that in *Phaseolus vulgaris* seeds lectins were associated with cytoplasmic sites. Similarly, cytochemical localization of lectins with FITC-labelled globulins also indicated that in the seeds of *Canavalia ensiformis* and *Phaseolus vulgaris* the lectins had a cytoplasmic location (Clarke *et al.*, 1975). However, these studies were not conducted with antibodies specific for the lectins; neither were there any proper sugar haptens used for the *Phaseolus* lectin. Furthermore the lack of fine enough resolution of fluorescent microscopy also made it difficult to draw firm conclusions about the precise location of lectins in the cytoplasm.

Subcellular fractionation studies, particularly when coupled to microscopy, provided further refinements in the location of the lectins. Thus by successive non-aqueous linear density gradient centrifugation (Youle and Huang, 1976) or by fractionation with glycerol (Tully and Beevers, 1976) of the endosperm of castor bean the lectins were found in the protein bodies. There is also substantial evidence to indicate that lectins are associated with the protein bodies in the cotyledonary cells of *Phaseolus vulgaris* (Barker *et al.*, 1976; Begbie, 1979; Manen and Miège, 1981). Similarly, in the seeds of *Datura stramonium* some of the lectin was in the protein bodies though a part was also associated with membranous structures (Kilpatrick *et al.*, 1979; Jefree and Yeoman, 1981). In more recent high-resolution electron microscopic localization studies with the gold method (Horisberger and Rosset, 1977) thin sections of the soya bean seed were marked with gold granules (12 nm in size) labelled with anti-soya bean lectin antibodies. A subsequent examination by transmission electron microscopy revealed that the lectin was uniformly distributed in most of the protein bodies of the cotyledon and the embryo axis (Horisberger and Vonlanthen, 1980). Immunocytochemical studies coupled with light- and electron-microcopy with the cotyledons of *Phaseolus vulgaris* seeds also suggested an association between the lectins and the protein bodies (Manen and Pusztai, unpublished). Furthermore, protein body ghosts obtained by the controlled rupture of isolated protein bodies from the cotyledons of *Phaseolus vulgaris* (Croy, 1977; Pusztai *et al.*, 1978; Begbie, 1979) and purified by sucrose-density centrifugation were also shown by SDS-PAGE to contain lectins. A fraction was isolated by preparative SDS-PAGE from purified ghosts which after the removal of SDS by phenol-borate, pH 8 partitioning (Pusztai, 1966) had a high haemagglutinating titre, an isoelectric point of pH 5·8 and gave a reaction of identity with anti-globulin lectin serum in double diffusion experiments (Pusztai *et al.*, 1979a). It was homogeneous by SDS-PAGE and by molecular sieve chromatography in 5 M guanidinium HCl. The results of both experiments indicated that the material had a subunit weight of 30 000. On the basis of the evidence obtained, this material was suggested to be identical with the seed lectin of pI = 5·8. There have also been a number of other reports suggesting the general presence of lectins in plant cell membranes. However, as lectins have a generally high reactivity, their binding to membranes may come about as an artefact during the rupture of the cells. Thus for example the lectin content of the mitochondrial inner membrane from castor bean (Bowles and Kauss, 1975) was later shown to be an artefact (Köhle and Kauss, 1979). Similarly, Manen and Miège (1981) found no lectins in their protein body membrane preparation. Another problem in this context was highlighted by the properties of the lectins associated with the membranes prepared from soya bean seeds at all stages of development. Most of these lectins were different from the seed lectins and required detergents for their solubilization (Bowles *et al.*, 1979). These results appeared to confirm earlier observations. For example, in mung bean hypocotyls lectin activity was found in the membranes of endoplasmic reticulum, Golgi-apparatus and plasma-membranes. A potential role for these carbohydrate-binding lectins in the synthesis and transport of cell wall components was also suggested (Bowles and Kauss, 1976). However, these lectins could not be absorbed and purified by affinity chromatography (on haemagglutination-inhibitory glycoproteins coupled to Sepharose) from their detergent-containing extracts. More recent studies indicated that lipid-type materials in such extracts were also excellent agglutinins and that this reaction could not be abolished by boiling the extracts, though the agglutination of red cells by these materials could be inhibited by such glycoproteins as fetuin or bovine submaxillary mucin (Pusztai and Grant, 1980; Tsivion and Sharon, 1981). These results again demonstrated that though haemagglutination is a convenient way for the measurement of lectins, the results without further corroborating evidence or purification of the lectins can be misleading. In fact, it is increasingly recognized that by perturbing the lipid bilayers of red cell membranes or synthetic liposomes (with or without glycolipids added) agglutination effects can be brought about by a number of compounds of hydrophobic character. The hydrophobicity of lectins, particularly that of concanavalin A, and the effects of hapten sugars on this character are being studied intensely (Boldt et al., 1977; Ochoa et al., 1979; Slama and Rando, 1980).

Extracts from cell walls of higher plants have also been reported to contain haemagglutinating substances. Thus Kauss and coworkers have, in the past seven years, described a number of lectins associated with plant cell walls (Kauss and Glaser, 1974; Kauss and Bowles, 1976; Haas et al., 1977). More recently they purified a galactose-specific haemagglutinating lectin from Vigna radiata hypocotyl cell wall preparations which, similar to the seed lectin from this plant (Hankins et al., 1980a and b) also had α-galactosidase activity (Haas et al., 1981). Because of its enzymatic properties, according to our present definition of lectins (Kocourek and Hořejši, 1981) this agglutinin should not be called a lectin. However, the authors presented some experimental evidence to suggest that the enzymatic and the agglutinating properties of the molecule could be separated by dissociation into subunits. Thus the native molecule could be regarded as a lectin combined with an enzyme. Unfortunately, as the haptenic sugar, galactose, does not enhance the elution of the lectin from the wall preparations or inhibit its binding to these, it is questionable if the lectin was located in the cell wall in vivo.

In conclusion, the association of seed lectins with the cellular storage organs, the protein bodies, is now well established experimentally. A part of these lectins are probably bound to, though not integral components of the single membranes surrounding these organelles. It also appears that other internal membranes, the plasma membrane and the cell walls may also bind lectins with various degrees of tightness. However, as some of these lectins are not released by their haptenic sugar and require dissociating conditions and/or detergents for their solubilization, the interpretation of the results and their relevance to in vivo conditions remains to be established.

C. BIOLOGICAL ROLE

Most of the proposed functions for the lectins in plants are based on their characteristic ability to bind carbohydrates. It is argued (Callow, 1975) that such a major property can hardly be fortuitous and must have some relevance in vivo. However this presupposes the existence of materials in the plant which can bind these lectins in vivo and that this binding has some unknown biological function in the plant. Such receptors or "lectin-binders" have already been obtained from pea and jack bean seeds at acidic pH values where the association of these binders to the endogenous lectins is at a minimum (Gansera et al., 1979). The amount of the lectin-binder proteins and the lectins in pea seeds was about equivalent and in the germinating jack bean the concanavalin A-binder was also found in the roots. Both these suggested some biological role for these receptor proteins. Receptors were also obtained from the seeds of soya bean and jack bean by affinity chromatography on columns containing immobilized endogeneous lectins (Bowles and Marcus, 1981). However, as in this instance no dissociation at low pH values was attempted and only trace amounts of these receptors were recovered, their biological significance remains yet to be proven.

Legume seeds also usually contain substantial amounts of sucrose and its α-galactosidic derivatives. As these are a part of the stored carbohydrate reserves of the seeds, they would be ideal candidates for fixation and transportation by the seed lectins as proposed by Ensgraber (1958). However, at least in the seeds of kidney bean there was no evidence for the binding of these oligosaccharides to the lectins (Begbie, 1979).

An alternative function for the lectins in plants is that they recognize and bind to external materials or organisms which contain those carbohydrates for which the lectins are specific. It is possible that surface-localized complementary macro-molecular mechanisms, where at least one of the complementary pair of molecules possesses lectin activity, may be the chemical basis of the various recognition processes between plants and external organisms. Such a recognition step may be an essential pre-requisite in host-parasite or host-symbiont relationships (Callow, 1975). There has been much work and speculation on the nature of the complementary materials from the organisms in these relationships. However their chemistry and biology is outside the scope of this review. Similarly, the rather speculative role envisaged for lectins as inhibitors of polygalacturonases and other enzymes secreted by these organisms will not be discussed here.

Some of the experimental evidence appears to support the suggestion that plant lectins are essential components of the initial and specific recognition step. Thus wheat germ lectin was shown to interact with fungal hyphae (Mirelman et al., 1975). Furthermore, lectins could inhibit fungal spore germination and interfere with normal growth (Barkai-Golan et al., 1978). These lectins are then suggested to function in plants as part of their protection system in combating the attacks by fungal pathogens. Similarly, the presence of lectin in the rhizosphere of young wheat plants might have a fungostatic effect (Mishkind et al., 1980). The active immobilization of saprophytic bacteria in kidney bean leaves was also suggested to be the result of a lectin-mediated binding, while the pathogens were not immobilized (Sing and Schroth, 1977). A similar role might be played by the potato lectin in fixing avirulent strains of pathogens (Sequeira and Graham, 1977). The experimental evidence for such a role for the lectins in the recognition step is not always clear cut. For example, with *Phytophthora infestans* both compatible and incompatible races were bound to the cell membranes of potato and mediated specifically by the potato lectin (Furnichi *et al.*, 1980). Similarly, there was no correlation between the agglutination activity and the pathogenicity of the new bacterial agglutinin recently shown to be present in soya bean (Fett and Sequeira, 1980). In a recent review of lectins and their role in host–pathogen specificity Sequeira (1978) concluded that the role of lectins in host-pathogen interaction is still a matter of conjecture. Binding of lectins to a carbohydrate constituent of a pathogen cannot be regarded as direct evidence for such a role. The recognition step has still not been experimentally linked to the "message" sent to the internal organelles of the cells which eventually will set up either an "accept" or "reject" response.

The host-symbiont relationship with *legume-Rhizobium* interaction has also been very intensively studied since the original observation by Hamblin and Kent (1973) that root hairs of *Phaseolus vulgaris*, just like its seed lectin, will bind erythrocytes. It has been suggested that surface-localized legume lectins provide receptor sites for the specific binding of *Rhizobium* prior to infection (Schmidt, 1979). However the evidence here, just as with the host–pathogen interaction, appears to be contradictory. For example, Bohlool and Schmidt (1974) found that FITC-labelled soya bean seed lectin reacted with 22 out of the 25 nodulating strains of bacteria, while it was unreactive with strains which did not nodulate soya beans. The *Rhizobium*-binding lectin in the seeds of both *Glycine max.* and *Glycine soya* was also present on the root surface at the site of the initial steps of the infection and was also capable of binding *Rhizobium japonicum* to the root (Stacey *et al.*, 1980). One of the best studied examples of *Rhizobium*-recognition proteins, trifoliin, was obtained from the seeds of white clover. The same protein could also be washed off with the haptenic sugar, 2-deoxy-D-glucose from clover roots. It was usually present in the highest concentrations at sites where *Rhizobium trifolii* bound to the roots. The lectin probably provided a specific bridge between the root cell wall and the *Rhizobium* (Dazzo and Brill, 1977; Dazzo *et al.*, 1978). On the other hand, it was established that several lines of *Glycine max.* lacked the 120 000-Dalton seed lectin but were still successfully nodulated by *Rhizobia* (Pull *et al.*, 1978). In the USDA *Glycine soya* collection 49% of the lines contained none of this lectin but were nodulated (Stahlhut *et al.*, 1981). In addition, no soluble lectin could be demonstrated by radioimmunoassay in 5-day old roots and hypocotyls of 20 soya bean lines in which the cotyledons did contain the 120 000-Dalton lectin (Su *et al.*, 1980). Furthermore some members of the cross inoculation group of *Rhizobium japonicum* could also nodulate other plants, including cowpea and peanut. Some of the rhizobia which normally nodulated these plants could also reciprocally nodulate soya bean, although these bacteria did not bind the soya bean seed lectin. Conversely, three of these bacteria strains, although they bound the soya bean seed lectin, did not nodulate the soya bean roots. Furthermore, cells of the soya bean and the peanut rhizobia did not bind the peanut seed lectin. These results indicated that there was no relationship between the ability of peanut and soya bean *Rhizobia* to nodulate the reciprocal host plant and their ability to bind to the lectin from the same plant (Pueppke *et al.*, 1980). Further conflicting evidence was obtained by Chen and Phillips (1976) who found that strains of *Rhizobia* which were incapable of infecting pea roots were bound to these roots just as well as the pea-nodulating *Rhizobium leguminosarum*. Thus simple attachment of *Rhizobium* to the legume root is not necessarily the basis of host-symbiont specificity (Wong, 1980; Dazzo and Hubbell, 1975). It is now suggested that after the initial attachment, a precisely timed and co-ordinated expression of specific plant and bacterial genes is essential for the establishment of an effective nitrogen-fixing nodule. In addition to leghaemoglobin which is an obligatory protein for symbiotic nitrogen fixation in legumes, a number of other "nodule-specific" host proteins (nodulins) have also been detected (Legocki and Verma, 1979; Auger and Verma, 1981). These proteins which were induced by *Rhizobia* following infection may be involved in symbiotic nitrogen fixation.

In most of these experiments, usually for convenience's sake, the lectins tested for their binding to the bacteria were from the seeds although there is no *a priori* reason to assume that if a lectin was produced by the root cells at all it should necessarily be identical with the seed lectin. However chemical evidence has more recently been published which suggests a close similarity, indeed identity between some of the lectins produced by the seeds and the roots of the same plant. Thus, pea root slime contained two glucose-binding proteins one of which had the same electrophoretic mobility as the glycosylated seed lectin (Kijne *et al.*, 1980). This lectin agglutinated a strain of *Rhizobium leguminosarum*. Gatehouse and Boulter (1980) also purified a lectin from pea roots which, though it was similar to the seed lectin by SDS-PAGE and immunochemical techniques, had a somewhat different sugar specificity. In these studies however it was not clear if adequate attention had been paid to the possible binding to the roots of seed lectins which are known to be released from the seeds during water uptake (Fountain *et al.*, 1977). This could not have happened in the case of the root lectin isolated from the nodulated roots of *Phaseolus vulgaris* cv. "Pinto III" (Pusztai and Grant, 1980; Pusztai *et al.*, 1982a). This variety of kidney bean was shown to lack type-1 and type-2 lectins but contained small amounts (0·03%) of a number of different type-3 lectins (Pusztai *et al.*, 1982b). The main root lectin component was purified by affinity chromatography on fetuin-Sepharose from roots which had first been washed with organic solvents to remove non-specific lipid-type agglutinins. The lectin contained subunits of 29–30 000 subunit weight and cross-reacted with both anti-type-1 and anti-type-2 lectin sera. However, it had little reactivity with the anti-Pinto seed lectin serum. The presence of other and minor lectin components was also demonstrated. These could be purified partially by affinity chromatography first on ε-N-caproyl-galactosamine-Sepharose 4B and then on pronase-treated rat stroma. These did not cross-react with any of the anti-seed lectin sera tested. Thus these results suggested that the root cells of "Pinto III" beans, when nodulated with *Rhizobia*, produced a number of lectins different from the "Pinto III" seed isolectins. As the concentration of these lectins was negligible in non-nodulated roots a potentially important and, possibly sequential, role for those lectins in the nodulation is envisaged (Pusztai *et al.*, 1982a).

In conclusion, there is a great deal of experimental evidence to suggest that the specificity of host-parasite and host-symbiont interactions are mediated by lectins. It appears however that the lectins involved in the specific recognition step are not necessarily identical with the normal seed lectin of the same plant. Further progress is to be expected when the physico-chemical properties of more root lectins are explored and the biochemical mechanism of the modulation of host genes by *Rhizobia* and the co-ordinated expression of both genes in successful symbiosis is unravelled.

Seeds of jack bean (Sumner and Howell, 1936) and soya bean (Liener and Pallansch, 1952) contain lectins in large concentrations. Similarly, in some varieties of *Phaseolus vulgaris* seed lectins were the second largest group of proteins comprising of 10 to 20% of the total protein (Pusztai *et al.*, 1979b). However seeds of a number of lines of these plants are also reported to have no measurable haemagglutinin or lectin content (Brücher, 1968; Pull *et al.*, 1978). For example, in the USDA *Glycine soya* collection 49% of the seeds contained no lectin (Stahlhut *et al.*, 1981). It appears that in the seeds of some plants, such as in the seeds of *Phaseolus vulgaris*, there is no significant selection pressure for the presence of lectins (Brücher *et al.*, 1969). However it was recently demonstrated (Pusztai *et al.*, 1982b) that a variety of *Phaseolus vulgaris* "Pinto III" which had been claimed to be lectin-free (Brücher, 1968) did in fact contain small amounts of a different dimeric lectin. Thus, according to Jaffè (1980), failure to detect lectins in seeds could be due to the methodology used for screening.

Since the early findings of von Eisler and von Portheim (1911) most observations have generally tended to indicate that lectins are synthesized during seed maturation and degraded during germination together with other reserve nutrients of the seeds (Howard *et al.*, 1972; Mialonier *et al.*, 1973; Rougé, 1976; Gracis and Rougé, 1977; Manen, 1978; Sun *et al.*, 1978; Bollini and Chrispeels, 1978). For example, in the maturing seeds of

Phaseolus vulgaris lectins are synthesized on the proliferating rough endoplasmic reticulum (Bollini and Chrispeels, 1978) and are eventually stored in the protein bodies together with other reserve proteins (Pusztai *et al.*, 1977, 1978; Begbie, 1979). Thus at least in the seeds of *Phaseolus vulgaris*, in addition to variability in content, both their ontogeny and cellular location suggests a reserve role for the lectin. However it is not clear if both basic types of *Phaseolus vulgaris* lectin subunits, E and L, fulfil the same role in the seeds. It has already been shown by immunochemical and other methods that both subunits are associated with the protein bodies and that one of the lectins, pI = 5·8, is bound to the membranes of the protein bodies and possibly to other internal membranes (Pusztai *et al.*, 1979a). However, the preliminary results which indicated that the two subunits are not synthesized synchronously during seed maturation needed further confirmation.

Both type-1 ("Processor") and type-3 ("Pinto III") beans were grown to full maturity under controlled conditions. Accumulation of seed dry matter, when plotted against (Fig. 4(A)(B)) stage of maturation (days after flowering, daf.) gave the usual S-shape curve (Öpik, 1968; Walbot *et al.*, 1972). Samples from each developmental stage were then extracted with 0·1 M borate buffer, pH 8·0. The extracts were dialysed against 0·033 M acetate, pH 5·0 and globulin-(insoluble part at pH 5·0) and albumin-type (soluble) proteins were separated and recovered (details of these experiments will be given elsewhere). The original samples, albumins, globulins and insoluble residues were then extracted with 2% SDS—0·1% 2-mercaptoethanol and run on SDS-PAGE (Figs. 5 and 6). Albumins and globulins from "Processor" bean were also dissolved in 0·9% NaCl, the solutions were adjusted to pH 7·0 and their lectin-content was measured by rocket immunoelectrophoresis with rabbit anti-E and anti-L-type antibodies. The results were calculated as mg lectin protein/seed of each type and plotted in Fig. 4(A). The results of both SDS-PAGE and rocket immunoelectrophoresis confirmed earlier findings for the "Processor" seed and showed that the synthesis of the globulin lectin, containing mainly the E-type subunit, preceded that of the albumin (L-type) lectin. Appreciable amounts of globulin lectins were already present in the early samples, while most of the albumin lectins were synthesized in the last third of the maturation time. During this last period more globulin lectin was also synthesized. The insoluble residues from the early samples of "Processor" seeds also contained appreciable amounts of proteins which could be solubilized only with 2% SDS—0·1% 2-mercaptoethanol or with 1% Triton X-100—0·1% 2-mercaptoethanol containing 6 M urea. One of these "structure-bound" proteins contained a subunit of 30 000 which was similar to that of the lectins. A component of these proteins also reacted with anti-lectin antibodies. The structure and properties of these proteins is now

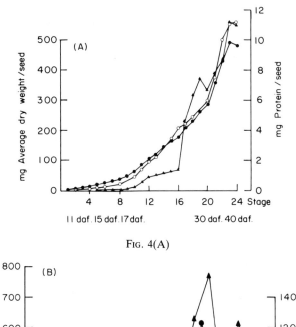

FIG. 4(A)

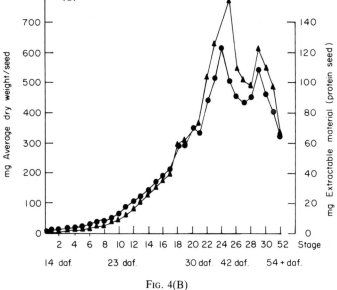

FIG. 4(B)

FIG. 4. Development of *Phaseolus vulgaris* (A) type-1 ("Processor") and (B) type-3 ("Pinto III") bean seeds. Collection and pooling of the seeds was done by size and days after flowering (daf) and average dry weight per seed (●——●) is plotted against stage of development. Globulin and albumin proteins were prepared from the seeds at each stage. In (A) the amounts of both L (▲——▲) and E (○——○) lectins were measured by rocket immuno-electrophoresis with specific antisera. In (B) only quantitative haemagglutination tests (results not given) were carried out on the extracted proteins (▲——▲).

under investigation. These studies gave further support to earlier sugges-
tions that as the globulin lectins (E-type) are started to be synthesized just
after the cell-division stage and that as they are partly membrane-bound,
these lectins might fulfil a structural, as yet unknown, role in the maturing
seed cells. The L-type lectin on the other hand is synthesized late and is
found in the protein bodies. It is difficult to envisage any other role for this
lectin component than that of a reserve protein.

Similarly detailed immunochemical studies have not yet been performed
with the "Pinto III" seeds. Nevertheless from SDS-PAGE experiments
(Fig. 6) the increase in total lectin content appears to follow the course of
protein accumulation. However the solubility of this type-3 lectin appears
to change with seed maturation. This lectin initially has the solubility
characteristics of an albumin-type protein. The amount of this reaches a
maximum before half-maturity stage. After this the lectin becomes increasingly less soluble in dilute aqueous buffers. The significance of this
change is not apparent.

As very few studies on the synthesis and degradation of other seed
isolectins have been carried out up to date, any generalization at this stage
is premature. In addition, the results of these studies are also occasionally
contradictory. For example, the two isolectins from the seeds of garden
pea behave differently on germination; PS I disappears more quickly than
PS II (Rougè, 1975). On the other hand, in *Lathyrus ochrus* L. both seed
isolectins are synthesized at the same rate on seed maturation. They are
also degraded equally on germination (Rougé and Pére, 1982). The various
lectins may thus fufil several distinct and not necessarily mutually exclusive
roles in the plants depending on their ontogeny, location in the plant,
concentration in the cells, specificity and other, as yet unknown factors.
One such attractive role which has been proposed for lectins is that they
may be selective agents for mitotic stimulation of callus cells or protoplasts.
Such a role, if established to be generally applicable, would be of great
significance. Some of the results obtained with the soya bean agglutinin on
callus cells appeared to support such a role (Howard *et al.*, 1977). However
the results with both root explants (Vasil and Hubbell, 1977) and with
isolated protoplasts (Larkin, 1978; Fenton and Labavitch, 1980) were
disappointing. On the other hand, the finding of some measure of
structural homology between several lectins from leguminous plants
(Foriers *et al.*, 1977; 1978; Richardson *et al.*, 1978; Bauman *et al.*, 1979)
and the conservation of immunochemical determinants (Howard *et al.*,
1979) suggests that during evolution, protein lectins with these conserved
portions of structure may have performed beneficial and similar functions
for the plants. The extensive immunochemical and other similarities
between some lectins and the widely occurring plant α-galactosidases,
some of which are also erythroagglutinating while some others are not,

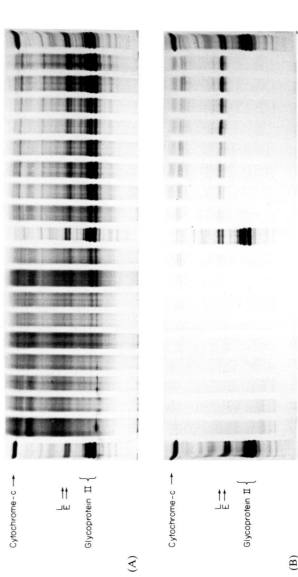

Cytochrome-c →

E ↑↑

Glycoprotein II ⎰⎱

(A)

Cytochrome-c →

E ↑↑

Glycoprotein II ⎰⎱

(B)

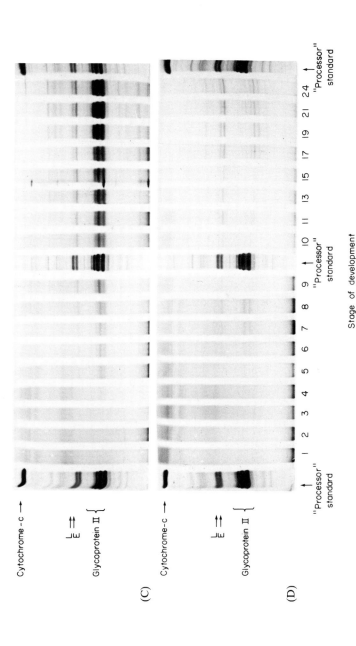

Cytochrome - c →

L →
M →

Glycoprotein II {

(C)

Cytochrome-c →

L →
M →

Glycoprotein II {

(D)

"Processor"
standard

1 2 3 4 5 6 7 8 9 10 11 13 15 17 19 21 24

Stage of development

"Processor"
standard

"Processor"
standard

Fig. 5. SDS-PAGE patterns of developing type-1 ("Processor") seeds of *Phaseolus vulgaris* of various stages of maturity: (A) seed meal (15 mg/ml extract); (B) albumins (2 mg/ml); (C) globulins (2 mg/ml) and (D) residue from aqueous, pH 8 extraction of seed meal (20 mg/ml). The materials were treated with 2% SDS-0·1% 2-mercaptoethanol, heated at 100° for 15 min and 10 μl of the resulting clear solutions were applied to the gels. Running conditions (Multiphor) as in Fig. 1. The position of standards and the stage of development are marked in the figure.

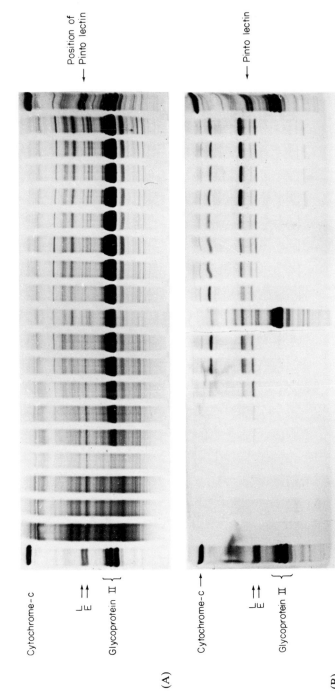

Position of
→ Pinto lectin

Cytochrome-c

L
E ↑↑

Glycoprotein II

(A)

Pinto lectin →

Cytochrome-c →

L
E ↑↑

Glycoprotein II

(B)

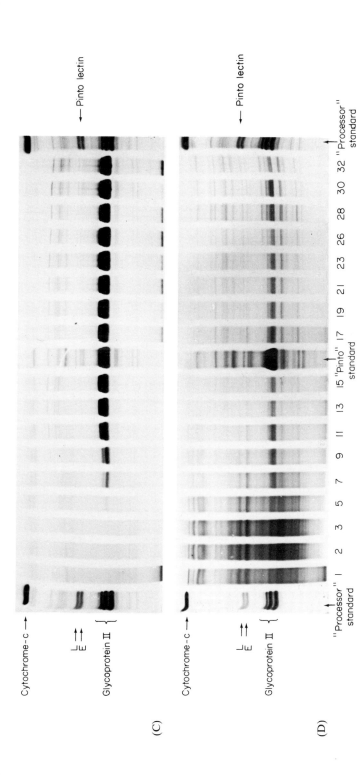

FIG. 6. SDS-PAGE patterns of developing type-3 ('Pinto III') seeds of *Phaseolus vulgaris* at various stages of maturity. The samples were applied as in Fig. 5: (A) seed meal (15 mg/ml extract); (B) albumins (3 mg/ml); (C) globulins (3 mg/ml) and (D) residue (30 mg/ml). Running conditions as in Fig. 1.

may indicate one such possible evolutionary relationship (Hankins *et al.*, 1979; 1980a, b). In this context the recent report of the identification of canavalin as an artificially proteolytically modified form of jack bean α-mannosidase is significant (McPherson and Smith, 1980). Since α-mannosidase and the mannose-specific oligosaccharide-binding protein, concanavalin A, are the two most abundant proteins of this seed, the view that these two proteins are simply storage proteins is no longer tenable (McPherson and Smith, 1980).

Callow (1975) in the last sentence of the introduction to her review wrote: "As will become apparent, there are large gaps in our knowledge of the biology of lectins, and where appropriate, areas for future experimentation will be outlined". Some of those areas have been extensively studied in the six years elapsed since her review and substantial progress has also been made during this time. However, the gaps are still there. Perhaps with future progress on a scale similar to that attained in these past six years, some of the gaps, hopefully, may soon disappear.

ACKNOWLEDGEMENT

We are grateful to Professor Lindsay Innes of National Vegetable Research Station (Wellesbourne, Warwick, England) for the samples of *Phaseolus* beans from their seed collection.

REFERENCES

Aub, J. C., Tieslau, C. and Lankester, A. (1963). *Proc. Natl. Acad. Sci. U.S.A.* **50**, 613–619.
Auger, S. and Verma, D. P. S. (1981). *Biochemistry* **20**, 1300–1306.
Barkai-Golan, R., Mirelman, D. and Sharon, N. (1978). *Arch. Microbiol.* **116**, 119–124.
Barker, R. D. J., Derbyshire, E., Yarwood, A. and Boulter, D. (1976). *Phytochemistry* **15**, 751–757.
Bauman, C., Rüdiger, H. and Strosberg, A. D. (1979). *FEBS Lett.* **102**, 216–218.
Begbie, R. (1979). *Planta* **147**, 103–110.
Bohlool, B. B. and Schmidt, E. L. (1974). *Science* **185**, 269–271.
Boldt, D. H., Speckart, S. F., Richards, R. L. and Alving, C. L. (1977). *Biochem. Biophys. Res. Commun.* **74**, 208–214.
Bollini, R. and Chrispeels, M. J. (1978). *Planta* **142**, 291–298.
Boyd, W. C. (1963). *Vox Sang.* **8**, 1–32.
Boyd, W. C. (1970). *Ann. NY. Acad. Sci.* **169**, 168–190.
Boyd, W. C. and Reguera, R. M. (1949). *J. Immunol.* **62**, 333–339.
Boyd, W. C. and Shapleigh, E. (1954). *Science* **119**, 419.
Boyd, W. C., Waszczenko-Zacharczenko, E. and Goldwasser, S. M. (1961). *Transfusion* **1**, 374–382.

Bowles, D. J. and Kauss, H. (1975). *Plant. Sci. Lett.* **4**, 411–418.
Bowles, D. J. and Kauss, H. (1976). *Biochim. Biophys. Acta* **443**, 360–374.
Bowles, D. J. and Marcus, S. (1981). *FEBS Lett.* **129**, 135–138.
Bowles, D. J., Lis, H. and Sharon, N. (1979). *Planta* **45**, 193–198.
Brücher, O. (1968). *Proc. Trop. Region Am. Soc. Hort. Sci.* **12**, 68–72.
Brücher, O., Wecksler, M., Levy, A., Palozzo, A. and Jaffé, W. G. (1969). *Phytochemistry* **8**, 1739–1743.
Callow, J. A. (1975). *Curr. Adv. Plant Sci.* **7**, 181–193.
Chen, A. P. T. and Phillips, D. A. (1976). *Physiol. Plant* **38**, 83–88.
Clarke, A. E., Knox, R. B. and Jermyn, M. A. (1975). *J. Cell. Sci.* **19**, 157–167.
Croy, R. R. D. (1977) Ph.D. Thesis, University of Aberdeen.
Dazzo, F. B. and Brill, W. J. (1977). *Appl. Environ. Microbiol.* **33**, 132–136.
Dazzo, F. B. and Hubbell, D. H. (1975). *Plant Soil* **43**, 717–722.
Dazzo, F. B., Yanke, W. E. and Brill, W. J. (1978). *Biochim. Biophys. Acta* **539**, 276–286.
Derbyshire, E., Yarwood, J. N., Neat, E. and Boulter, D. (1976). *New Phytol.* **76**, 283–288.
Eisler, von M. and Portheim, von L. (1911). *Ber. Dtsch. Bot. Ges.* **29**, 419–430.
Ensgraber, A. (1958). *Ber. Dtsch. Bot. Ges.* **71**, 349–361.
Etzler, M. E. and Borrebaeck, C. (1980). *Biochem. Biophys. Res. Commun.* **96**, 92–97.
Farnes, P., Barker, B. E., Brownhill, L. E. and Fanger, H. (1964). *Lancet* **2**, 1100–1102.
Felsted, R. L., Leavitt, R. D., Chen, C., Bachur, N. R. and Dale, R. M. K. (1981). *Biochim. Biophys. Acta* **668**, 132–140.
Fenton, C. A. L. and Labavitch, J. M. (1980). *Physiol. Plant.* **49**, 393–397.
Fett, W. F. and Sequeira, L. (1980). *Plant Physiol.* **66**, 853–858.
Foriers, A., Wuilmart, C., Sharon, N. and Strosberg, A. D. (1977). *Biochem. Biophys. Res. Commun.* **72**, 980–986.
Foriers, A., de Neve, R., Kanarek, L. and Strosberg, A. D. (1978). *Proc. Natl. Acad. Sci. U.S.A.* **75**, 1136–1139.
Fountain, D. W., Foard, D. E., Replogle, W. D. and Yang, W. K. (1977). *Science* **197**, 1185–1187.
Furnichi, N., Tomiyama, K. and Doke, N. (1980). *Physiol Plant Pathol.* **16**, 249–256.
Gansera, R., Schurz, H. and Rüdiger, H. (1979). *Hoppe-Seyler's Z. Physiol. Chem.* **360**, 1579–1585.
Gatehouse, J. A. and Boulter, D. (1980). *Physiol. Plant* **49**, 437–442.
Gold, E. R. and Balding, P. (1975). *In* "Receptor-Specific Proteins. Plant and Animal Lectins", pp. 151–236. Excerpta Medica, Amsterdam.
Goldstein, I. J. and Hayes, C. E. (1978). *Adv. Carbohydr. Chem. Biochem.* (R. S. Tipson and D. Horton, eds), Vol. 35, pp. 127–340. Academic Press, New York.
Gracis, J. P. and Rougé, P. (1977) *Bull. Soc. Bot. Fr.* 124, 301–306.
Haas, D., Frey, R. and Kauss, H. (1977). *Hoppe-Seyler's Z. Physiol. Chem.* **358**, 1210–1211.
Haas, D., Frey, R. Thiesen, M. and Kauss, H. (1981). *Planta* **151**, 490–496.
Hamblin, J. and Kent, S. P. (1973). *Nature (London), New Biol.* **245**, 28–30.
Hankins, C. N., Kindinger, J. I. and Shannon, L. M. (1979). *Plant Physiol.* **64**, 104–107.

Hankins, C. N., Kindinger, J. I. and Shannon, L. M. (1980a). *Plant Physiol.* **65,** 618–622.
Hankins, C. N., Kindinger, J. I. and Shannon, L. M. (1980b). *Plant Physiol.* **66,** 375–378.
Horisberger, M. and Rosset, J. (1977). *J. Histochem. Cytochem.* **25,** 295–305.
Horisberger, M. and Vonlanthen, M. (1980). *Histochemistry* **65,** 181–186.
Howard, I. K., Sage, H. J. and Horton, C. B. (1972). *Arch. Biochem. Biophys.* **149,** 323–326.
Howard, J., Shannon, L., Oki, L. and Murashige, T. (1977). *Exp. Cell. Res.* **107,** 448–450.
Howard, J., Kindinger, J. and Shannon, L. M. (1979). *Arch. Biochem. Biophys.* **192,** 457–465.
Hutchinson, J. (1964). *In* "The Genera of Flowering Plants.", Vol. 1. Clarendon Press, Oxford.
Jaffé, W. G. (1980). *In* "Toxic Constituents of Plant Foodstuffs" (I. E. Liener, ed.) 2nd edition, pp. 73–102. Academic Press, New York.
Jeffree, C. E. and Yeoman, M. M. (1981). *New Phytol.* **87,** 463–471.
Jermyn, M. A. and Yeow, Y. M. (1975). *Aust. J. Plant Physiol.* **2,** 501–531.
Kauss, H. and Bowles, D. J. (1976). *Planta* **130,** 169–174.
Kauss, H. and Glaser, C. (1974). *FEBS Lett.* **45,** 304–307.
Kijne, J. W., Van der Schaal, I. E. M. and De Vries, G. E. (1980). *Plant. Sci. Lett.* **18,** 65–74.
Kilpatrick, D. C., Yeoman, M. N. and Gould, A. R. (1979). *Biochem. J.* **184,** 215–219.
Kloz, J. and Klozova, E. (1974). *Biol. Plant.* **16,** 290–300.
Kocourek, J. and Hořejši, V. (1981). *Nature* **290,** 188.
Köhle, H. and Kauss, H. (1979). *Biochem. J.* **184,** 721–723.
Krüpe, M. (1956). Blutgruppen spezifische pflanzliche Eiweiskörper (Phytoagglutinine). Enke, Stuttgart.
Laemmli, U. K. (1970). *Nature* **227,** 680–685.
Landsteiner, K. and Raubitschek, H. (1908). *Zbl. Bakt. I. Abt.* **45,** 660–667.
Larkin, P. J. (1978). *Plant Physiol.* **61,** 626–629.
Legocki, R. P. and Verma, D. P. S. (1979). *Science* **205,** 190–193.
Liener, I. E. and Pallansch, M. J. (1952) *J. Biol. Chem.* **197,** 29–36.
Lis, H. and Sharon, N. (1977) *In* "The Antigens" (M. Sela, ed.), Vol. IV, pp. 429–529. Academic Press, New York.
Lis, H. and Sharon, N. (1981). *In* "The Biochemistry of Plants. A Comprehensive Treatise". (P. K. Stumpf and E. E. Conn, editors-in-chief; A. Marcus, ed.), Vol. 6, pp. 371–447. Academic Press, New York.
Manen, J. F. (1978). Thése Doct. és-Sc. No. 1866, Genéve.
Manen, J. F. and Miége, M. N. (1981). *Physiol. Veg.* **19,** 45–58.
Marinkovich, V. A. (1964). *J. Immunol.* **93,** 732–741.
Mäkela, O. (1957). *Ann. Med. Exp. Biol. Fenn.* **35,** Supplement 11.
McPherson, A. and Smith, S. C. (1980). *Phytochemistry* **19,** 957–959.
Mialonier, G., Privat, J. P., Monsigny, M., Kahlem, G. and Durand, R. (1973). *Physiol. Veg.* **11,** 519–537.
Mirelman, D., Galun, L., Sharon, N. and Lotan, R. (1975). *Nature* **256,** 414–416.
Mishkind, M., Keegstra, K. and Palevitz, B. A. (1980). *Plant Physiol.* **66,** 950–955.
Nowell, P. (1960). *Cancer Res.* **20,** 462–466.
Ochoa, J. L., Kristiansen, T. and Påhlman, S. (1979). *Biochim. Biophys. Acta* **577,** 102–109.

Öpik, H. (1968). *J. Exp. Bot.* **19**, 64–76.

Pueppke, S. G. (1979). *Plant Physiol.* **64**, 575–580.

Pueppke, S. G. Freund, T. G., Schulz, B. C. and Friedman, H. P. (1980). *Can. J. Microbiol.* **26**, 1489–1497.

Pull, S. P., Pueppke, S. G., Hymowitz, T. and Orf, J. H. (1978). *Science* **200**, 1277–1279.

Pusztai, A. (1966). *Biochem. J.* **99**, 93–101.

Pusztai, A. (1980). *Rep. Rowett Inst.* **36**, 110–118.

Pusztai, A. and Grant, G. (1980). *Abstr. 13th FEBS Meeting* 52–P41, p. 44.

Pusztai, A. and Stewart, J. C. (1978). *Biochim. Biophys. Acta* **536**, 38–49.

Pusztai, A. and Watt, W. B. (1974). *Biochim. Biophys. Acta* **365**, 57–71.

Pusztai, A., Croy, R. R. D., Grant, G. and Watt, W. B. (1977). *New Phytol.* **79**, 61–71.

Pusztai, A., Stewart, J. C. and Watt, W. B. (1978). *Plant Sci. Lett.* **12**, 9–15.

Pusztai, A., Croy, R. R. D., Stewart, J. C. and Watt, W. B. (1979a). *New Phytol.* **83**, 371–378.

Pusztai, A., Clarke, E. M. W., King, T. P. and Stewart, J. C. (1979b). *J. Sci. Food Agric.* **30**, 843–848.

Pusztai, A., Grant, G. and Stewart, J. C. (1982a). *In* "Lectins. Biology, Biochemistry, Clinical Biochemistry (T. C. Bøq-Hansen, ed.) Vol. 2. Walter de Gruyter, Berlin, New York (in preparation).

Pusztai, A., Grant, G. and Stewart, J. C. (1982b). *Biochim. Biophys. Acta* (accepted).

Renkonen, K. O. (1948). *Ann. Med. Exp. Biol. Fenn.* **26**, 66–72.

Richardson, C., Behnke, W. D., Freisheim, J. H. and Blumenthal, K. M. (1978). *Biochim. Biophys. Acta* **537**, 310–319.

Rougé, P. (1975). *C.R. Acad. Sci. Paris* **280**, 2105–2108.

Rougé, P. (1976). *CR. Acad. Sci. Paris* **282**, 621–623.

Rougé, P. and Pére, D. (1982). *In* "Lectins. Biology, Biochemistry, Clinical Biochemistry". (T. C. Bøq-Hansen, ed.) Vol. 2. Walter de Gruyter, Berlin, New York (in preparation).

Saint-Paul, M. (1961). *Transfusion* **4**, 3–37.

Schmidt, E. L. (1979). *Ann. Rev. Microbiol.* **33**, 355–376.

Sequeira, L. (1978). *Ann. Rev. Phytopathol.* **16**, 453–481.

Sequeira, L. and Graham, T. L. (1977). *Physiol. Plant Pathol.* **11**, 43–54.

Sharon, N. and Lis, H. (1972). *Science* **177**, 949–959.

Sing, V. O. and Schroth, M. N. (1977). *Science* **197**, 759–761.

Slama, J. S. and Rando, R. R. (1980). *Biochemistry* **19**, 4595–4600.

Stacey, G., Paau, A. S. and Brill, W. J. (1980). *Plant Physiol.* **66**, 609–614.

Stahlhut, R. W., Hymowitz, T. and Orf. J. H. (1981). *Crop Sci.* **21**, 110–112.

Stillmark, H. (1888). *Inaug. Dis., Dorpat.*

Su, L. C., Pueppke, S. G. and Friedman, H. P. (1980). *Biochim. Biophys. Acta* **629**, 292–304.

Sumner, J. B. and Howell, S. F. (1936). *J. Bacteriol.* **32**, 227–237.

Sun, S. M., Mutschler, M. A., Bliss, F. A. and Hall, T. C. (1978). *Plant Physiol.* **61**, 918–923.

Talbot, C. F. and Etzler, M. E. (1978). *Biochemistry* **17**, 1474–1479.

Tobiška, J. (1959). *Z. Immunforsch.* **117**, 190–196.

Tobiška, J. (1964). Die Phytohäemagglutinine. Akademie Verlag. Berlin.

Toms, G. C. and Western, A. (1971). *In* "Chemotaxonomy of the Leguminosae". (J. B. Harborne, D. Boulter and B. L. Turner, eds.). Phytohaemagglutinins. pp. 367–462. Academic Press, London.

Tsivion, Y. and Sharon, N. (1981). *Biochim. Biophys. Acta* **642**, 336–344.
Tully, R. E. and Beevers, H. (1976). *Plant Physiol.* **58**, 710–716.
Vasil, I. K. and Hubbell, D. H. (1977). *Z. Pflanzenphysiol.* **84S**, 349–353.
Walbot, V., Clutter, M. and Sussex, I. M. (1972). *Phytomorphology* **22**, 59–68.
Watkins, W. M. and Morgan, W. T. J. (1952). *Nature* **169**, 825–826.
Wong, P. P. (1980). *Plant Physiol.* **65**, 1049–1052.
Yachnin, S. and Svenson, R. H. (1972). *Immunology* **22**, 871–883.
Youle, R. J. and Huang, A. H. C. (1976). *Plant Physiol.* **58**, 703–709.

CHAPTER 4

Allergens in Oilseeds

R. L. ORY AND A. A. SEKUL

Southern Regional Research Center, USDA–SEA, AR P.O. Box 19687
New Orleans, USA

I. INTRODUCTION

Von Pirquet (1906) introduced the term "allergie" to describe the altered capacity of a human to react to a second injection of horse serum. Since then, all types of hypersensitivity in humans have been called "allergies" to identify the body's response to any allergen-antibody reaction that causes release of the chemical mediators of hypersensitivity (Austin, 1965). Allergens are found in both animal and plant materials consumed by people but those in foods have attracted the most attention. Allergens are ordinarily harmless, but certain sensitized people react with the onset of symptoms such as flushing of the face, skin disorders, respiratory problems, and gastrointestinal disturbances. Allergens are generally large molecular weight compounds, non-dialyzable, and are most often identified with the protein moiety of the material.

There are three basic types of allergy: atopic, delayed, and anaphylactic (Spies, 1974). Atopic allergies are triggered by the specific reaction of the allergen with skin-sensitizing antibodies, homocytotropic antibodies or immunoglobulin-E (IgE). The symptoms appear within a few minutes and continue for up to an hour after exposure. Delayed allergies are triggered by a specific reaction between the allergen and small lymphocytes; the symptoms may not appear until a few hours up to 96 hours after exposure. Anaphylactic allergies are the worst. Like atopic allergies, they are also triggered by reaction of allergen with mast cell bound IgE, but they appear within a few seconds or minutes after exposure and can cause violent, sometimes fatal symptoms. Both atopic and anaphylactic reactions are immediate types of reactivity.

This report concentrates on allergens in oilseeds, principally those in castor beans, peanuts, and cottonseed, their properties, characterization, and methods used for their analysis.

II. GENERAL FOOD ALLERGIES

The most common offenders among food allergens are cow's milk, chocolate, cola beverages (made from Kola nuts), corn, eggs, the pea family of legumes (chiefly peanuts, which are not nuts but are legumes), citrus fruits, tomato, wheat and other small grains, cinnamon, artificial food colors (Speer, 1976). The large family of legumes is an important source of both foods and food allergens. Extracts of peanuts often cross-react with various beans and peas; especially soybeans. General symptoms displayed by persons who have adverse reactions to food allergens are headache, shock, nausea, sneezing, wheeze in breathing, vomiting, skin rash, redness, swelling, diarrhea, eczema, abdominal pains, urticaria, and in severe cases, convulsive seizures or loss of consciousness (Bock *et al.*, 1978).

Cereal grains include many staple foods, some of which have been reported to produce allergic responses in humans. Wheat, rye, barley, corn, oats, and rice are the primary cereal grains and of these, wheat is the most frequent offender (Ferguson, 1976). Wheat gluten hypersensitivity (coeliac disease) produces an adverse response in individuals who are sensitive to wheat gluten. There are two primary theories on the cause of coeliac disease; that it is caused (a) by a deficiency of peptidase or (b) by hypersensitivity to the gluten. The general conclusion however, is that, in most patients the precipitating factor and exact composition of the allergenic fraction are still unknown.

The two most reliable methods for clinical diagnosis of hypersensitivity to ingested foods are the skin puncture, or skin prick test, (SPT) and the radioallergosorbent test (RAST). These were reliable for diagnosis of allergy to peas, codfish, tree nuts, peanuts, and eggwhite (Aas, 1978). Positive SPT and RAST responses were rather common for wheat or wheat flour but only partly reliable for allergy to cow's milk, soybeans and white beans. There is also a report of strong positive SPT and RAST response to pineapple in a pharmaceutical worker who had handled bromelain during his duties (Baur and Fruhmann, 1979).

III. Oilseed Allergens

Oilseed proteins are generally of two types; metabolic (cytoplasmic) and storage. Legumes and oilseeds contain large amounts of storage proteins, most of which are globulins that are soluble in dilute salt or buffer and insoluble in water. The metabolic fraction contains water-soluble albumins, enzymes, and other non-storage proteins. Legumes also contain glycoproteins which include toxicants, trypsin inhibitors, lectins, and allergens, which can affect nutritional value of the protein. Many of these are frequently found with the storage globulins.

The classical isolation scheme for chemical characterization of the cottonseed allergens by Spies and coworkers (Spies, et al., 1941), soon became the prototype for subsequent isolation of allergens from other oilseeds and nuts. The nomenclature adopted for these listed the primary cottonseed allergen as CS–1A, the secondary allergen as CS–1B, etc. As allergens from other seeds were isolated by this same procedure they were also listed as CB–1A and CB–1B (castor bean seed), PN–1A and PN–1B (peanut), etc. The CS–1A isolation procedure is based upon solubility of the allergens in water and 25% ethyl alcohol but not in 75% ethyl alcohol. The 1–A allergens are stable in boiling water and not precipitated by basic lead acetate, which is used to separate other antigens from them. Excess lead acetate is important to obtain the primary allergen free of other allergens and antigens that precipitate as lead salts. The secondary allergen (1–B) is isolated by a milder procedure that avoids heat and requires adjusting the pH of the solution to 9·6. Suspended protein in the original water extract is precipitated by lowering the pH to 4·7 with 50% acetic acid, centrifuging and discarding the precipitate. The pH is readjusted to 5·6 with NaOH and the (1–A) procedure is then continued to obtain the (1–B). Allergens of some of the principal oilseeds and tree nuts are shown in Table I. Except for flaxseed, castor beans, and cottonseed, yields (dry weight of recovered material) of the allergens are less than 1%. Allergens in seeds of oil-bearing plants appear to be present in higher concentrations than those in tree nuts and the two legumes, peanuts and soybeans. Nitrogen contents range from 11 to 18% but carbohydrate moieties show very wide variations, from 3 to 39%. Peanuts and the tree nuts are eaten in many countries but peanuts appear to be more allergenic than tree nuts, possibly because they are consumed by more people in the United States. Cottonseed is considered a potential source of protein for humans and, like peanuts, is generating much research on the proteins and allergens. Castor beans, because of potent allergens in the seed, are not a promising source of protein for food or feed use but the oil is industrially important. Allergens of these seeds will be discussed in more detail.

TABLE I

Properties of allergens in some oilseeds and tree nuts

Seed	Isolation method[a]	Yield %	% N	% CHO
Almonds	1A	0·46	16·9	12·6
Brazil nuts	1A	0·81	17·6	6·9
Castor beans	1A	1·76	18·4	3·1
	1C	0·33	17·1	8·0
Cottonseed	1A	1·38	12·1	36·4
Filberts	1A	0·30	17·1	10·5
	H1B	0·26	18·8	6·6
Flaxseed	1A	1·98	11·3	39·4
Peanuts	1A	0·07	15·4	16·7
	1B	0·18	15·2	10·6
Soybeans	1B	0·10	13·3	20·4

[a] Isolated by method of Spies et al. (1951).

A. PEANUTS (*Arachis hypogaea* L.)

Peanut (groundnut) is one of the major oilseeds grown in tropical and subtropical countries. It is the most extraordinary legume because of its manner of growth. The flower produces a long stalk (peg) that turns downward and grows until it penetrates the ground for several centimeters. The pods of seeds develop at the tips of these pegs underground; thus the name groundnut. Peanuts have been reported to contain various adventitious components, such as hemagglutinins (Dechary et al., 1970), anti-hemophilia factors (Frampton and Boudreaux, 1963), trypsin inhibitors (Ory and Neucere, 1971), and allergens, but Spies et al. (1951) were probably the first to characterize the peanut allergens. These allergens were not prominent nor especially harmful at that time but in recent years, reports of allergic reactions to peanuts have appeared (Gillespie et al., 1976; Sachs et al., 1981). One possible cause of this could be the improved breeding of new varieties for larger seed size, better flavor, increased insect/disease resistance, etc., with no attention to the possible effects on allergen contents. As noted in Table I, the total amounts of both peanut allergens, 1–A and 1–B, are only 0·25%. Peanuts and soybeans yield immunologically atypical fractions by the Spies et al., (1951) procedure. Using crossed immunoelectrophoresis to study specific binding of patient serum IgE to allergens of soybean, Nyrup et al. (1981) identified 14 allergens in extracts of defatted soybean flour.

PN–1A was antigenic but exhibited a specificity different from that of any other component of peanut extracts, suggesting that possible structural changes may have occurred.during isolation. PN–1B, isolated by a milder procedure, was serologically identical to another component in the unfractionated peanut extract, but PN–1A and PN–1B are serologically different. Since antigenicity of PN–1B is destroyed by heat it is unlikely that this antigen is a precursor of the antigen(s) in PN–1A. It is therefore apparent that persons who are allergic to roasted peanuts (or peanut butter) are sensitive to PN–1A or some other heat-stable allergen. It cannot be PN-1B, which is heat-labile. There have been several reports of clinical demonstrations of allergy to nuts (including peanuts) by RAST (Gillespie et al., 1976; Sachs et al., 1981).

B. COTTONSEED (*Gossypium* spp.)

Cotton has been grown for centuries as a source of textile fiber but since the twentieth century, the cotton seed has been recognized as a source of edible oil and protein for animals and humans. With the proposed introduction of cottonseed protein into human diets, investigations on the metabolic and storage proteins and on other materials associated with the proteins, such as allergens, were initiated. The classical isolation procedure for CS–1A, the most potent allergen of cottonseed, by Spies et al. (1941), has been the basis for isolation of the allergen for these studies.

Cottonseed meal, which contains the protein and allergens, is a by-product of the cotton textile fiber industry. It constitutes 60% of the harvested fiber-free seed. Subcellular structure of the seed is typical of oilseeds; the cotyledons contain the major deposits of storage oil and protein. Storage proteins, about 70% of the total proteins, are located within protein bodies (aleurone grains) that contain one or more globoids, the subcellular site of phytic acid deposits (Lui and Altschul, 1967).

The cottonseed allergen, CS–1A, was purified by the Spies et al. (1951) method, to determine the chemical nature of the material. CS–1A is immunologically distinct from other allergens in the seed and it is the only one not precipitated by lead acetate. If CS–1A is refluxed for 4 hours in 0·1N HCl, a new allergen, designated CS–51R by Spies and coworkers, is produced. These were early recognized as glycoproteins but an active carbohydrate-free allergen, CS–60C, was isolated from CS–1A by Spies and Umberger (1942). This is similar to CB–65A, which is isolated from CB–1A, the castor bean primary allergen (Spies *et al.*, 1944).

Cottonseed contains three major classes of proteins; 2S, 5S, and 9S, in about equal amounts. The 5S and 9S proteins are typical globulin storage proteins. The 2S proteins are albumins but are also classified as storage proteins based upon their amino acid composition, developmental properties, and high amounts in the seed (Youle and Huang, 1979). By combining SDS gel electrophoresis, immuno-cross reactivity tests and amino acid compositions, they showed that 2S albumins contained the cottonseed allergen, CS–1A (Fig. 1). The 2S albumins cross-reacted with

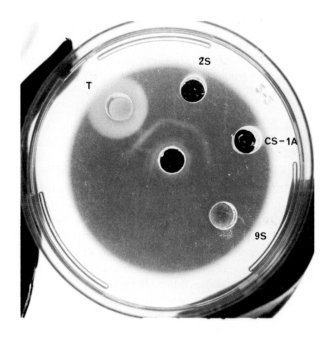

FIG. 1. Double diffusion of antibodies to cottonseed allergen, CS-1A, against cottonseed proteins and allergens. Center well, antibodies to CS-1A; well T, total proteins of cottonseed; 2S, cottonseed 2S proteins; CS-1A, cottonseed allergens; 9S, cottonseed 9S proteins. (Reproduced by courtesy of A.H.C. Huang.)

cottonseed allergens in double diffusion tests against antibodies to CS–1A. They showed a reaction of identity between total proteins, 2S proteins, and the allergen but not between the allergen and 9S proteins, thereby localizing the allergen in the albumin storage protein fraction. By gel electrophoresis of proteins from germinating seeds, the 2S proteins and the CS–1A allergen disappeared completely after 2 days germination.

Byssinosis, an occupational disease of textile mill workers, is believed to be caused by inhalation of something in the mill dust. The exact cause is still not known, but it is generally agreed that exposure to cotton textile mill dust is responsible for the acute respiratory reaction experienced by some cotton mill workers (Rooke, 1981). Although the symptoms are very much like those of allergenic reactions, byssinosis is still not confirmed as an allergen/antibody type disorder.

Butcher et al., (1979) surveyed workers in cottonseed processing plants by SPT with cottonseed and cotton dust extracts and by RAST of blood samples, to detect specific immunoglobulin-E antibodies (IgE). There was 83% agreement between RAST and skin testing with cottonseed extracts and 93% agreement between RAST and skin testing with cotton dust extracts, suggesting that cottonseed products are capable of stimulating an IgE response in exposed workers.

In our studies on the causative agent of cotton byssinosis, we extracted cotton textile mill dust and cotton bract (small leaf-like tissues that surround the developing cotton bolls) and produced antibodies to the extracted antigens in rabbits (Sekul and Ory, 1978, 1979a). Antibodies to cotton dust and bract, when reacted with either dust or bract antigens, formed precipitin lines that fused in a reaction of identity indicating the presence of bract antigens in textile mill dust. A white glycoprotein isolated from the dust was shown to contain the active dust/bract antigen(s) (Sekul and Ory, 1979b). This active antigen was present in cotton stem, leaf, burr, and cotton gin trash but it was not present in house dust, clean cotton fibers, cotton seed hulls or cottonseed protein extracts (Sekul and Ory, 1981). Thus, antigens found in cotton bract or textile mill dust are not derived from cottonseed tissues, which contain the CS–1A allergen. The purified bract and dust antigens were also shown to activate the alternative complement pathway in a manner similar to inulin and asbestos dust (Wilson et al., 1980). Kamat et al. (1979) reported that blood serum immunoglobulins from cotton mill workers in India (some with byssinosis) and those from normal persons showed differences in the IgG and IgE levels. Normal sera had IgG levels of 1400 mg and IgE of 3000 mg, whereas blood sera of byssinotics averaged 1850 mg IgG and 6050 mg IgE. Although it seems apparent that something in cotton dust is the probable cause of the increased IgG and IgE levels, this disorder is still not classified as an allergic response. These results raise a question, however, of whether cotton plants may have some allergen-like materials in plant tissues other than those identified in the seed.

C. CASTOR BEAN (*Ricinus communis* L.)

Castor beans (seeds) are not grown as a source of edible oil or protein because the deoiled meal contains several allergens, but they are an industrially important source of oil. The seed contains approximately 50% oil and 18% protein. The deoiled meal, called castor cake or pomace, contains 36–40% protein, but also contains the toxalbumin, ricin, the CB-1A series of allergens, and the alkaloid, ricinine. It is used almost exclusively as fertilizer.

1. Ricin

Ricin is an extremely potent phytotoxin, the third most potent toxin known. It is not included as one of the castor CB–1A series of allergens, though many animals (rabbits, mice, rats, goats) can be immunized to produce antibodies to ricin. Grabar and Koutseff (1934a,b) confirmed the difference between ricin and the classical allergens, although they strongly resemble each other. They suggested that the allergen of castor seed be called "ricin allergen." The allergic symptoms are specific to man only, not to animals. The symptoms of ricin toxicity, however, are similar to those for the allergens.

Ricin has been the subject of as many investigations as the CB–1A allergens. Mourgue *et al.* (1958) separated the phosphate buffer-soluble castor bean proteins by paper electrophoresis and identified two toxic fractions. They referred to ricin as a toxalbumin. Janssen (1964) reported a molecular weight of 40 000 for ricin and showed that it was not digested by proteases. This explains why ricin is so toxic when ingested; it is not destroyed in the gastrointestinal tract. Ishiguro *et al.* (1964) purified ricin-D free of proteolytic and hemagglutinating activities by ion exchange chromatography and reported the molecular weight as 60 000.

2. CB–1A Series of Allergens

Spies and Coulson (1943) employed the CS–1A procedure to isolate the classical CB–1A from castor beans. About 1·8% of a non-toxic but allergenic protein-polysaccharide, the CB–1A, was prepared. CB–1A and CS–1A are remarkably similar in composition, chemical, and immunological properties, but they are different in allergenic specificity and yields. CB–1A is the more potent and is present in slightly higher concentration (Table I). Spies *et al.* (1944) subsequently isolated the less potent CB–65A (analogous to CS–60C), a carbohydrate-free protein allergen, from the

CB–1A. CB–65A retained both the allergenic and the antigenic specificities of CB–1A.

Using paper electrophoresis, Layton *et al.* (1962) separated CB–65A into 5 distinct bands, compared to 7 bands for CB–1A. Layton *et al.* (1969) also examined cross-reactivity of castor allergens with other species of Euphorbiaceae (Table II). Humans allergic to castor allergens reacted with

TABLE II

Allergenic cross reactivity of different species of Euphorbiaceae to castor bean allergens[a]

Species	No. patients reacting, skin test
Ricinus communis L.	50
Euphorbia lagascae Spreng	23
E. poinsettia Willd.	13
E. paralias L.	30
E. salicifolia L.	24
E. terracina L.	25
Chrozophora hierosolymitana Spreng	46
Cnidoscolus angustidens Torr.	20
Pedilanthus macrocarpus Benth	44

[a] From Layton *et al.* (1969).

antigen extracts of species within the same botanical family. To explain this, they proposed that certain cross-reactive tropical species must have developed in an isolation that began at the time of separation of Africa and South America some 100 to 150 million years ago. The morphological and immunochemical similarities between Old World *R. communis* and the American genera, *Cnidoscolus* and *Jatropha*, are so close as to tempt one to consider them as different species within the same genus. In a more recent investigation of reports that coffee beans (and coffee-generated dust) produced allergic symptoms in coffee bean handlers, Lehrer and Karr (1980) examined blood of 6 workers with allergic symptoms to dust by RAST for specific IgE antibodies. Five control subjects did not react but RAST indices ranged from 3–15 for green coffee bean extracts and 28–60 for castor bean. Closer investigation revealed that castor bean allergens were the actual cause of the reactions in coffee bean handlers. Burlap sacks used to ship the coffee beans had been used earlier to ship castor beans.

Several attempts have been made to inactivate the allergens in castor meal (pomace) so that the deoiled meal could be fed to animals. All methods destroyed the allergens but the remaining protein had no nutritive value.

3. Characterization/Localization of CB–1A

Early studies on protein bodies of seeds suggested these particles as sites for protein storage only, but later studies have localized albumins and some enzymes there. Proteins exist in both crystalline and amorphous states. Sobolev *et al.* (1972), assuming that only proteins in aleurone grains are true storage proteins, studied the synthesis of storage proteins in ripening castor seeds. They proposed that albumins and many enzymes are localized in the amorphous protein zones, such as those described in barley (Ory and Henningsen, 1969; Tronier and Ory, 1970). The crystalloids are believed to be the main storage protein. They concluded that the crystalloids appear in early ripening endosperms and that storage proteins are totally absent during early ripening. They did not examine the proteins for albumins or allergens.

Daussant *et al.* (1976) and Youle and Huang (1976, 1978a), characterized the changes in allergens and proteins of germinating castor beans. Daussant *et al.* (1976) identified CB–1A in the total protein extract of mature ungerminated seeds. Immunoelectrophoresis was employed to separate and identify allergen fractions and to follow their changes during seed germination. CB–1A was shown to contain a major antigen and at least two minor antigens (Fig. 2). Although major storage proteins changed significantly upon germination, CB-1A was still unchanged after 6 days germination and showed only slight degradation after 10 days. There was a slight anodic shift in the CB–1A but it was still antigenically intact. Youle and Huang (1976) fractionated castor bean protein bodies into membranes, a protein matrix, a protein crystalloid, and several phytin globoids. Ricin and the hemagglutinins were localized in the matrix proteins. After 5 days germination, 3 major disc gel electrophoresis bands of matrix proteins did not change much but a major band of crystalloid protein disappeared rapidly. They concluded that the crystalloid is a storage protein. (This could be major protein F in Fig. 2.) Youle and Huang (1978a) then isolated the albumin storage proteins from castor bean protein bodies and found that 40% of the total proteins consist of closely related albumins (2S fraction). These disappeared rapidly after the first day of germination as do storage globulins. The 2S albumins are present in the protein bodies, they degrade rapidly upon germination, and have an amino acid composition very similar to those of storage proteins. Subsequently, Youle and Huang (1978b) proposed that the CB–1A allergens were low molecular weight storage albumins. Total proteins of ungerminated castor beans were separated into matrix proteins, lectins (4–6S), 11S globulins, and 2S storage albumins. Pure CB–1A allergen was used to prepare antibodies in rabbits. Double diffusion showed a reaction of identity between the 2S albumins, CB–1A, matrix proteins, and a total

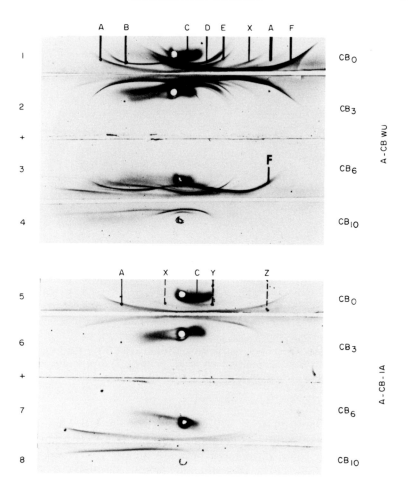

FIG. 2. Immunoelectrophoresis of CB-1A in ungerminated and germinated castor beans. A-CBWU, antibodies to total proteins; A-CB-1A, antibodies to CB-1A; CB_0, CB_3, CB_6, CB_{10} are total protein extracts of ungerminated seeds, 3-, 6-, and 10-days germinated seeds, respectively. Immunoelectrophoresis conditions: 200 V, 22°C (room temp.) for 2 hr, 1·5% Ionagar in 0·025 M veronal buffer (pH 8·6).

protein extract (Fig. 3). This localized the CB–1A allergens with the 2S albumins of protein bodies, in the matrix proteins. Gel electrophoresis of the 2S albumins separated them into several proteins with a molecular weight around 11 000 which is similar to that for CB–1A reported by Spies and Coulson (1943).

Although the CB–1A allergen is associated with 2S storage albumins in the protein bodies matrix, and storage proteins are metabolized during

94   R. L. ORY AND A. A. SEKUL

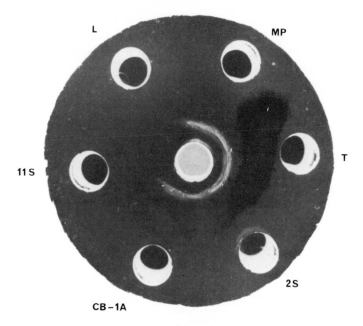

Fɪɢ. 3. Double diffusion of antibodies to castor bean allergen, CB-1A, against castor bean proteins and allergens. Center well, antibodies to CB-1A; well CB-1A, CB-1A allergen; 11S, 11S globulins; L, lectins; MP, protein body matrix proteins; T, total seed proteins; 2S, 2S albumins. (Reproduced by courtesy of A.H.C. Huang.)

germination, Daussant *et al.* (1976) found that the CB–1A itself is not degraded. Only after 10 days germination does CB–1A show any changes. There are also some differences between amino acid compositions of Youle and Huang's (1978a) 2S albumin fraction and Spies *et al.*'s (1951) classical CB–1A. Thus, while the two are associated in the same organelle, the CB–1A is not a true storage protein in the traditional sense.

4. Cross-Reaction of CB–1A and Cotton Dust Antigens

Layton *et al.* (1969) found cross-reactivity between antibodies to castor bean allergens and antigen extracts from other Euphorbiaceae. In our earlier studies on cotton dust/bract antigens, we found antigenic similarities and cross-reactions between cotton dust/bract antibodies and antigens isolated by the same procedure from flax, sisal, hemp, and jute (Sekul and Ory, 1979b). All cross-reacted with cotton dust antibodies and all are believed to induce byssinotic symptoms in susceptible textile mill workers, even though these other textile fiber-producing plants are of different botanical families than cotton. We also examined cotton dust antibodies and antigens from cotton dust and castor bean allergen, CB–1A, for cross-reactions. The CB–1A was first adsorbed on a DEAE-cellulose column from water. Continued elution with water produced a small amount of material, CB–1A (1). Elution with 10% NaCl produced a second fraction, CB–1A (2). Results in Fig. 4 show that there was some cross-reactivity between cotton dust antibodies and both the NaCl-eluted CB–1A (2) and total castor bean antigens, but not with the water-eluted fraction from CB–1A.

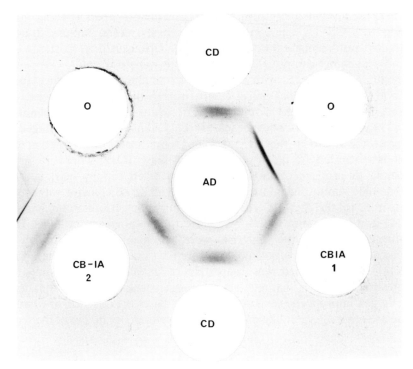

FIG. 4. Double diffusion of cotton dust antibodies and antigens extracted from castor beans and cottonseeds. AD, antibodies to cotton dust antigen; CD, cotton dust antigens; 0, total proteins from ungerminated castor beans; CB-1A (1), water-soluble fraction from DEAE-cellulose chromatographed CB-1A; CB-1A (2), NaCl fraction from DEAE-cellulose.

It, therefore, appears that certain small glycoproteins isolated from textile fiber-producing plants (i.e. cotton, flax, sisal, jute, hemp) by the same procedure (Sekul and Ory, 1979a,b), and allergens isolated from castor seeds can cross-react in immunodiffusion tests. These plants have been implicated in disorders that are clinically quite different; i.e. allergy to castor bean allergens and ricin; byssinosis of textile mill workers.

5. Other CB–1A-like Allergens

During investigations on the lipase of castor beans, a heat-stable protein cofactor for the enzyme was separated by incubation at pH 5·2 for 25 min, then raising to pH 6·0 and centrifuging (Ory et al., 1967). Its size and unusual heat stability was similar to that of the CB–1A allergen and prompted a chemical/biochemical comparison of the lipase cofactor and the allergen (Ory, 1969). Both are small glycoproteins stable to boiling temperature. By gel electrophoresis, both appeared to have the same protein bands. By high voltage paper electrophoresis, the protein cofactor of the lipase produced an electrophoretogram almost superimposable upon that of the CB–1A allergen. In lipolysis tests, crude buffer-extracted allergens could replace the lipase cofactor. Ultracentrifuge patterns and amino acid compositions of the two materials were also similar. When the first rabbit was injected with a solution of the lipase cofactor to prepare antibodies for immunochemical comparison to the CB–1A allergen, the animal died within an hour, reminiscent of ricin toxicity. This could not be ricin, however, since castor beans are homogenized and centrifuged to separate the lipase in the fat pad spherosomal membranes fraction and ricin is discarded in the lower aqueous phase (Ory, 1969). In other attempts to produce antibodies, dosage was reduced by half, but the rabbits died soon after booster injections. The lipase cofactor was finally compared to the CB–1A allergen by immunoelectrophoretic analysis (IEA) using antibodies to CB–1A and to total castor bean proteins. IEA confirmed that the two materials were not identical even though all chemical comparisons suggested they were the same and the potency of the cofactor when injected was suggestive of an allergen or toxin. It is interesting, however, that a cofactor for an enzyme could be so similar to an allergen in properties and potency and it raises the question of whether there are other allergen-like materials in subcellular organelles of the seed.

IV. CONCLUDING REMARKS

All of the oilseed allergens isolated by the classical CS–1A procedure of Spies et al. (1941) yield a heat stable primary allergen and a secondary (1–B) allergen. There are similarities in chemical properties of these allergens and some cross-reactions between cottonseed and castor bean, castor bean and Euphorbiaceae (Layton et al., 1969); but there are also some pertinent differences in their composition. Amino acid contents are not uniform between the different allergen fractions (Spies et al., 1951). (No data is available for PN–1A because yields are too low to obtain workable quantities.) The only similarities are in arginine contents of

CB–1A (27·1%) and CS–1A (27·5%) compared to 10·8% in PN–1B, and in glutamic acid contents of CB–1A (21·1) and (CS–1A) (20·5) against 7·9% in PN–1B. Glycine content in PN–1B (29·4) is 5–10 times higher than that in CB–1A (5.9) and CS–1A (3·0). It is apparent that neither amino acids, % nitrogen, nor % carbohydrate of allergens isolated by the CS–1A procedure are consistent enough to be correlated with allergenic properties. It is noteworthy, however, that Youle and Huang (1978b), using current methods, separated the castor bean 2S albumins and allergen into several proteins and obtained a molecular weight of 11 000, similar to the value for CB–1A obtained by Spies and Coulson (1943) using the classical chemical methods.

Cross-reactions that occur during immunodiffusion of certain glycoproteins isolated from cotton plant tissues and castor bean allergens suggest similarities between the two antigenic materials. The clinical disorders caused by these antigens are quite different so that neither chemical analysis nor immunodiffusion tests alone can be used to confirm a relationship between chemical compositions and the disorders. It is possible that these allergens contain a similar basic amino acid sequence in the protein moiety that is associated with allergic reactions in general (or byssinotic symptoms). However, sequence studies of the amino acids in highly purified allergen fractions would be needed to verify this. This might also be applicable to a sequence analysis of amino acids in the lipase cofactor (Ory, 1969), to determine how the cofactor is so similar to the CB–1A allergen in properties and potency.

In recent years, more sensitive techniques have been introduced for identification of allergens in crude or purified fractions, based upon specific reactions with IgE, with and without radioisotopes. One technique mentioned earlier is the RAST assay. This has been employed to diagnose allergy to various foods, including peanuts (Aas, 1978; Baur and Fruhman, 1979; Gillespie et al., 1976; Sachs et al., 1981), to survey sensitivity to cottonseed and cotton dust extracts in workers in a cottonseed processing plant (Butcher et al., 1979) and to coffee and castor bean antigens in coffee bean handlers (Lehrer and Karr, 1980). Also, immunoelectrophoretic methods have been modified and combined with radioisotopically labelled IgE to specifically identify allergenic compounds in semipurified or crude fractions. Weeke and Løwenstein (1973) identified allergens in a timothy pollen extract by crossed immunoelectrophoresis. The proteins are separated by electrophoresis and submitted to a second dimensional immunolectrophoresis using immune serum specific for these proteins raised in rabbits. After suitable washing to remove excess proteins, the gel on the plate is immersed in blood serum overnight to allow binding of allergens to IgE, washed again, then incubated once more in a solution containing ^{125}I-antihuman IgE. ^{125}I-IgE binds only to peaks of the allergen-IgE complexes. After the final washing, autoradiography is performed on the plates to specifically identify the allergens in the total mixture. Weeke and Løwenstein (1973) observed 6 peaks in timothy pollen extract but, of these, only 2 peaks appeared in autoradiographs when serum from two patients with hay fever were employed as the source of blood serum.

Nyrup *et al.* (1981) applied the methods of Løwenstein (1978) to the identification of 14 allergen in an extract of soybean flour. Although crossed immunoelectrophoresis with or without labelled IgE has not been employed yet for investigation of peanuts, cottonseed and castor bean allergens, results of Nyrup *et al.* on soybeans indicate that this technique should be applicable to other seeds. For successful application of the method, the allergens must be present in the extract being examined, the development of antibodies in an animal must be possible, and the patient IgE must bind to the allergens on the plate, even though the allergens are already fixed in the immunoprecipitate (Weeke and Løwenstein, 1973). If these three conditions are present, it should be posssible to employ this method to identify small quantities of allergen not detectable by the older methods.

REFERENCES

Aas, K. (1978). *Clin. Allergy* **8,** 39–50
Austin, K. F. (1965). *In* "Immunological Diseases" (M. Samter, ed.), pp. 221–225. Little, Brown and Co., Boston, Massachusetts.
Baur, X., and Fruhmann, G. (1979). *Clin. Allergy* **9,** 443–450.
Bock, S. A., Lee, W-Y., Remingo, L. K., and May, C. D. (1978). *J. Allergy Clin. Immunol.* **62,** 327–334.
Butcher, B. T., Lehrer, S. B., O'Neil, C. E., Hughes, J. M. Salvaggio, J. E., and Weill, H. (1979). *J. Allergy Clin. Immunol.* **63,** 213 (abstract).
Daussant, J., Ory, R. L. and Layton, L. L. (1976). *J. Agric. Food Chem.* **24,** 103–107.
Dechary, J. M., Leonard, G. L., and Corkern, S. (1970). *Lloydia* **33,** 270–274.
Ferguson, A. (1976). *J. Hum. Nutr.* **30,** 193–201.
Frampton, V. L., and Boudreaux, H. B. (1963). *Econ. Bot.* **17,** 312–316.
Gillespie, D. N., Nakajima, S., and Gleich, G. J. (1976). *J. Allergy Clin. Immunol.* **57,** 302–309.
Grabar, P., and Koutseff, A. (1934a). *C. R. Soc. Biol.* **117,** 700–701.
Grabar, P., and Koutseff, A. (1934b). *C. R. Soc. Biol.* **117,** 702–704.
Ishiguro, M., Takahashi, T., Funatsu, G., Hayashi, K., and Funatsu, M. (1964). *J. Biochem. (Tokyo)* **55,** 587–592.
Janssen, C. (1964). *Bull. Soc. Chim. Biol.* **46,** 317–322.
Kamat, S. R., Taskar, S. P., Iyer, E. R., and Naik, M. (1979). *J. Soc. Occup. Med.* **29,** 102–106.
Layton, L. L., Greene, F. C., DeEds, F., and Green, T. W. (1962). *Am. J. Hyg.* **75,** 282–286.

Layton, L. L., Panzani, R., and Von Helms, L. T. (1969). *Folia Allergol.* **16**, 204–212.

Lehrer, S. B., and Karr, R. M. (1980). *Food Sci. Technol.* **12**, 2C58 abstract.

Løwenstein, H. (1978). *Prog. Allergy* **25**, 1–62.

Lui, N. S., and Altschul, A. M. (1967). *Arch. Biochem. Biophys.* **121**, 678–684.

Mourgue, M., Baret, R., Reynaud, J., and Bellini, J. (1958). *Bull. Soc. Chim. Biol.* **40**, 1453–1463.

Nyrup, A., Valentin-Hansen, L., and Hejgaard, J. (1981). International Symposium: Seed Proteins, The Phytochemical Society of Europe Meeting, Versailles, France, September 22–24, abstract L–9.

Ory, R. L. (1969). *Lipids* **4**, 177–185.

Ory, R. L., and Henningsen, K. W. (1969). *Plant Physiol.* **44**, 1488–1498.

Ory, R. L., and Neucere, N. J. (1971). *J. Am. Peanut Res. Educ. Assoc.* **3**, 57–62.

Ory, R. L, Kircher, H. W., and Altschul, A. M. (1967). *Biochim. Biophys. Acta.* **147**, 200–207.

Rooke, G. B. (1981). *Text. Res. J.* **51**, 168–173.

Sachs, M. I., Jones, R. T., and Yunginger, J. W. (1981). *J. Allergy Clin. Immunol.* **67**, 27–34.

Sekul, A. A., and Ory, R. L. (1978). *Fed. Proc., Fed. Am. Soc. Exp. Biol.* **37**, 604 abst.

Sekul, A. A., and Ory, R. L. (1979a) *Text. Res. J.* **49**, 523–525.

Sekul, A. A. and Ory, R. L. (1979b). Proc. Beltwide Cotton Prod. Res. Conf., Special Session on Cotton Dust, 1979, Phoenix, Arizona, 30–32.

Sekul, A. A., and Ory, R. L. (1981). *J. Am. Oil Chem. Soc.* **58**, 690–692.

Sobolev, A. M., Suvorov, V. I., Safonova, M. P., and Prokof'ev, A. A. (1972). *Sov. Plant Physiol.* **19**, 894–899.

Speer, F. (1976). *Am. Fam. Physician* **13**, 106–112.

Spies, J. R. (1974). *J. Agric. Food Chem.* **22**, 30–36.

Spies, J. R., and Coulson, E. J. (1943). *J. Am. Chem. Soc.* **65**, 1720–1725.

Spies, J. R., and Umberger, E. J. (1942). *J. Am. Chem. Soc.* **64**, 1889–1891.

Spies, J. R., Bernton, H. S., and Stevens, H. (1941). *J. Am. Chem. Soc.* **63**, 2163–2169.

Spies, J. R., Coulson, E. J., Chambers, D. C., Bernton, H. S., and Stevens, H. (1944). *J. Am. Chem. Soc.* **66**, 748–753.

Spies, J. R., Coulson, E. J., Chambers, D. C., Bernton, H. S., Stevens, H., and Shimp, J. H. (1951). *J. Am. Chem. Soc.* **73**, 3995–4001.

Tronier, B., and Ory, R. L. (1970). *Cereal Chem.* **47**, 464–471.

von Pirquet, C. (1906). *Muench. Med. Woechenschr.* **53**, 1457.

Weeke, B., and Løwenstein, H. (1973). *Scand. J. Immunol.* **2**, supplement 1, 149–153.

Wilson, M. R., Sekul, A. A., Ory, R. L., Salvaggio, J. E., and Lehrer, S. B. (1980). *Clin. Allergy* **10**, 303–308.

Youle, R. J., and Huang, A. H. C. (1976). *Plant Physiol.* **58**, 703–709.

Youle, R. J., and Huang, A. H. C. (1978a). *Plant Physiol.* **61**, 13–16.

Youle, R. J., and Huang, A. H. C. (1978b). *Plant Physiol.* **61**, 1040–1042.

Youle, R. J., and Huang, A. H. C. (1979). *J. Agric. Food Chem.* **27**, 500–503.

CHAPTER 5

Immunochemistry of Seed Proteins

J. DAUSSANT AND A. SKAKOUN

Laboratoire de Physiologie des Organes Végétaux C.N.R.S. 4 ter, Route des Gardes, 92190 Meudon, France

I. Introduction

Studies on seed proteins cover various aspects including identification, characterization, modifications during the seed life, localization and regulation. These studies concern proteins with known biological activity such as enzymes, inhibitors, phytohaemagglutinins as well as large protein groups generally referred to as "storage proteins". In the last two decades and particularly in the last few years, immunochemical methods have been increasingly used in different aspects of these studies. The uniqueness of the immunochemical approach is based on the specificity of the reaction between the biological reactive used in these methods and the proteins to be analyzed. Moreover, the numerous developments carried out on the immunochemical methods provide many possibilities of application.

There is a large amount of literature concerned with immunochemical methods, and reviews dealing with their application to studies on plant proteins have been already published (Daussant, 1975; Daussant *et al.*, 1977; Daussant and Skakoun, 1981; Knox and Clarke, 1978; Manteuffel, 1982). Therefore this report aims at presenting a survey of the main immunochemical methods useful in studies on seed proteins taking into account recent developments in the methodology and also present trends in research concerning seed proteins. The principles of the main groups of these methods will be summarized and the sort of information which can be obtained will be underlined. Difficulties inherent to the application of this methodology to seed protein studies will be mentioned.

II. CHARACTERISTICS OF THE ANTIGEN–ANTIBODY REACTION

Immunochemical methods involve the use of a biological reagent, the immune serum. Its active constituents are the antibodies. They are produced by the immune system of higher vertebrates, a system which recognizes and then rejects foreign constituents which may have invaded the organism. The system is well developed both at the cellular and at the molecular levels and may include non specific and specific responses against foreign constituents. The formation of antibodies is part of the specific response of higher vertebrates.

The injection of one protein, called antigen, into a higher vertebrate induces the formation of antibodies with the following characteristics:

(1) The antibodies can specifically react *in vitro* with the protein which induced their formation.

(2) The injection of one protein results in the formation of a family of antibodies. These antibodies are each specific for a distinct limited area on the protein surface called the antigenic determinant. One protein molecule is thus immunochemically characterized by several distinct antigenic determinants.

(3) There are several classes of antibodies. The main class in mammals consists of immunoglobulins G (IgG). They are formed by two identical heavy chains bound together by disulphide bonds and by two identical light chains each bound to one heavy chain by a disulphide bond. The IgG differ from other immunoglobulins by the type of their heavy chains. The IgG are involved in the methods

dealt with in this report. On each antibody molecule there are two antigen binding sites each situated between one light and one heavy chain in the N terminal part of these chains. The two antigen binding sites of one antibody molecule are specific for the same antigenic determinant. Thus, one antibody molecule will react with two identical antigenic determinants generally situated on two distinct antigen molecules.

(4) An immune serum which contains antibodies against one single protein only is referred to as a monospecific immune serum. When a mixture of proteins is used for immunization, the immune serum obtained contains several families of antibodies each of which is specific for one protein. In this case, the immune serum can be referred to as a polyspecific immune serum.

The specific reaction between antigens and antibodies is used in different types of methods each of which includes various techniques. These methods can be grouped as follows:

(1) Methods based on precipitation between antigens and antibodies following the formation of the immunocomplex. This reaction can be carried out in solution or in gels.

(2) Methods of passive agglutination which involve the use of particles (red cells or latex particles for instance) coated with the antigens.

(3) Methods of immunohistology in which antibodies are labelled with a marker (such as a fluorescent dye, ferritin or enzymes) and which are used for localizing antigens using light or electron microscopy.

(4) Methods involving the immobilization of either antigens or antibodies on an insoluble support or on a matrix.

(5) Methods combining immobilization of antigens or antibodies with the labelling of one of the reactants with radioactive molecules (radio immunoassay) or with enzymes (enzyme linked immunosorbent assays).

(6) Methods involving the property of the antibodies to bind to a series of seric constituents once the antibodies are bound to the corresponding antigen (methods of complement fixation).

III. Analysis of Solubilized Proteins

This section will deal mainly with the application of the classical methods based on the reaction of precipitation carried out either in test tubes or in gels and with the more recent methods involving the enzyme linked immunosorbent assays.

A. METHOD OF ANTIGEN–ANTIBODY SPECIFIC PRECIPITATION

One characteristic of the precipitation of the immunocomplexes formed in solution between proteins and their corresponding antibodies will briefly be outlined here.

When adding increasing amounts of one antigen to a constant amount of the corresponding antibodies an immunoprecipitate forms. This reaction is usually carried out by exposing the antigen–antibody mixtures at 37°C for 30 minutes, then overnight or for several days at 4°C. The precipitates are washed and then measured. The precipitate first increases proportionally to the antigen added, then remains the same and finally decreases. The maximum of precipitation is called equivalence zone. In excess of antigen, part of the immuncomplexes is soluble.

The curve of precipitation is used for titrating one protein in a protein mixture and the ascending part of the curve where there is no excess of antigen is used. In this application the use of a monospecific immune serum is essential.

The classical reaction of precipitation became, more recently, an essential complementary means to the analysis of one protein species among the products synthesized in an acellular system. The specific precipitation aims at separating one specific protein from all other proteins synthesized. Three procedures for this separation are shown in Fig. 1.

In the single antibody precipitation procedure, carrier antigen and

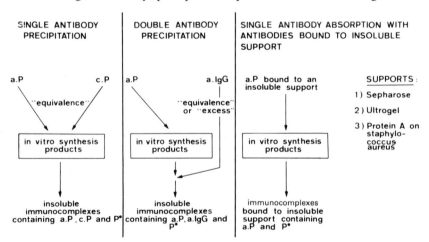

FIG. 1. Use of antibodies for separating one protein from the products synthesized in an acellular system. P* labelled protein P, one of the products synthesized in the *in vitro* system; c.P purified protein P used as carrier; a.P rabbit antiprotein P immune serum (generally the IgG fraction of the immune serum or the antibodies anti-protein P separated from the IgG fraction); a.IgG goat or sheep immunoglobulin fraction of an immune serum specific for rabbit immunoglobulins G.

antibodies are added in proportions corresponding to the equivalence zone in order to get the maximum amount of precipitate (the precipitate includes the newly synthesized antigen, the quantity of which is very small compared to the amount of the carrier antigen). This procedure was used in studies concerning viral proteins synthesized by particles in beans which had been infected by the virus (Higgins *et al.*, 1976a), storage proteins in peas (Higgins and Spencer, 1977; Croy *et al.*, 1980a,b) and storage proteins in beans (Sun *et al.*, 1978).

For double antibody precipitation, rabbit antiprotein antibodies are first added and form complexes with the protein. Thereafter goat anti rabbit IgG are added in proportions corresponding to the equivalence zone in order to guarantee the precipitation of all rabbit IgG, particularly those complexed with the antigen. This procedure was used in studies on barley and wheat α-amylases (Higgins *et al.*, 1976b; Okita *et al.*, 1979), on maize zein (Wienand and Feix, 1978), on storage bean or field bean protein (Hall *et al.*, 1978; Püchel *et al.*, 1979), on glyoxysomal castor bean proteins (Lynne and Lord, 1979).

The third type of procedure reported on Fig. 1 does not involve a specific immunoprecipitation but the use of immobilized antibodies on insoluble supports. In this system the support is kept in suspension, the antigen will absorb on the antibodies bound to the support. The immobilization of the antibodies may reduce the specific binding capacity of the antibodies because some parts of the antibody molecules close to one antibody site may bind to the support. This is probably one cause of the very low absorption capacity of the antibodies bound to Sepharose which was underlined by Croy *et al.* (1980a) in a study on pea storage proteins. In contrast, the interaction between the protein A purified from the cell wall of *Staphylococcus aureus* does not involve the antigen binding site of the immunoglobulin G (Forsgren and Sjöquist, 1966; Kronvall and Frommel, 1970), thus, the immobilized antibodies remain very active. The whole killed bacterium was used as such for absorbing the immunocomplex from the *in vitro* protein system in studies on pea storage proteins (Bollini and Chrispeels, 1979) and on castor bean glyoxysomal proteins (Bowden-Bonnett and Lord, 1979). In this procedure the absorption capacity of the protein A is higher for the antigen–antibody complexes than for the free antibodies, and the absorbed immunocomplex cannot be displaced by large amounts of extra free antibodies or other IgG molecules (Kessler, 1975). Purified protein A immobilized on a support can be also used.

The use of antibodies in the *in vitro* protein synthesis systems constitutes an intermediary step only. However, it represents a unique means of getting information concerning the synthesis regulation of one protein. Aspects of the immunochemical identification of proteins synthesized *in vitro* will be discussed at the end of this chapter. Difficulties inherent to the application of these techniques were underlined by Evans *et al.* (1979) and Croy *et al.* (1980a): the use of the double antibody precipitation resulted in the modification of the electrophoretic patterns caused by the presence of excessive amounts of IgG molecules; the single antibody precipitation as well as the double antibody precipitation involving immobilized goat anti rabbit IgG antibodies resulted in considerable non specific binding. In order to prevent coprecipitation or non specific absorption, particularly when microsomal membranes are present in the *in vitro* synthesis systems, the precipitation is generally carried out in the presence of detergents (0·1% to 1% Triton x 100, sometimes together with 1% deoxycholate, or 0·1% SDS).

B. METHODS OF ANTIGEN–ANTIBODY SPECIFIC PRECIPITATION IN GELS

In the late forties new techniques were developed in which the reaction of precipitation was carried out in gels. The pioneers for this new methodology were Oudin (1946) for single diffusion, Ouchterlony (1949) for double diffusion and Grabar and Williams (1953) for immunoelectrophoretic analysis. Four basic methods from which a large number of techniques were developed are shown on Fig. 2. In contrast to the method of precipitation in test tubes, the techniques of precipitation in gels are particularly appropriate for distinguishing from each other the different precipitates obtained by a polyspecific immune serum and the corresponding antigens.

The techniques of immunoprecipitation in gel can be combined with other biochemical preparative or analytical methods such as electrophoresis of different types (Smyth *et al.*, 1977 for review). There is indeed a great number of variations of these techniques and they have found many applications (Axelsen *et al.*, 1973; Axelsen, 1975; Kaminski, 1979 for reviews). Moreover, when the antigen is an enzyme, it is generally possible to characterize the activity on the corresponding precipitin band in the gel (Uriel, 1971). The methods of precipitation in gel have already been extensively used in seed protein analyses and there are recent reviews concerning their application to studies on plant proteins (Daussant, 1975; Manteuffel, 1982), soybean proteins (Catsimpoolas, 1977), barley seed proteins (Daussant, 1977), wheat proteins (Ewart, 1977), peanut proteins (Neucere, 1977), plant agglutinin (Jaffé, 1977), proteinase inhibitors (Ryan, 1977), oil seed allergens (Spies, 1977), plant and seed enzymes (Daussant *et al.*, 1977; Daussant and Skakoun, 1981; Manteuffel, 1982). Therefore a few specific points concerning studies on characterization and on physiological and molecular genetic aspects will be dealt with here.

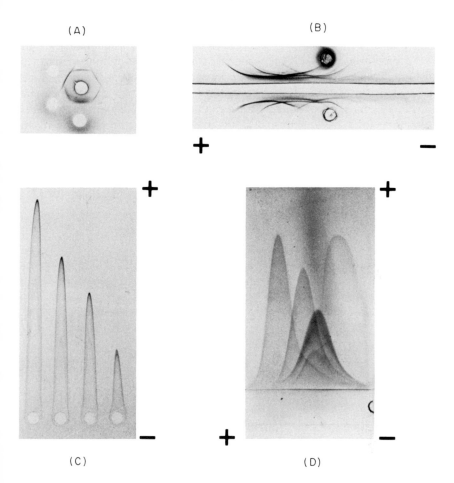

(A)   (B)

+   −

+   +

−   +   −

(C)   (D)

FIG. 2. Basic methods of immunoprecipitation in gel.

(A) Double diffusion according to Ouchterlony (1949). The central well is filled with the immune serum. The outer wells are filled with the protein solutions. Antigens and antibodies diffuse, meet each other and form precipitin arcs.

(B) Immunoelectrophoretic analysis according to Grabar and Williams (1953). Antigens are first separated by electrophoresis. After electrophoresis a trough is cut in the gel parallel to the axis of electrophoretic migration and filled with the immune serum. Antigens and antibodies diffuse, meet and form precipitin arcs.

(C) Electroimmunodiffusion, or, rocket-immunoelectrophoresis (Laurell, 1966). The antigen solutions are deposited in wells cut in a gel containing antibodies. Electrophoresis is carried out at a pH at which the antibodies do not move or if so only slightly. The antigen move into the gel and form immunocomplexes first in an excess of antigen. During the migration, due to electrophoresis and also diffusion, the immuno complexes become richer in antibodies and when their proportion (Ag/Ab) is optimal they form precipitates which occur as peaks (rockets). The surface of the peaks is linearly related to the amount of antigen deposited in the wells (Clarke and Freeman, 1967).

(D) Crossed immuno electrophoresis (Ressler, 1960; Laurell, 1965). Antigens are separated by electrophoresis before being submitted to electroimmunodiffusion.

1. Characterization Studies

These methods constitute one criterion for testing the homogeneity of a purified protein. Usually the purified fraction is tested by using an immune serum specific for the constituents of the crude extract from which the protein was purified or specific for a putative contaminant. This test was used for biologically active proteins such as a lectin prepared from *Bandeiraea simplicifolia* seeds (Hayes and Goldstein, 1974), a mung bean inhibitor (Chrispeels and Baumgartner, 1978) or for storage proteins such as a glycoprotein purified from *Phaseolus aureus* (Pusztai and Watt, 1970), vicilin, legumin, narbonin from *Vicia* (Manteuffel and Scholz, 1975; Schlesier *et al.*, 1978).

The antigens identified with an immune serum may each be composed of several constituents differing in charge or molecular weight. Numerous examples of this sort exist concerning enzymes (Daussant *et al.*, 1977; Daussant and Skakoun, 1981 for reviews). This is also the case for other biologically active proteins such as lectins (Petryniak *et al.*, 1977). Seed storage proteins also present this characteristic as reported for vicilin and legumin from *Vicia faba* (Scholz and Manteuffel, 1975; Millerd *et al.*, 1978; Thomson *et al.*, 1978). Concerning pea, no antigenic differences were found between two constituents of vicilin with different electrophoretic mobilities. Furthermore, no antigenic differences were found between the constituents which were bound to Concanavalin A and those which did not bind (Davey and Dudman, 1979). The storage proteins of wheat and barley (the prolamins) exhibit large electrophoretic polymorphism but they correspond to only a few distinct antigens (Ewart, 1977; Kling, 1975; Laurière, 1981). Thus, the antigenic definition of seed proteins appears less efficient for establishing differences between constituents than other tools, particularly electrophoresis. However, the antigenic identity or similarity shown between constituents separated by other tools indicates that there are structural relationships between them. Gradual antigenic similarities which can be established between proteins are used as markers in evolutionary and taxonomic studies because they indicate a more or less closed structural relationship between the proteins (see chapter 6). One particularly interesting example is provided by a study showing common antigenical structures between lectins from different legumes and particularly with the lectin from *Vigna radiata* which shows a glycolytic hydrolase activity: the results indicate that most of the legume lectins are evolutionarily closely related proteins and they suggested furthermore that lectins are enzymes which have lost their activity during evolution but have kept their specificity for their substrate (Hankins *et al.*, 1979).

Conversely, in some cases, constituents which are not distinguished by electrophoresis or by their molecular weight may appear antigenically distinct. This is the case for instance for enzyme constituents (Daussant and Skakoun, 1981), for one constituent of the vicilin of *Vica faba* which is not distinguished from the legumin by polyacrylamide gel electrophoresis and which was shown antigenically unrelated to the legumin by comparing this fraction and the legumin fraction by using crossed immunoelectrophoresis (Scholz and Manteuffel, 1975). In those cases the antigenic definition can solve a difficult problem of identification showing that the two constituents are structurally quite different.

In order to illustrate these two sorts of results, Fig. 3 shows the combination of pore gradient polyacrylamide gel electrophoresis (Margolis and Kenwick, 1968), a non denaturing analysis for determining the molecular weight of globular proteins, with the crossed immunoelectrophoresis (Daussant and Skakoun, 1975). On Fig. 3(A), the two constituents of the main globulin fraction purified from sunflower seeds (Baudet and Mossé, 1977) appear as two oligomeric forms of the same constituent. The antigenic identity obtained between the two constituents confirms that hypothesis. In contrast, two barley malate dehydrogenases which have very close molecular sizes (Fig. 3(B.V.)) correspond to two distinct antigens (the two precipitin bands cut each other as shown on Fig. 3(BIV) and can therefore be considered as two distinct molecular species).

It is worth noting that immunoelectrophoretic analysis was proposed as a basis for the nomenclature of barley proteins (European Brewery Convention, 1967). Crossed immunoelectrophoresis which incorporates a quantitative dimension is still found to be a proper means of identifying seed constituents as shown by Millerd *et al.* (1978) and Guldager (1978) in studies on *Vicia faba* and pea seed proteins. The immunochemical approach is still useful for solving certain identification problems such as the identification of the allergen in castor bean to the low molecular weight storage protein of these seeds (Youle and Huang, 1978). The immunochemical identification of a particular trypsin inhibitor in soybean provided a unique means of quantitating this constituent in a crude extract containing many other inhibitors (Freed and Ryan, 1978).

A practical problem of identification of seed proteins is the detection and quantitative evaluation of seed proteins in food products which may have been heat denatured. Among the many techniques assayed for this purpose, immunoprecipitation and particularly immunoprecipitation in gel have been used (see for recent reviews Baudner, 1978; Daussant, 1981; Llewellyn, 1979; Olsman and Hitchcock, 1980).

2. Physiological Studies

In recent studies, qualitative and quantitative information was obtained by using crossed immunoelectrophoresis with polyspecific immune sera or

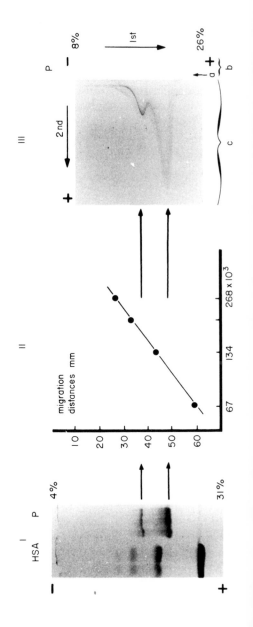

(A)

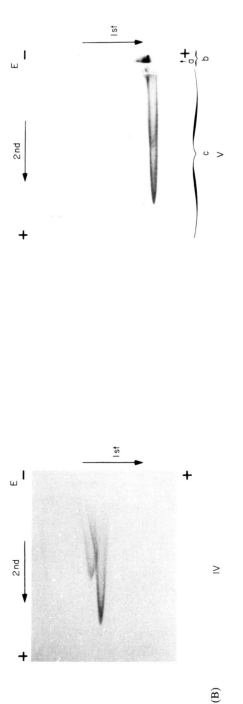

(B)

IV

FIG. 3. Combination of pore gradient polyacrylamide gel electrophoresis (PGAGE) with crossed immunoelectrophoresis.

(A) Purified sunflower globulin. Amidoblack staining. I: Long duration PGAGE (110 v, 36 hours, 4°C). HSA, human serum albumin; P, purified globulin. The migration distance of each of the constituents becomes dependant on their molecular size only when the electrophoresis duration is sufficiently long (35 hours for HSA and this globulin). II: Relationship between HSA oligomer migration distances and the logarithm of their molecular weights. Taking this scale as reference, the globulin constituents can be estimated at 110 000 and 200 000 daltons. III: Crossed agarose gel electrophoresis. PGA gel slab (a) was taken after the first dimension electrophoresis (1st) of P and embedded in agarose gel (b). The second dimension electrophoresis (2nd) was carried out in agarose gel containing the antiglobulin immune serum (c) under 1·5 v/cm for 13 hours at 20°C. The two globulin constituents form a continuous precipitin line.

(B) Barley malate dehydrogenases (MDH). The barley extract (E) was submitted to the first electrophoresis (1st) in PGA gel. a, b, as for (A) III. An immune serum anti-barley proteins was included in the agarose gel (c) for the second dimension immunoelectrophoresis (2nd). MDH characterization. IV: After a short duration electrophoresis in the first dimension (5 hours), two MDH forms were detected. Antigenic differences between them can easily be seen: the precipitin bands cross each other. V: After a long duration electrophoresis (51 hours) the migration distances were solely dependent on the molecular weights. The two MDH forms do not differ by their migration distances.

rocket immunoelectrophoresis with monospecific immune sera. The information concerns the sequential appearance, the accumulation and the decrease of each of the immunochemically defined constituents as well as the changes in their electrophoretic migration. A precise quantitative evolution of the different storage proteins can be followed during seed development, ripening and germination by using purified protein preparations and by establishing the proper reference scales. Thus, it is possible to evaluate the quantitative importance of each of these proteins in relation to each other and to the total protein amount in the seeds. These techniques were used in recent studies concerning storage proteins in *Vicia faba* (Lichtenfeld *et al.*, 1979; Manteuffel *et al.*, 1976), *Phaseolus vulgaris* (Mutschler, 1977; Mutschler *et al.*, 1980; Sun *et al.*, 1978) and pea (Guldager, 1978; Millerd *et al.*, 1978) (see also Chapter 3 for lectins). The methods were also used to evaluate the effect of deficiency of certain minerals during plant growth on the production of each of the protein species (Randall *et al.*, 1979; Millerd *et al.*, 1979). They were used for studying the capacity of *Phaseolus vulgaris* cotyledons to produce the storage proteins when they were detached at different steps of their development and placed in *in vitro* culture (Millerd *et al.*, 1975). Proteins of developing mitochondria at the onset of germination in maize (Ivanov and Khavkin, 1976) and the trypsin inhibitor of soybean during germination (Freed and Ryan, 1978) were also investigated by using these methods.

Important information can be obtained by the detection of precursor forms of the proteins, if these forms are recognized by the antibodies specific for the "mature" protein. This was shown in studies on enzymes (Daussant *et al.*, 1977; Daussant and Skakoun, 1981 for reviews). An example of the sort was reported concerning two proteins of pea during seed development. The proteins were detected with only part of their antigenic structure at an early step of seed development, and gradually these forms vanished as the complete forms increased (Millerd *et al.*, 1978). The immunochemical approach was used recently for detecting incomplete or functionally inactive proteins in several studies concerning molecular genetics; when an organism is deprived of enzymatic activity this may be caused by modification of the protein structure or by a lack of synthesis of active protein. That was the case for a maize mutant which lacked a catalase isoenzyme (Tsaftaris and Scandalios, 1981) and several soybean inbred lines which lacked a lectin (Su *et al.*, 1980). In both cases the absence of function was related to the absence of antigenic protein. The same problem appeared with certain inbred lines of rye in which the β-amylase activity was very low and amounted to 1 to 3% of the activity found in other inbred lines. This activity was shown to be displayed by the same single antigen in the normal and in enzyme deficient inbred lines. It was found that the very small activity in the β-amylase deficient inbred lines corresponded to a similar small amount of the β-amylase antigen. In these three studies the results indicate that a defect in the production of the functional protein was the cause of the lack or of the drastically reduced activity. The last example suggests that regulatory genes are involved in this abnormal production of protein (Daussant *et al.*, 1981).

The combination of immunoelectrophoresis or crossed immunoelectrophoresis and *in vivo* or *in vitro* incorporation of labelled amino acids into proteins is of particular interest in these studies as shown by the following examples: the *in vivo* incorporation of labelled amino acids during germination of wheat seeds (Daussant and Corvazier, 1970) combined with immunoelectrophoresis showed that some of the proteins already existing in mature seeds, but not all of them, continue to be synthesized upon germination. New protein types were also synthesized, namely two distinct antigenic α-amylases. However, the results showed that strong modifications in the electrophoretic mobility observed for the β-amylases were to be ascribed essentially to post translational events. This combination was used to investigate the rate of synthesis of vicilin and legumin in the cotyledons of *Vicia faba* seeds taken at different stages of their development. The cotyledons were exposed in conditions where tritiated amino acids could be incorporated into proteins. The radioactivity was measured on the specific precipitates obtained from the extracts by using respectively the mono specific anti vicilin or anti legumin immune sera. The results showed that the rate of vicilin synthesis is higher than the rate of legumin synthesis at early stages of maturation, whereas the reverse situation appeared with seeds taken at later stages of maturation (Neumann and Weber, 1978). The incorporation of labelled sulphur amino acids during pea maturation led to the identification of the antigens containing sulphur amino acids (several albumins and the legumin but not the vicilin and the phytohaemagglutinin) (Guldager, 1978). An interesting use of crossed immunoelectrophoresis for analysing proteins synthesized *in vitro* should be mentioned because it does not involve the denaturation of the immunocomplex by SDS polyacrylamide gel electrophoresis and does not need the use of monospecific immune sera. Carrier proteins are added to the *in vitro* system in quantities which give a good immunoelectropherogram. After autoradiography it is possible to identify the precipitin bands which are radioactive, and consequently the antigens which have been synthesized *in vitro*. With this approach Püchel *et al.* (1979) showed that the globulins of *Vicia faba* are preferentially produced on membrane bound polysomes and that the poly A containing RNA of developing seeds includes the mRNA for both vicilin and legumin.

C. ENZYME LINKED IMMUNOSORBENT ASSAY, ELISA

Other well developed techniques for analyzing proteins in solution have a much higher sensitivity than the techniques of immunoprecipitation. That is the case for the passive agglutination (Stavitsky, 1977; Litwin and Bozicevich, 1977 for reviews), for the micro complement fixation (Levine, 1973 for review) or for the radio immuno-assays (Hunter, 1973 for review). The relatively recent ELISA technique (Engvall and Perlmann, 1971) is used increasingly in protein analysis and is now being applied to seed protein studies. There are several variations to the method (Voller *et al.*, 1979 for review), two of which are shown in Fig. 4. The methods are characterized by a high degree of sensitivity and because of this characteristic they were used for testing the absence of any cross reactivity between vicilin and legumin with each of the corresponding immune sera. Less than 10 ng storage proteins could be detected (Craig *et al.*, 1980). ELISA can be used for titrating one protein in a mixture as exemplified in a study on the *in vivo* and *in vitro* initiation of legumin synthesis in immature embryos of pea. Nanogram amounts of the storage protein could be detected (Domoney *et al.*, 1980). The method was also proposed for detecting soybean proteins in meat products although it does not solve the problem of immunochemical quantitation of proteins which may be denatured to different degrees (Hitchcock *et al.*, 1981). Its application underlined a practical aspect of the method which was found very convenient for carrying out a large number of inexpensive determinations.

IV. Localization

An important aspect of seed protein studies is their tissue and subcellular localization. The use of antibodies monospecific for one protein provides a unique means of localizing one particular protein species which does not show any specific characteristic other than the antigenic one. The principles of these methods applied to enzymes have been reviewed (Coombs and Franks, 1969; Wachsmuth, 1976) and the use of antibodies in studies on plants by electron microscopy have been described also (Knox and Clarke, 1978). Three immunohistological techniques are diagrammatically represented in Fig. 5.

In these techniques, antibodies are covalently bound to a label such as a fluorescent molecule (Coons, 1956), ferritin (Singer, 1959), an enzyme (Avrameas, 1970) (Fig. 5(A),(B)) or also they become specifically labelled by an enzyme in the course of the reactions involved in the procedures

(Wachsmuth, 1976 for review) (Fig. 5(C)). When an enzyme is used as a label, a reaction of enzymatic characterization is carried out; fluorescent molecules or enzymes can be used with light microscopy and ferritin or enzymes with electron microscopy.

The procedure of protein fixation in tissues should neither alter the antigenic structure of the protein nor prevent the antibodies from reaching the antigens. These questions and also the choice of the proper label are discussed in the reviews on the immunohistology of enzymes (Wachsmuth, 1976) and of plant proteins (Knox and Clarke, 1978).

Autofluorescence may be an obstacle to the use of the immunofluorescence techniques. As far as seeds are concerned, the use of 2·5% glutaraldehyde for the fixation of the bean cotyledon tissues apparently did not result in non specific fluorescence (Graham and Gunning, 1970). Autofluorescence appeared, however, when the wheat seed tissue was prepared with 3% glutaraldehyde. Nevertheless, this kind of autofluorescence could be distinguished from the fluorescence due to labelled antibodies (Barlow, 1973). In pea and in other seeds, the fixation of the tissues was found to create too high a level of autofluorescence and 4% paraformaldehyde was therefore used instead of glutaraldehyde (Craig et al., 1979, 1980).

The earliest application of immunofluorescence in seeds (by using a direct technique) established the sequential appearance of vicilin (first) and legumin in the developing bean cotyledons and showed that most protein bodies contained both the vicilin and the legumin. One question, however, remained unanswered: a few protein bodies remained unlabelled by the fluorescent antibodies. This could be due either to the existence of protein bodies containing albumins instead of globulins, or to the limitation of the immunohistological technique, the globulins being in these protein bodies inaccessible to the antibodies (Graham and Gunning, 1970). Similar results were also obtained for peas during their ripening by using an indirect immunofluorescence technique (Craig et al., 1979, 1980). Vicilin and legumin appeared as deposits in vacuoles which became filled by the deposits during ripening. These deposits were then named protein bodies. Furthermore, the results indirectly showed that both globulins were stored in the same protein deposits during development. Immunofluorescence appeared to be more sensitive than the immunoelectrophoretic technique since vicilin and legumin could be detected in this study at an early stage of development whereas they were not detected at the same stage by using immunoelectrophoresis in an earlier study (Millerd et al., 1975). The immunohistological approach appeared particularly appropriate for checking experiments which suggested an association of a high proportion of the water soluble, non-gluten forming proteins, with starch granules in mature wheat endosperm cells (Barlow, 1973). Antibodies

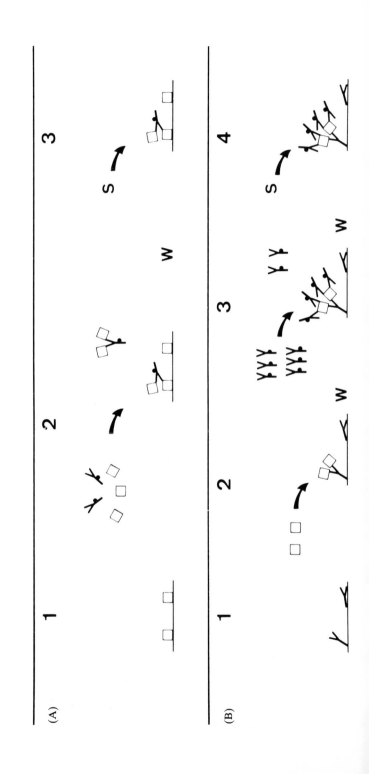

□ antigen

Y antibody

Y antibody labelled with an enzyme

W washing

S substrate

FIG. 4. Diagrammatic representation of two ELISA techniques.

(A) 1. Antigens are immobilized. 2. Standard amounts of enzyme labelled antibodies are added together with the free antigen (either reference samples or unknown samples). There is competition between the free and the immobilized antigens to bind to the antibodies. 3. After washing to eliminate unabsorbed materials, the substrate is added. The activity is maximum when no free antigen is added in step 2 and decreases proportionally to the amount of free antigen added. NB. A variation of this technique consists in using unlabelled antibodies at step 2 and then to add commercially available goat anti rabbit immuno globulin G antibodies which are labelled. They will react with the rabbit IgG specifically bound to the immobilized antigens. After washing, the substrate is added and the enzymatic activity is evaluated.

(B) 1. Antibodies are immobilized. 2. Free antigens are added (reference samples or unknown samples). 3. After washing, labelled antibodies are added which will bind to the specifically attached antigens. 4. After washing the substrate is added. The activity level depends on the amount of antigen added in step 2. NB. As for technique A, a variation of technique B avoids the use of labelled antibodies but involves a further step in which commercially available labelled goat antibodies specific for the rabbit IgG are used.

Another version of this technique involves the use of labelled antigen in step 2 instead of labelled antibodies in step 3. In step 2, a standard amount of free labelled antigen is added together with unlabelled antigen (reference samples or unknown samples). There is competition between unlabelled and labelled antigens to react with the attached antibodies. After washing substrate is added and the activity measured. In the absence of unlabelled antigen the activity is maximum. The activity decreases proportionally to the amount of unlabelled antigen added.

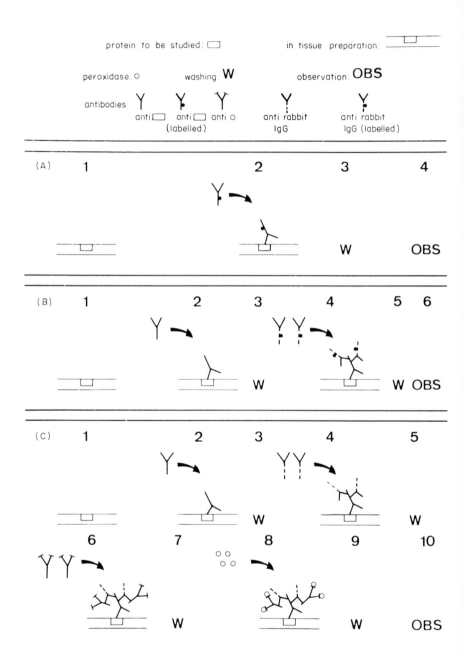

protein to be studied: ☐   in tissue preparation:

peroxidase: O   washing: W   observation: OBS

antibodies: Y Y Y   Y   Y
anti☐ anti☐ anti O   anti rabbit   anti rabbit
(labelled)   IgG   IgG (labelled)

(A)   1   2   3   4

☐   ☐   W   OBS

(B)   1   2   3   4   5 6

☐   ☐   W   ☐   W OBS

(C)   1   2   3   4   5

☐   ☐   W   ☐   W

6   7   8   9   10

☐   W   ☐   W   OBS

Fig. 5. Diagrammatic representation of three histochemical techniques.
(A) Direct technique. In this technique the antibodies specific for the protein to be studied
are labelled. 1. Proteins are fixed in the tissue. 2. The tissue is treated with the immune serum
(generally with the IgG fraction of the immune serum or with the antiprotein antibodies

specific for the proteins in an aqueous wheat extract were used for the indirect technique. The results indicated that the soluble proteins were found in high concentrations around each starch granule. Many enzymes, including β-amylase, are present in the aqueous extract and the results suggest that β-amylase could be localized on the starch granules. An immunohistological approach using anti β-amylase antibodies (in the indirect technique) showed that this was the case for rice (Okamoto and Akazawa, 1979). Furthermore, enzyme histological detection with starch as substrate failed to detect the enzyme on similar slices of rice seeds. The results therefore indicated that the β-amylase found in the rice seeds was inactive and corresponded to the "latent" or "bound" β-amylase. Different aspects of immunohistochemistry of enzymes particularly for the localization of inactive, precursory of isoenzymatic forms have been reported in recent reviews (Knox and Clarke, 1978; Daussant and Skakoun, 1981). One example concerning enzymes will be given here. It provides new information on the mutual importance of the scutellum and the aleurone layer in the production of α-amylase during the germination of whole barley grain (Gibbons, 1980a and b). Earlier studies on barley reviewed by Briggs (1978) and by Palmer (1980) indicated that the aleurone layer was dominant in this production. However, the indirect immunofluorescence technique carried out on slices of whole seeds taken at different stages of germination showed that the major α-amylase isozyme was localized in the epithelial part of the scutellum less than one day after soaking. It was then found spread out into the endosperm. The enzyme was detected in the aleurone layer only after several days of germination. This study (Gibbons, 1980a and b) underlined the important part played by the scutellum in the production of α-amylase, a fact that

separated from the IgG fraction). One or several antibodies bind to each antigen. 3. Washing in order to eliminate proteins unspecifically bound to the tissue preparation. 4. Observation with the light or the electron microscope depending on the label used on the antibodies.

(B) Indirect technique. In this technique the anti protein antibodies are not labelled. However, a further step is added which involves the use of antibodies specific for the IgG of the animal species used for preparing the antiprotein immune serum (generally rabbits). The anti rabbit IgG antibodies are prepared in goats or sheep and are conjugated with a fluorescent molecule or with an enzyme (generally peroxidase or alkaline phosphatase). These are commercially available. Several antibodies bind to each IgG molecule which were previously specifically fixed.

(C) Indirect technique with uncovalently bound label. In contrast to the two preceeding techniques which involve a label covalently bound to antibodies, this technique needs no covalent binding but the use of three antibody specificities and of one enzyme (peroxidase for example). The first five steps in C are the same as the first five steps in B with the difference that the unlabelled goat anti rabbit IgG antibodies are used here. 6. Addition of rabbit antiperoxidase antibodies—these antibodies will bind to the free antibody site of the goat antirabbit IgG already specifically bound to the antiprotein antibodies preparation. 7. Washing. 8. Addition of peroxidase. 9. Washing. 10. Observation.

seems to have been overlooked in preceding studies. However, in this immunohistological approach, the presence of α-amylase in the aleurone layer may have been underestimated for reasons inherent to the technique: seed tissues may present differences in their permeability towards the antibodies especially at the beginning of germination. Thus, the antigen may be easily detected in the scutellum but only with difficulty in the aleurone layer. For example, this difficulty was encountered with pumpkin seeds. It was impossible to detect by immunofluorescence a globulin present in the protein bodies by using antibodies specific for this globulin (Hara and Matsubara, 1980).

An immunochemical approach different from the immunohistological one can be used for localizing proteins in tissues or subcellular particles. It consists in analyzing the proteins extracted from tissue or subcellular preparations by using one of the numerous immunochemical techniques based on specific precipitation in gel as illustrated by the following examples. The protein composition of protein bodies in barley seeds was studied by using the immunoelectrophoretic analysis of proteins extracted from subcellular preparations. Hordein was found to be confined inside the protein bodies whereas the fine structure and the membrane surrounding the protein bodies contained several other antigens (Tronier et al., 1971). Quantitative immunoelectrophoretic techniques were also used for characterizing protein bodies and their membrane as shown in a study on the protein bodies from mature and germinating *Vicia faba* seeds (Weber et al., 1978, 1979). Vicilin, legumin and a lectin were shown to be aleurins and the aleurone membrane did not contain any of these storage proteins (particularly the lectin). In a study on the role of the endoplasmic reticulum in glyoxysome formation in castor been seed endosperm it was shown that the immune serum anti glyoxysomal proteins strongly reacted with the proteins of the endoplasmic reticulum membranes prepared from the endosperm at an early stage of germination when the glyoxysomes were being produced. This result supported the view that glyoxysomes are derived directly from the endoplasmic reticulum (Gonzales and Beevers, 1976). The approach has been used also for enzymes (Daussant and Skakoun, 1981 for review). This approach and the immunohistochemical technique were both used for localizing a trypsin inhibitor in mung bean cotyledon which was found associated with the cytoplasm (Chrispeels and Baumgartner, 1978).

V. IMMUNOABSORPTION

Immunoabsorption represents a means of eliminating from a protein solution a given antigen by using the corresponding antibodies. Conversely, immunoabsorption can be adapted for removing from an immune serum certain antibodies by using the corresponding antigen.

A. IMMUNOPRECIPITATION

Immunoabsorption was generally carried out by using the reaction of precipitation for preparative or analytical purposes in seed protein analysis. In order to get an immune serum specific for the new antigens appearing in the peanut cotyledons during germination, an immune serum specific for the proteins extracted from the cotyledons of germinated seeds was used. Proteins extracted from the cotyledons of ungerminated seeds were added to the immune serum. So, antibodies specific for antigens common to ungerminated and germinated seeds were removed by precipitation. The supernatant solution contained antibodies against new types of antigens appearing during germination only (Daussant *et al.*, 1969). The immunoabsorption was used for obtaining immune sera specific for either legumin or vicilin starting with one immune serum prepared by immunizing with a mixture of both bean proteins. Either vicilin or legumin were added to the immune serum in order to eliminate by precipitation all the corresponding antibodies (Graham and Gunning, 1970). Another type of immunoabsorption was used in order to get an immune serum monospecific for the legumin from *Vicia faba*. A purified preparation of the legumin was used for immunization. However, it contained vicilin as a contaminant. The contaminant was precipitated by adding small amounts of the antivicilin immune serum to the preparation of the purified legumin. After elimination of the specific precipitate, the supernatant was injected and monospecific immune sera were obtained (Scholz *et al.*, 1974). Castor bean allergen was identified as a low molecular weight storage albumin by using immunoabsorption as a complementary means of investigation. The immunoglobulin G fraction of the serum specific for the allergen was added to the albumin storage protein fraction and the precipitate formed was discarded (a similar amount of the IgG fraction from the serum of a non immunized rabbit was added to an aliquot of the albumin fraction). The comparison between polyacrylamide gel electropherograms of the treated and untreated aliquots showed that the albumin constituent was specifically removed by the action of the antiallergen antibodies (Youle and Huang, 1978).

B. IMMUNOAFFINITY CHROMATOGRAPHY

Methods of immobilizing biologically active proteins have considerably improved the technique of immunoabsorption and its possible applications. The principle and use of the immunoaffinity chromatography, in which either antigen or antibodies are bound to a support (generally Sepharose or Ultrogel) as well as the absorption and desorption procedures involved have been reported (Lowe and Dean, 1974). The two ways of using immunoaffinity chromatography by immobilizing either the antigen or the antibodies are diagramatically represented in Fig. 6. The

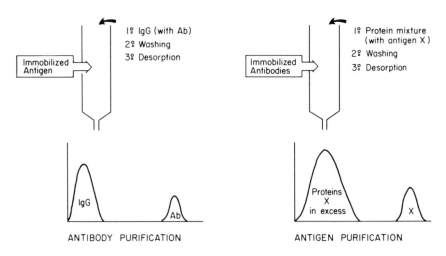

FIG. 6. Diagrammatic representation of immunoaffinity chromatography for the purification of antibodies or antigens. Ab, antibodies specific for the immobilized antigen; X, antigen which reacts with the immobilized antibodies.

solution is poured on to the column under conditions of minimum non specific binding by using a nearly neutral pH and sodium chloride in the solution. After washing, desorption is carried out by using low pH or chaotropic reagents. In order to prevent exposure of the desorbed proteins to the drastic conditions of desorption for too long a period of time, the desorbed fractions are immediately buffered at neutral pH. The column can be equilibrated at the initial pH and so can be used many times.

1. Use of Immobilized Antigens

The technique was used in seed protein studies mainly for purifying the antibodies from the IgG fraction of the immune sera. Use of the antibody fraction instead of the total IgG fraction in immunohistological techniques or in the immune precipitation of proteins in an acellular protein synthesis system indeed improves the quality of the results. Moreover, this procedure proved to be useful for obtaining monospecific antibodies from immune sera which were not monospecific.

An early application of this technique involved the use of crossed linked antigens by glutaraldehyde (Graham and Gunning, 1970). More recently, the immobilization of purified antigens on a support was used for purifying the corresponding antibodies in a number of studies. They concern antibodies specific for pea legumin and vicilin (Casey, 1979a; Croy et al., 1980a; Spencer et al., 1980), for vicilin and "Gl protein" from Phaseolus vulgaris (Bollini and Chrispeels, 1979; Sun et al., 1978), for vicilin from Vicia faba (Püchel et al., 1979), for the peptidohydrolase from mung beans (Baumgartner et al., 1978) and for barley β-amylase (Okamoto and Akazawa, 1979). Moreover, this procedure constitutes a powerful means for obtaining monospecific antibodies from an immune serum which is not monospecific, provided that the protein preparation used for immunoaffinity chromatography is highly purified (Okamoto and Akazawa, 1979). Based on this application of immunoaffinity chromatography, monospecific antibodies can now be obtained by using much reduced amounts of highly purified antigens: the crude protein is used for immunization and use of the highly purified protein is restricted to the immunoaffinity chromatography. This approach was defined and used for preparing antibodies monospecific for pea legumin (Casey, 1979a). The same approach was used for obtaining antibodies monospecific respectively for legumin and for vicilin, starting with an immune serum prepared by immunizing with the whole extract of protein bodies from pea cotyledons (Craig et al., 1980).

In particular cases, when a known contamining constituent is very difficult to separate from the antigen fraction to be injected, the corresponding antibodies may appear in the immune serum. These unwanted antibodies can be eliminated from the immunoglobulin fraction by passage of this fraction through a column containing the immobilized contaminant. That was used for eliminating the anti vicilin parasite antibodies contained in an immune serum specific for the legumin (Domoney et al., 1980; Evans et al., 1979).

2. Use of Immobilized Antibodies

This type of application is hampered by the fact that many proteins or enzymes are denatured by the drastic conditions of desorption normally used.

However, that is not important in cases when the desorbed proteins are analyzed by SDS polyacrylamide gel electrophoresis. The procedure was used for separating on one hand the entire protein body proteins, on the other hand the legumin from the whole proteins extracted from pea cotyledons at different stages of development (Spencer et al., 1980). This procedure was also used for separating proteins synthesized in an acellular protein synthesis system (Higgins and Spencer, 1981). The same conditions of desorption (3 M KSCN, pH 8) were also used for purifying the legumin from pea. These did not alter its antigenic structure (Casey, 1979a). Based on the cross reactivity between legumin from a range of *Pisum* types, the immunoaffinity chromatography was used for purifying the protein from different pea type seeds. In these experiments, the immunoaffinity column prepared in the preceding study was used. Purified legumin from different pea types were then compared by SDS polyacrylamide gel electrophoresis for their subunit composition (Casey, 1979b).

When an active protein has to be purified, the conditions of desorption have to be adapted to this protein. That was the case for bean α-mannosidase. The enzyme exists in two electrophoretically distinct forms. One of these forms was purified and the corresponding antibodies prepared. As the two forms were found to be closely related antigenically, immunoaffinity chromatography was used to purify the two forms together and subsequently they were separated from each other by isoelectric focusing (Paus, 1976).

However, the conditions of desorption are still an obstacle to the general application of immunoaffinity chromatography to the purification of biologically active proteins. Many studies have been carried out on methods for avoiding these drastic conditions: they involve the association of organic solvents with basic media (Andersson *et al.*, 1979), they proposed the use of antibodies which have less affinity with heterologous antigens (Erikson and Steers, 1970: Ruoslahti, 1978) or the chemical modification of immobilized antibodies (Murphy *et al.*, 1976). Electrophoretic desorption of antigens has also been proposed (Brown *et al.*, 1977; Morgan *et al.*, 1978). Recently, gentle and simpler procedures using distilled water (Vidal *et al.*, 1980) or the combination of distilled water with an interruption in the desorption (Bureau and Daussant, 1981) succeeded in desorbing enzymes without altering either their enzymatic activity or their antigenic specificity. Immunoaffinity chromatography will probably be increasingly used for purifying proteins and enzymes provided that gentle desorption procedures of general use are developed.

VI. Concluding Remarks

In the preceding sections some aspects of the use of antibodies in seed protein studies were mentioned, such as the need for monospecific antibodies in certain methods, the use of antibodies for recognizing precursory or denatured forms of proteins, the use of chaotropic reagents and of detergents for solving particular problems.

A. CONTROL OF THE MONOSPECIFICITY OF AN IMMUNE SERUM

Usually the monospecificity of an immune serum is tested by using double diffusion, immunoelectrophoretic analysis or crossed immunoelectrophoresis which have a certain degree of sensitivity. However, the immune serum may be used in techniques having a higher sensitivity for instance in *in vitro* protein synthesis where proteins are detected by their radioactivity or in immunohistological studies which involve the use of labelled antibodies. The problem was well illustrated in a study concerning the regulation of pea storage proteins, particularly the legumin. The monospecificity of the anti legumin immune serum was checked by immunoelectrophoretic analysis. However, the SDS polyacrylamide gel electrophoresis of the proteins immunoprecipitated from the acellular synthesis system indicated the presence of constituents migrating like vicilin constituents. The additional bands in the electropherogram suggested the presence of vicilin constituents precipitated by parasite anti vicilin antibodies in the anti legumin immune serum. Indeed when the IgG fraction of the immune serum was immuno absorbed on to a column containing vicilin, these parasite antibodies were eliminated and the treated fraction no longer precipitated proteins which migrated like the vicilin constituents after SDS polyacrylamide gel electrophoresis (Evans *et al.*, 1979). This example clearly showed that the monospecificity of the immune serum controlled by the classical techniques of specific precipitation in gels does not guarantee that the immune serum does not contain parasite antibodies. Their presence can be detected by more sensitive techniques.

An example of checking the monospecificity of an immune serum at the appropriate level of sensitivity was given in a study of *Phaseolus vulgaris*. An immune serum specific for vicilin was prepared in order to precipitate the protein from an *in vitro* protein synthesis system. The monospecificity was checked by double diffusion, immunoelectrophoretic analysis and also by SDS PAG electrophoresis of radioactively labelled antigens. The proteins were labelled *in vivo* by using ^{35}S methionine. The immune serum was used to precipitate vicilin from the cotyledon extract and the precipitate analysed by SDS PAG electrophoresis and visualized by auto fluorography. Only vicilin constituents were detected (Bollini and Chrispeels, 1979). This control of the monospecificity of the immune serum was particularly important in regard to some results of this study: under certain conditions of *in vitro* synthesis, additional bands were observed on the electropherogram. They corresponded to constituents smaller in size than the vicilin constituents. Since the antibodies were checked to be vicilin monospecific as mentioned before, the constituents could not be due to the appearance of constituents reacting with parasite antibodies in the serum.

Therefore, the presence of these constituents must be due to the presence of early termination products of the vicilin cross reacting with the anti complete vicilin antibodies (Bollini and Chrispeels, 1979).

Controlling the monospecificity of an immune serum means also controlling the absence of cross reactivity of the immune serum with other proteins. That was underlined in a study on the immunohistological localization of pea vicilin and legumin. The ELISA technique was used to control at an appropriate level of sensitivity the absence of cross reaction between vicilin and legumin which were previously purified by immunoaffinity chromatography (Craig et al., 1980).

B. IMMUNOCHEMICAL RECOGNITION OF MODIFIED ANTIGENS

The ability of antibodies to recognize to a certain extent the corresponding antigen when the antigen is present in a "modified" or incomplete form has been discussed in several studies.

Examples of inactive enzymes being recognized by anti active enzyme antibodies were previously reviewed (Arnon, 1973; Daussant et al., 1977). They indicated that most, generally all or part of the anti active enzyme antibodies, react with the inactive form of the enzyme, whatever the cause of the inactivation (inhibited enzyme, apoenzyme, mutated enzyme, enzyme precursors). This seems to apply also to other proteins such as a putative precursory form of the pea legumin which cross reacted with part of the anti legumin antibodies and which vanished as complete legumin appeared (Millerd et al., 1978).

The capacity of recognizing modified antigens is particularly important in the immunochemical separation of a protein from the products of in vitro protein synthesis. These proteins can be different from the "mature" protein either because the extra signal peptide is present or because some post translational modifications like glycosylation are lacking. As far as seed proteins are concerned, the references reported in Section III.A. indicate that generally the anti "mature" protein antibodies react with the in vitro synthesized proteins even when they have the extra signal peptide. It has been suggested even that early termination products could be recognized by the antibodies specific for the "mature" protein (Bollini and Chrispeels, 1979; Higgins et al., 1976b). However, it seems that there are some exceptions illustrated by conflicting results obtained in a study on pea globulin cell free synthesis programmed by seed polysomes. Tryptic peptide mapping of in vitro synthesized products indicated that a large proportion of the synthesized proteins were globulins. However, the antiglobulin antibodies precipitated only a small proportion of these globulins. The authors suggested that post translational events—including glycosylation—were necessary for the completion of the antigenic determinants (Higgins and Spencer, 1977). The phytohaemagglutinin (PHA) from *Phaseolus vulgaris* was separated from an *in vitro* protein synthesis sytem by using its ability to bind to thyroglobulin. However, the antibodies which react strongly with PHA failed to react with the protein synthesized *in vitro*. The results strongly suggest that the antigenic determinants are formed during post translational events and particularly that the addition of the polysaccharide moiety plays a major part in the formation of the antigenic structure (Bollini and Chrispeels, 1979). In contrast antibodies anti Gl globulin from *Phaseolus vulgaris* reacted well with the cell free synthesis system under conditions where glycosylation did not take place. Moreover, the immunochemically separated globulin contained only two of its three subunits. That shows the recognition capacity of the antibodies in this case (Hall *et al.*, 1978; Sun *et al.*, 1978).

A prerequisite to any immunochemical quantitative determination of one antigen is to establish that the antibodies used in the technique react identically with the protein from the different sources. The quantification of denatured proteins from different sources, by using a quantitative immunochemical technique is therefore a difficult task. The same problem arises in studies concerning the identification and quantitation of seed proteins in food products (see for recent reviews Baudner, 1978; Daussant, 1981; Llewellyn, 1979; Olsman and Hitchcock, 1980). To avoid the difficulty that antigens exposed to different sorts and degrees of denaturation will not react with the same intensity with the "standard antibodies" another approach can be proposed. This is the immunoaffinity chromatography used in unsaturated conditions. Here the immunochemical step is used only for separating the antigens from the protein mixture. A chemical quantitative determination would then be carried out on the desorbed fraction. However, care must be taken to avoid leakage of the antibodies from the support during desorption (Pekonen *et al.*, 1980; Dahl, 1980).

Structural relationships between proteins which have been differently modified during evolution can be evidenced by using immunochemical approaches (see Chapters 3 and 6 of this book). The structural relationships between constituents which are apparently evolutionary or functionally not related have also been evidenced by the existence of a certain degree of cross reactivity (Julliard *et al.*, 1980).

C. USE OF DETERGENTS AND UREA

Ionic and non ionic detergents are used for solubilizing membrane proteins (Helenius and Simons, 1975 for review), the non ionic detergents being less efficient than the ionic ones (Schäfer-Nielsen and Bjerrum, 1975). However, the detergents may cause difficulties in immunochemical analysis (Bjerrum et al., 1980 for review). One of these difficulties is the inhibition of the antigen–antibody precipitation reaction, the degree of which depends on each antigen. For example, the concentrations of sodium dodecyl sulphate (SDS) which result in 50% inhibition of the immunoprecipitation carried out in test tubes range from 0·02 to 0·27% depending on the antigen (Crumpton and Parkhouse, 1972; Dimitriadis, 1979). Nevertheless these studies and several others (Lee and Baden, 1977; Lee et al., 1978a) show that for a few proteins at least, the immunochemical analysis can be carried out in test tubes or in gel with SDS in a concentration not exceeding 0·1% in the analysis medium. In contrast to SDS, non ionic detergents such as Triton X 100 generally do not prevent antigen antibody precipitation even when used at 2% concentration (Bjerrum and Lundhall, 1973; Crumpton and Parkhouse, 1972; Dimitriadis, 1979; Bjerrum et al., 1980 for review). The milder effect of non ionic detergents in comparison to the effect of SDS on proteins led to the use of the non ionic detergents in conjunction with SDS in order to overcome the difficulties caused by SDS in immunochemical techniques. Several techniques have been proposed which combine the use of SDS in a first step with the use of non ionic detergents in a second one (Bjerrum et al., 1980; Chua and Blomberg, 1979; Converse and Papermaster, 1975; Dimitriadis, 1979; Lee et al., 1978b; Schäfer-Nielsen and Bjerrum, 1975). Although these studies show that conditions can be obtained for the use of detergents in immunochemical analysis, certain proteins are so sensitive to SDS or even to the non ionic detergents that they irreversibly lose their reactivity with the antibodies as shown by Menke et al. (1975) for a thylakoïd protein solubilized by SDS and by Shäfer-Nielsen and Bjerrum (1975) for two lipoproteins denatured by non ionic detergents.

Another difficulty with SDS is the formation of unspecific precipitates due to the reaction of the detergent, particularly with seric proteins (Lee et al., 1978b). This difficulty can be overcome by using the IgG fraction instead of the whole immune serum (Bjerrum et al., 1980 for review).

In addition to the solubilization of membrane proteins, the detergents are used in low concentrations in several immunochemical techniques in order to prevent non specific protein interactions as already mentioned.

Cereal storage proteins, the prolamins are usually solubilized in acid pH or in urea, conditions which makes it difficult to investigate them by immunochemical techniques. Urea 8 M was used in immunoaffinity chromatography for desorbing antigens. However, lower concentrations of urea were used in immunochemical studies on prolamins carried out in test tubes or in gel. The prolamins dissolved in high concentrations of urea were generally analyzed by the double diffusion test without urea in the agar gel. In these conditions an ill defined concentration gradient of urea in the gel develops, a small part of the prolamins which remains soluble in this gradient diffuses and the small urea concentration does not prevent the immunoprecipitation appearing (Ewart, 1977; Kling, 1975). However, in these conditions it is difficult to obtain quantitative data because an unknown—and small—proportion of the prolamins diffuses in the gel. Although some indications were reported showing that the antigen antibody precipitation is perceptible in gel containing 2 M, 3 M and even 4 M urea (Lee and Baden, 1977; Laurière, 1981) the problem of immunochemical analysis of prolamins in the presence of urea remains to be further investigated.

ACKNOWLEDGEMENT

The authors thank Mrs Renate Daussant for her help in writing the English text.

REFERENCES

Andersson, K. K., Benyamin, Y., Douzou, P. and Balny, C. (1979). *J. Immunol. Methods* **25**, 375–381.
Arnon, R. (1973). *In* "The Antigens" (Sela, M. ed.), Vol. 1, pp. 88–159. Academic Press, London and New York.
Avrameas, S. (1970). *Int. Rev. Cytol* **27**, 349–385.
Axelsen, N. H. (1975). *Scand. J. Immunol.* **2** (S2), 1–230.
Axelsen, N. H., Krøll, J. and Weeke, B. (1973). *Scand. J. Immunol.* **2** (S1), 1–169.
Barlow, K. K. (1973). *Experientia* **29**(1), 229–231.
Baudet, J. and Mossé, J. (1977). *J. Am. Oil. Chem. Soc.* **54**, 82–86.
Baudner, S. (1978). *Getreide, Mehl Brot* **32**, 330–337.
Baumgartner, B., Tokuyasu, K. T. and Chrispeels, M. J. (1978). *J. Cell. Biol.* **79**, 10–19.
Bjerrum O. J. and Lundahl, P. (1973). *Scand. J. Immunol.* **2** (S1), 139–143.
Bjerrum, O. J., Ramlau, J., Bock, E. and Bøg-Hansen, T. C. (1980). *In* "Methods for membrane receptors, characterization and purification" (Jacobs, S. and Cuatrecasas, P. eds.), pp. 117–156. Chapman and Hall, London.
Bollini, R. and Chrispeels, M. J. (1979). *Planta* **146**, 487–501.
Bowden-Bonnett, L. and Lord, J. M. (1979). *Plant Physiol.* **63**, 769–773.
Briggs, D. E (1978). "Barley". Chapman and Hall, London.
Brown, P. J., Leyland, M. J., Keenan, J. P. and Dean, P. D. G. (1977). *FEBS Lett.* **83**, 256–259.
Bureau, D. and Daussant, J. (1981). *J. Immunol. Methods* **41**, 387–392.
Casey, R. (1979a). *Biochem. J.* **177**, 509–520.
Casey, R. (1979b). *Heredity* **43** (2), 265–272.
Catsimpoolas, N. (1977). *In* "Immunological Aspects of Foods". (N. Catsimpoolas, ed.), pp. 37–59. The Avi Publishing Company, Inc. Westport, Connecticut.
Chrispeels, M. J. and Baumgartner, B. (1978). *Plant Physiol.* **61**, 617–623.

Chua, N. H. and Blomberg, F. (1979). *J. Biol. Chem.* **254**, 215–223.
Clarke, H. G. M. and Freeman, T. (1967). *In* "A quantitative immuno-electrophoresis method" (H. Peeters, ed.), pp. 503–509. *Prot. Biol. Fluids*, **14**, Elsevier, Amsterdam.
Converse, C. A. and Papermaster, D. S. (1975). *Science* **189**, 469–472.
Coombs, R. R. A. and Franks, D. (1969). *Prog. Allergy*, **13**, 174–272.
Coons, A. H. (1956). *Int. Rev. Cytol.* **5**, 1–23.
Craig, S., Goodchild, D. J. and Millerd, A. (1979). *J. Histochem. Cytochem.* **27**, 1312–1316.
Craig, S., Millerd, A. and Goodchild, D. J. (1980). *Aust. J. Plant Physiol.* **7**, 339–351.
Croy, R. R. D., Gatehouse, J. A., Evans, I. M. and Boulter, D. (1980a). *Planta* **148**, 49–56.
Croy, R. R. D., Gatehouse, J. A., Evans, I. M. and Boulter, D. (1980b). *Planta*, **148**, 57–63.
Crumpton, M. J. and Parkhouse, R. M. E. (1972). *FEBS Lett.* **22**, 210–212.
Dahl, D. (1980). *Biochim. Biophys. Acta* **622**, 9–17.
Daussant, J. (1975). *In* "The Chemistry and Biochemistry of Plant Proteins". (J. B. Harborne and C. F. Van Sumere, eds.), pp. 31–69. Academic Press, London and New York.
Daussant, J. (1977). *In* "Immunological Aspects of Foods" (N. Catsimpoolas, ed.), pp. 60–86. The Avi Publishing Company, Inc., Westport, Connecticut.
Daussant, J. (1981). *Ernährung* **5**, (2), 75–80.
Daussant, J. and Corvazier, P. (1970). *FEBS Lett.* **7** (2), 191–194.
Daussant, J. and Skakoun, A. (1975). *J. Immunol. Methods* **7**, 39–46.
Daussant, J. and Skakoun, A. (1981). *In* "Isoenzymes: Current topics in biological and medical research" (M. C. Rattazi, J. C. Scandalios, G. S. Whitt, eds), Vol. 5 pp. 175–218. Alan R. Liss Inc. Scientific, Medical and Scholarly, Publications, New York.
Daussant, J., Neucere, N. J. and Conkerton, E J. (1969). *Plant Physiol.* **44** (4), 480–484.
Daussant, J., Lauriere, C., Carfantan, N. and Skakoun, A. (1977). *In* "Regulation of Enzyme Synthesis and Activity in Higher Plants" (H. Smith, ed.), pp. 197–223. Academic Press. London and New York.
Daussant, J., Zbaszyniak, B., Sadowski, J. and Wiatroszak, I. (1981). *Planta* **151**, 176–179.
Davey, R. A. and Dudman, W. F. (1979). *Aust. J. Plant. Physiol.*, **6**, 437–447.
Dimitriadis, G. J. (1979). *Anal. Biochem.* **98**, 445–451.
Domoney, C., Davies, D. R. and Casey, R. (1980). *Planta* **149**, 454–460.
Engvall, E. and Perlmann, P. (1971). *Immunochemistry* **8**, 871–874.
Erikson, R. P. and Steers, E. (1970). *Arch. Biochem. Biophys.* **137**, 399–408.
European Brewery Convention (1967). "Barley Protein Committee". *J. Inst. Brew.*, **73**, 381–386.
Evans, I. M., Croy, R. R. D., Hutchinson, P., Boulter, D., Payne, P. I. and Gordon, M. E. (1979). *Planta*, **144**, 455–462.
Ewart, J. A. D. (1977). *In* "Immunological Aspects of Foods". (N. Catsimpoolas, ed.), pp. 87–116. The Avi Publishing Company Inc., Westport, Connecticut.
Forsgren, A. and Sjöquist, J. (1966). *J. Immunol.* **97**, 822–827.
Freed, R. C. and Ryan, D. S. (1978). *J. Food Sci.* **43** (4), 1316–1319.
Gibbons, G. C. (1980a). *Cereal Res. Commun.* **8** (1), 87–96.
Gibbons, G. C. (1980b). *Carlsberg Res. Commun.* **45**, 177–184.

Gonzales, E. and Beevers, H. (1976). *Plant Physiol.* **57**, 406–409.
Grabar, P. and Williams, C. A. (1953). *Biochim. Biophys. Acta*, **10**, 193–194.
Graham, T. A. and Gunning, B. E. S. (1970). *Nature*, **228**, 81–82.
Guldager, P. (1978). *Theor. Appl. Genet.* **53**, 241–250.
Hall, T. C., Ma, Y., Buchbinder, B. U. Pyne sun, S. M. and Bliss, F. A. (1978). *Proc. Natl. Acad. Sci. U.S.A.* **75** (7), 3196–3200.
Hankins, C. N., Kindiger, J. I. and Shannon, L. (1979). *Plant Physiol.* **64**, 104–107.
Hara, I. and Matsubara, H. (1980). *Plant Cell. Physiol.* **21** (2), 247–254.
Hayes, C. E. and Goldstein, I. J. (1974). *J. Biol. Chem.* **249**, 1904–1914.
Helenius, A. and Simons, K. (1975). *Biochim. Biophys. Acta* **415**, 29–79.
Higgins, T. J. V. and Spencer, D. (1977). *Plant Physiol.* **60**, 655–661.
Higgins, T. J. V. and Spencer, D. (1981). *Plant Physiol.* **67**, 205–211.
Higgings, T. J. V., Goodwin, P. R. and Whitefield, P. R. (1976a). *Virolog.* **71**, 486–497.
Higgins, T. J. V., Zwar, J. A. and Jacobsen, J. V. (1976b). *Nature* **260**, 166–169.
Hitchcock, C. H. S., Bailey, F. J., Crimes, A. A., Dean, D. A. G. and Davis, P. J. (1981). *J. Sci. Fd. Agric.* **32**, 157–165.
Hunter, W. M. (1973). *In* "Immunochemistry" (D. M. Weir, ed.), Vol. 1, pp. 17.1–17.36. Blackwell Scientific Publications, Oxford, London, Edinburgh, Melbourne.
Ivanov, V. N. and Khavkin, E. E. (1976). *FEBS Lett.* **65**, (3), 383–385.
Jaffé, W. G. (1977). *In* "Immunological Aspects of Foods" (N. Catsimpoolas, ed.), pp. 170–181. The Avi Publishing Company, Inc., Westport, Connecticut.
Julliard, J. H., Shibasaki, T., Ling, N. and Guillemin, R. (1980). *Science* **208**, 183–185.
Kaminski, M. (1979). "La Pratique de l'Immunoélectrophorèse". (Masson, ed.), Paris.
Kessler, S. W. (1975). *J. Immunol.* **115**, 1617–1624.
Kling, H. (1975). *Z. Pflanzenphysiol* **76** (2), 155–162.
Knox, R. B. and Clarke, A. E. (1978). *In* "Electron Microscopy and Cytochemistry of Plant Cells" (J. L. Hall, ed.), pp. 149–185. Elsevier, North-Holland, Biochemical Press.
Kronvall, G. and Frommel, D. (1970). *Immunochemistry* **7**, 124–127.
Laurell, C. B. (1965). *Anal. Biochem.* **10**, 358–361.
Laurell, C. B. (1966). *Anal. Biochem.* **15**, 45–52.
Laurière, M. (1981). Thèse de 3ème cycle, Université de Paris.
Lee, L. D. and Baden, H. P. (1977). *J. Immunol. Methods* **18**, 381–385.
Lee, L. D., Baden, H. P. and Cheng, C. K. (1978a). *J. Immunol. Methods* **24**, 155–162.
Lee, L. T., Deas, J. E. and Howe, C. (1978b). *J. Immunol. Methods* **19**, 69–75.
Levine, L. (1973). *In* "Immunochemistry" (D. M. Weir, ed.), Vol. 1, pp. 22.1–22.8. Blackwell Scientific Publications, Oxford.
Lichtenfeld, C., Manteuffel, R., Müntz, K., Neumann, D., Scholz, G. and Weber, E. (1979). *Biochem. Physiol. Pflanz.* **174**, 255–274.
Litwin, S. D. and Bozicevich, J. (1977). *In* "Immunology and Immunochemistry". (C. A. Williams and M. W. Chase, eds), Vol. IV, pp. 115–125. Academic Press, London and New York.
Llewellyn, J. W. (1979). *Int. Flavours Fd. Additive* **10**, 115–128.
Lowe, C. R. and Dean, P. D. G. (1974). "Affinity chromatography". Wiley and Sons, London.
Lynne, M. and Lord, J. (1979). *Plant Physiol.* **64**, 630–634.

Manteuffel, R. (1982). *In* "Encycl. Plant Phys: Nucleic Acids and Proteins in Plants" (D. Boulter and B. Parthier, eds.) Vol 14A, pp. 459–503. Springer Verlag, Berlin.

Manteuffel, R. and Scholz, G. (1975). Biochem. Physiol. Pflanzen, **168**, 277–285.

Manteuffel, R., Müntz, K., Püchel, M. and Scholz, G. (1976). *Biochem. Physiol. Pflanz.* **169**, 595–605.

Margolis, J. and Kenwick, K. G. (1968). *Anal. Biochem.* **25**, 347–362.

Menke, W., Koenig, F., Radunz, A. and Schmid, G. H. (1975). *FEBS Lett.* **49**, 372–375.

Millerd, A., Spencer, D., Dudman, W. F. and Stiller, M. (1975). *Aust. J. Plant Physiol.* **2**, 51–59.

Millerd, A., Thomson, J. A. and Schroeder, H. E. (1978). *Aust. J. Plant Physiol.* **5**, 519–534.

Millerd, A., Thomson, J. A. and Randall, P. J. (1979). *Planta* **146**, 463–466.

Morgan, M. R. A., Brown, P. J., Leyland, M. J. and Dean, P. D. G. (1978). *FEBS Lett.* **87**, 239–243.

Murphy, R. F., Imam, A., Hugues, A. E., Mac Gucken, M. J., Buchanan, K. D., Conlon, J. M. and Elmore, D. T. (1976). *Biochim. Biophys. Acta* **420**, 87–96.

Mutschler, M. A. (1977). *Plant Physiol.* **59**, 112.

Mutschler, M. A., Bliss, F. A. and Hall, T. C. (1980). *Plant Physiol.* **65**, 627–630.

Neucere, N. J. (1977). *In* "Immunological Aspects of Foods". (N. Catsimpoolas, ed.), pp. 117–151. The Avi Publishing Company, Inc., Westport, Connecticut.

Neumann, D. and Weber, E. (1978). *Biochem. Physiol. Pflanz.* **173**, 167–180.

Okamoto, K. and Akazawa, T. (1979). *Plant Physiol.* **64**, 337–340.

Okita, T. W., Decaleya, R. and Rappaport, L. (1979). *Plant Physiol.*, **63**, 195–200.

Olsman, W. J. and Hitchcock, C. (1980). *In* "Development in food analysis techniques" (R. D. King, ed.), Vol. 2, pp. 225–260. Applied Science Publishers, London.

Oudin, J. (1946). *C. R. Acad. Sci. (Paris)* **222**, 115–116.

Ouchterlony, O. (1949). *Acta Pathol. Microbiol. Scand.* **26**, 507–515.

Palmer, G. H. (1980). *In* "Cereals for Food and Beverages. Recent Progress in Cereal Chemistry and Technology" (G. E. Inglett and L. Munck, eds.), pp. 301–338. Academic Press, London and New York.

Paus, E. (1976). *FEBS Lett.* **72**, 39–42.

Pekonen, F., Williams, D. M. and Weintraub, B. D. (1980). *Endocrinology* **106** (5), 1327–1332.

Petryniak, J., Pereira, M. E. A. and Kabat, E. A. (1977). *Arch. Biochem. Biophys.* **178**, 118–134.

Püchel, M., Müntz, K., Parthier, B., Aurich, O., Bassüner, R., Manteuffel, R. and Schmidt, P. (1979). *Eur. J. Biochem.* **96**, 321–329.

Pusztai, A. and Watt, W. B. (1970). *Biochim. Biophys. Acta* **207**, 413–431.

Randall, P. J., Thomson, J. A. and Schroeder, H. E. (1979). *Aust. J. Plant Physiol.* **6**, 11–24.

Ressler, N. (1960). *Clin. Chim. Acta* **5**, 795–800.

Ruoslahti, E. (1978). *J. Immunol.* **121**, 1687–1690.

Ryan, C. A. (1977). *In* "Immunological Aspects of Foods" (N. Catsimpoolas, ed.), pp. 182–198. The Avi Publishing Company, Inc., Westport, Connecticut.

Schäfer-Nielsen, C. and Bjerrum, O. J. (1975). *Scand. J. Immunol.* **4** (S2), 73–80.

Schlesier, B., Manteuffel, R., Rudolph, A. and Behlke, J. (1978). *Biochem. Physiol. Pflanz.* **173**, 420–428.

Scholz, G. and Manteuffel, R. (1975). *Die Nahrung* **19**, 823–828.

Scholz, G., Richter, J. and Manteuffel, R. (1974). *Biochem. Physiol. Pflanz.* **166**, 163–172.

Singer, S. J. (1959). *Nature* **183**, 1523–1524.

Smyth, C. J., Söderholm, J. and Wadström, T. (1977). LKB Application, Note 269.

Spencer, D., Higgins, T. J. V., Button, S. C. and Davey, R. A. (1980). *Plant Physiol.* **66**, 510–515.

Spies, J. R. (1977). *In* "Immunological Aspects of Foods" (N. Catsimpoolas, ed.), pp. 317–371. The Avi Publishing Company, Inc. Westport, Connecticut.

Stavistcky, A. B. (1977). *In* "Immunology and Immunochemistry". (C. A. Williams and M. W. Chase, eds), Vol. IV, pp. 30–41. Academic Press, London and New York.

Su, L. C., Pueppke, S. G. and Friedman, H. P. (1980). *Biochim. Biophys. Acta* **629**, 292–304.

Sun, S. M., Mutschler, M. A., Bliss, F. A. and Hall, T. C. (1978). *Plant Physiol.* **61**, 918–923.

Thomson, J. A., Schroeder, H. E. and Dudman, W. F. (1978) *Aust. J. Plant Physiol.* **5**, 263–279.

Tronier, B., Ory, R. L. and Henningsen, K. W. (1971). *Phytochem.* **10**, 1207–1211.

Tsaftaris, A. S. and Scandalios, J. G. (1981). *Mol. Gen. Genet.* **181** (2), 158–163.

Uriel, J. (1971). *In* "Methods in Immunology and Immunochemistry". (C. A. Williams and M. W. Chase, eds.) Vol. 3, pp. 294–321. Academic Press, London and New York.

Vidal, J., Godbillon, G. and Gadal, P. (1980). *FEBS Lett.* **118** (1), 31–34.

Voller, A., Bidwell, D. E. and Bartlett, A. (1979). Dynatech. Europe, Borough House, Guernsey, Great Britain.

Wachsmuth, E. D. (1976). *Histochem. J.* **8**, 253–270.

Weber, E., Manteuffel, R. and Neumann, D. (1978). *Biochem. Physiol. Pflanz.* **172**, 597–614.

Weber, E., Süss, K. H., Neumann, D. and Manteuffel, R. (1979). *Biochem. Physiol. Pflanz.* **174**, 139–150.

Wienand, U. and Feix, G. (1978). *Eur. J. Biochem.* **92**, 605–611.

Youle, R. J. and Huang, A. H. C. (1978). *Plant Physiol.* **61**, 1040–1042.

CHAPTER 6

The Use of Seed Proteins in Taxonomy and Phylogeny

J. G. VAUGHAN

Department of Biology, Queen Elizabeth College, London

I. Introduction

Taxonomy is the oldest of the biological disciplines and, although it can be defined in various ways, in practice it is the science that deals with classification and identification. There are, of course, many publications that discuss the aims and scope of plant taxonomy. The recent book by Stace (1980) surveys the subject in its modern sense in a very readable and professional manner.

There are broadly speaking three types of classification: artificial, natural (phenetic) and phylogenetic. The artificial system is one based on a small number of characters and is often accredited to the eighteenth century scientist Linnaeus. Natural classifications are constructed from as many characters as possible. The phylogenetic system is said to be based on evolutionary relationships. Evidence for such a system should really be derived from: (a) an observation of the evolutionary process, (b) a recreation of the process, (c) fossils. As far as the seed plants are concerned, such evidence is sadly lacking and, as a consequence, the published systems of phylogeny are largely subjective.

Taxonomic characters and information are derived from various sources:

structure (external and internal), chromosomes, breeding systems, ecology and geography, chemistry.

The use of chemical characters in taxonomy is not new but certainly in recent times chemotaxonomy has become most popular. There are a number of reasons for this state of affairs. The taxonomist likes to use so-called "objective" characters and chemical characters fit well into this category. Modern methods of chemical analysis, such as chromatography and electrophoresis, are relatively cheap, rapid and can process small quantities of materials.

Cronquist (1980) has recently surveyed the contributions of chemistry to plant taxonomy. It is his opinion that, so far, secondary metabolites (alkaloids, betalains, flavonoids and many others) have been taxonomically the most useful although he feels that proteins may hold the most promise. Over the years there has been much discussion concerning the relative advantages and disadvantages of studies of "macromolecules" (proteins and nucleic acids) and "micromolecules" (secondary metabolites) in taxonomic investigations (Vaughan, 1975). Modern instrumentation provides exact identification of secondary metabolites but these compounds are not necessarily universally distributed. On the other hand, proteins and nucleic acids are universally distributed but, as will be discussed later, the establishment of homologies is not always easy. It is also usually pointed out that "macromolecules" either form part of the genome or are its immediate products.

For a long time taxonomists have shown interest in protein investigations. The statement by Gibbs (1963) that "it seems probable that each kind of living organism has its own set of proteins: that the proteins of nearly related species are nearly alike; that those of more distantly related ones are unlike" is attractive to the taxonomist. The present chapter deals specifically with seed proteins and the two groups of seed-bearing plants are the gymnosperms and angiosperms. Taxonomic studies of these groups, as well as all other groups of plants, would generally make reference to the range of plant organs available—a taxonomy based on seeds alone would constitute a special purpose system. It is therefore important to consider the particular significance of seed proteins.

If the chemical taxonomist were to parallel the morphologist then he would investigate the proteins of all the organs of the taxa under review because the protein patterns of the organs of one taxon do vary (e.g. Kloz et. al., 1960). However, the technical effort required for such an operation would be enormous. This being the case, chemotaxonomic investigations have often been restricted to one plant organ and, in the majority of cases, the seed has been utilized. The use of seed proteins in taxonomy and phylogeny has recently been reviewed by Miège (1975) and Ladizinsky and Hymowitz (1979). Both accounts list a considerable number of taxa that have now been investigated. There are various reasons for the particular use of seed proteins. Seed material or powder is relatively easy to handle as regards protein extraction but, rather more important, the seed may be regarded as a fixed physiological state. In taxonomic studies, it is critical to compare organs at the same stage of development and this applies to chemistry as well as morphology. In this sense, the seed, and its proteins, may be regarded as a "conservative" unit, little affected by the environment, geographic origin, seasonal fluctuations and chromosomal rearrangements (Ladizinsky and Hymowitz, 1979). On the other hand, comparison of leaf proteins for taxonomic purposes can pose difficulties. Although the majority of protein investigations in taxonomy have involved seeds, there are remarkably few controlled experiments on record that have investigated the presumed "conservative" nature of seed proteins (Smith, 1976; Ladizinsky and Hymowitz, 1979).

As will be described later in the chapter, most of the published work on seed proteins and taxonomy has concerned relationships and possible phylogeny. However, as stated earlier, the taxonomist is also concerned with identification and, as far as seeds are concerned, this can be important from the applied point of view in agriculture and industry. Such identification is normally carried out on macroscopic and microscopic features but is only usually effective down to the species level. For reasons of quality control, cultivar identification is now required and, in recent times, seed protein patterns have been utilized for this purpose.

The use of seed proteins in taxonomy will be considered under the headings of serology, lectins, gel electrophoresis, amino acid sequence data, industrial and agricultural applications.

II. SEROLOGY

Of the various protein techniques that have been used in taxonomy, serology is the oldest and may be traced back, at least as far as animal systematics are concerned, to the end of the last century (Fairbrothers, 1977). The use of serology in taxonomy is based on the situation that if foreign proteins (antigens) are injected into the bloodstream of a higher animal then the antibodies (immunoglobins) are produced in response. The term antiserum is applied to blood containing antibodies, after the red cells have been removed. If antigens (homologous) are brought into contact with their antibodies then precipitation occurs but precipitation will also occur with nearly related antigens (heterologous). It is assumed that the degree of precipitation in the cross-reaction is an indication of the degree of similarity between the homologous and heterologous antigens and hence taxonomic affinity.

As far as plant serotaxonomy is concerned, the early work was carried out by Mez and his associates in Königsberg, Germany, from 1911–36 (for a summary of this work, see Chester, 1937). Mez embarked on his work because he was aware of the taxonomic difficulties encountered by the morphologist. He regarded serology as dealing with proteins specific to plant taxa although he did not consider the technique as a panacea for all the difficulties of taxonomy. Mez investigated many plant groups and a range of plant organs but, mainly, he used seeds.

To compare the cross-reactions of plant antigens, Mez employed the total precipitin test in liquid medium. Cross-reactions were estimated on the degree of precipitation as measured by dilution series of antigen extracts, precipitate weight and the time taken for the precipitate to appear. Although Mez and his associates published many papers, no great changes were recommended as regards the system of classification already in existence. His findings are summarized in the "Stammbaum" (Mez and Ziegenspeck, 1926). This was meant to be an evolutionary or phylogenetic tree but it must be pointed out that an evolutionary concept based on protein relationships is no more valid than one based on morphological data unless the criteria used to evaluate evolution are utilized. However, Mez must be credited with initiating an exciting area of chemotaxonomic research.

Mez was subjected to a great deal of criticism by the contemporary plant taxonomists but, fortunately, his general line of approach stimulated the zoologist Boyden to initiate serological studies of animal systematics at Rutgers University, USA (Boyden and De Falco, 1943). On the technical side, Boyden introduced two innovations. The estimation of turbidity in the precipitin reaction can be somewhat subjective but Boyden used the Libby photronreflectometer, a type of densitometer, for this purpose. He also showed that in a homologous antigen extract/antiserum system the amount of precipitate with increasing dilutions of the antigen extract (or antiserum) rose from zero to a maximum and then fell to zero. Heterologous antigen extracts might give the same type of curve but the peak was often at a different point to that of the homologous extract. Consequently, at certain dilutions, the rather bizarre situation of the heterologous extract giving a stronger reaction than the homologous extract was obtained. Boyden, however, dealt with this situation by regarding the area under the homologous curve as 100% and estimating the areas under the heterologous curves as proportions of this to give a taxonomic index.

At Rutgers University, Boyden's methods were adopted by the botanist M. A. Johnson and used to investigate problems of taxonomy in the families Magnoliaceae, Cucurbitaceae, Ranunculaceae, Solanaceae and the Gramineae. Seed proteins were used as antigens in these investigations. A good example was the work carried out into the taxonomy of the

Magnoliaceae (Johnson, 1954). At that time, *Illicium* was included in the family together with genera such as *Magnolia* and *Liriodendron*. Johnson showed that the seed proteins of *Illicium* were serologically inactive against the other members of the Magnoliaceae investigated (Table I). It is interesting to note that in modern classifications (e.g. Takhtajan, 1969), *Illicium* is included in a separate family, namely the Illiciaceae. As far as can be made out, the photronreflectometer technique has only been used by workers at Rutgers University (Fairbrothers, 1977; Lee and Fairbrothers, 1978).

TABLE I
Serological relationship of *Illicium* (Johnson, 1954)

Species			Homologous area	Heterologous area	% Het. Area Homol. Area
Liriodendron	*tulipifera*	L.	731		
Michelia	*champaca*	L.		158	21·6
Magnolia	*virginiana*	L.		153	20·9
Talauma	*candollei*	Blume.		104	14·2
Illicium	*floridanum*	Ellis.		0	0

The antiserum has been made to *Liriodendron* seed proteins. Seed proteins of *Illicium* show no cross reaction.

The next phase in phytoserology was initiated by a paper published by Gell *et al.* (1960) on the taxonomy of *Solanum* species. In this work gel diffusion methods were used rather then the total precipitin reaction. Tuber proteins were used to produce antisera. This situation is rather analogous to the use of seed proteins because the plant tuber is a fixed developmental stage. The gel diffusion method separates the antigen–antibody systems and taxonomic assessment may therefore be made on numbers of identical and non-identical antigens. It is possible that in the total precipitin reaction a high degree of cross-reaction may be the result of one antigen. In their work, Gell *et al.* used the techniques of Ouchterlony double diffusion (Crowle, 1961) and immunoelectrophoresis (Grabar and Burtin, 1964). The first technique separates the antigen–antibody systems on simple diffusion, the second technique initially resolves the proteins by electrophoresis and then utilizes an antiserum to form precipitin arcs. Gell *et al.* also utilized the procedure of absorption or adsorption where the heterologous antigen extract is added to the antiserum made to the homologous extract. The common antibodies are precipitated and then discarded so that the absorbed antiserum should only contain antibodies specific to the homologous extract. If the absorbed antiserum is analysed by gel diffusion method then the antigens specific to the homologous source may be quantified.

To illustrate the use of gel diffusion methods with seed proteins in taxonomy, the investigation of *Brassica campestris*, *B. oleracea* and *B. nigra* is presented (Vaughan *et al.*, 1966; Vaughan, 1975). Antisera were prepared to the seed proteins of each of the three species. The reaction of each antiserum was observed against the homologous and heterologous extracts in double diffusion and immunoelectrophoresis. Also, each antiserum was absorbed separately with each of the heterologous extracts and the absorbed antiserum analysed by means of double diffusion and immunoelectrophoresis. The results of this work indicated a clear serological distinction between the three *Brassica* species and also that *campestris* and *oleracea* are closer to each other than either is to *nigra*. This relationship is in absolute agreement with that suggested on morphological grounds by Schulz (1919).

Most of the recent phytoserological investigations have utilized gel diffusion methods and have considered taxonomic problems at all levels of the hierarchy. Reviews of the situation have been presented by Fairbrothers (1969) and Fairbrothers *et al.* (1975). It might be said that serology has made an impact on classical taxonomy when reference is made to the technique in an accepted classification. One example of this is the utilization of serological data in taxonomic problems pertaining to the Cornaceae and Nyssaceae (Cronquist, 1968).

It would appear that most phytoserological studies have employed mixtures of proteins from seeds. Recently, Jensen (1973) and Jensen and Penner (1980) have used single seed proteins (globulins) in taxonomic studies of the Ranunculaceae, Leguminosae and Berberidaceae. Jensen has claimed that reactions of partial identity using the double diffusion technique gives information on numbers of antigenic determinants with taxonomic implications. The use of purified plant proteins might well appeal to the biochemist.

One of the most active areas in contemporary plant (and animal) systematics is numerical taxonomy, that is the interpretation of data on a mathematical basis and sometimes involving the use of a computer. It might be argued that the serological investigations so far described are examples of numerical taxonomy. Cristofilini (1980) has presented a modern review of the subject in terms of plant serology.

As indicated, the serological methods used by plant taxonomists have been mainly the precipitin reaction, gel double diffusion and immunoelectrophoresis. Other methods, such as quantitative immunoelectrophoresis with bidimensional electrophoresis and microcomplement fixation (Daussant, 1975) have been applied to plant materials but have not been widely used in taxonomy although Fairbrothers (1980) has used the "rocket" technique.

Some recent serological investigations have concerned *Salicornia* (Cristofilini and Chiapella, 1970); *Triticum* (Aniol, 1974); Flagellariaceae and other families (Lee *et al.*, 1975); *Galeopsis* (Houts and Hillebrand, 1976); *Casuarina* (El-Lakany *et al.*, 1977); *Coffea* (Lee, 1977); Rubiaceae (Lee and Fairbrothers, 1978); *Lupinus* (Nowacki and Jaworski, 1978); *Acacia* (El Tinay and Karamalla, 1979); *Medicago, Melilotus, Trigonella* (Simon, 1979); *Aralidium* (Fairbrothers, 1980).

To assess the value of serological methods for analysing the seed protein pattern in plant taxonomy, it should be pointed out that there are disadvantages and advantages. From a practical point of view, animal house facilities are not always available to the taxonomist. Also, if one accepts the thesis that no serotaxonomic investigation is complete without full reciprocal analysis then a large number of antisera might be required. In addition, the technique deals only with the immunochemical properties of proteins and it might be that certain proteins give identical serological reactions but show dissimilarities on other biochemical criteria. However, the absorption method has the powerful advantage of indicating proteins specific to a taxon and serology is probably more effective for investigating the relationships between rather distant taxa than methods of gel electrophoresis to be described later in the chapter.

III. LECTINS

The potential use of lectins in plant taxonomy has recently been discussed by Toms (1981). He defined a lectin as a plant or animal protein (usually a glycoprotein), not know to be an antibody, that combines specifically with an antigen to produce a phenomenon resembling an immunological reaction. Lectins are able to agglutinate erythrocytes and have been investigated because they may show blood group specificity although there are other biological applications.

Legume seeds are a rich source of lectins, in excess of 1500 species of the family have now been investigated. An early and comprehensive study of legume seed lectins was that of Mäkelä (1957) but no new taxonomic information emerged. Toms (1981) has discussed the situation in great detail and has pointed out that there are lectin differences between varieties of a species and that lectin specificities may vary with seed age. It would appear that at present seed lectins are of very limited use in taxonomic work although Foriers *et al.* (1980), using modern methods for lectin analysis, are of the opinion that such work is important in chemotaxonomy (Fig. 1). The natural functions of lectins are not known but they may fulfil the roles of defence against pathogenic bacteria and/or binding agents for symbiotic bacteria. Possibly when their natural functions are more fully understood, then the taxonomic implications of lectins will

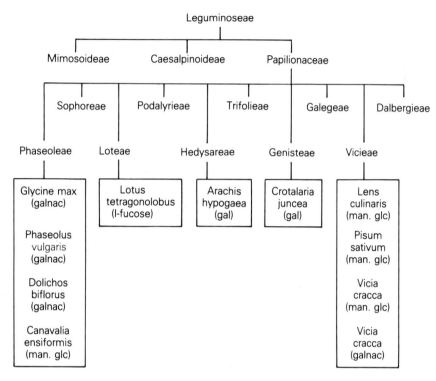

Fig. 1. Lectins and the taxonomy of some leguminous plants. The sugar specificities of the lectins are in parenthesis (Foriers *et al.*, 1980).

become clearer. However, because of the important biological applications of lectins, wide surveys of plant groups are to be encouraged.

IV. GEL ELECTROPHORESIS

Taxonomic studies using the separation of seed proteins by gel elec- trophoresis are now numerous and, although various gels have been utilized, polyacrylamide seems to have been the most popular. An early exposition of the techniques was given by Boulter *et al.* (1966).

A number of the investigations have employed general protein staining and taxonomic problems at various levels of the hierarchy have been considered. An early paper by Vaughan *et al.* (1966) considered the relationships between the three species—*Brassica campestris, B. oleracea* and *B. nigra.* This was combined with a serological study (as already described) and the results of the polyacrylamide gel electrophoresis, both for the globulins and albumins, were in agreement with the serological work. In this work, protein homology was based on identical Rp (Rf) values although this may or may not be the case because Mies and Hymowitz (1973) stated that bands with the same migration rate in profiles of two species do not necessarily represent similar proteins. Another technique that has been utilized for determining the similarity of proteins separated in electrophoresis is to create an artificial mixture of proteins from seeds of the different taxa and subject this to electrophoresis when different proteins will form separate bands (Johnson and Thein, 1970).

Some studies have concentrated on the seed globulins. MacKenzie and Blakely (1972) purified the seed globulins of some *Brassica* taxa, and, on their amino acid composition, supported the findings of Vaughan *et al.* (1966). Jackson *et al.* (1967) investigated the globulins of some legume genera by a fingerpinting technique (peptides separated by tryptic digestion) and found very similar patterns for members of the Vicieae but very different to the Phaseoleae examined. Boulter *et al.* (1967) studied some legume tribes and analysed the globulin patterns in polyacrylamide gel electrophoresis. A correlation with the tribal status was found. Przybylska *et al.* (1979) used sodium-dodecyl-sulphate (SDS) and urea electrophoresis for an examination of the legumin and vicilin fractions of *Pisum* taxa.

In the *Brassica* work quoted, a small number of species was considered and taxonomic relationships were based on the presence and absence of seed globulins and albumins as indicated by Rp values. No complicated mathematical deductions were involved. If, however, an investigation deals with a reasonable number of taxa, then interpretation becomes more difficult. Vaughan and Denford (1968) investigated nine *Brassica* and *Sinapis* taxa. In this situation, it was necessary to consider the reciprocal relationships between the protein spectra of all taxa and to analyse the results with a % similarity equation. A full understanding was only obtained when the % similarity results were converted into a three-dimensional model. Studies using the % similarity and similar techniques have also been carried out on *Suaeda* (Ungar and Boucard, 1974), *Acer saccharum* (Ziegenfus and Clarkson, 1971), *Sorghum* (Schechter and de Wet, 1975), *Lasthenia* (Altosar *et al.*, 1974).

A number of the taxonomic investigations involving general protein staining of seed proteins have utilized rods or sticks of polyacrylamide gel. This sometimes presents a difficulty for the taxonomist because assessment of Rp values of protein bands of differing thicknesses is not always easy nor is the assessment of homology between the protein spectra of different taxa. The situation is improved if seed proteins are subjected to electrophoresis in a slab (Stegemann, 1975).

An excellent example of the application of the technique under discussion to a problem of plant evolution is the work of Johnson and Hall (1965) and Johnson (1972) with the hexaploid bread wheat (*Triticum aestivum*).

The technique used for comparing protein patterns was that of estimating the optical density values of the bands in the gels and using this information to arrive at a correlation coefficient. They showed that *T. aestivum* (AABBDD) and *T. dicoccum* (AABB) possess almost all the proteins of the A genome as shown in the diploid *T. monococcum* (AA). Morphological and cytological evidence has suggested that the D genome came from *Aegilops squarrosa*. Johnson (1972) mixed the seed proteins of *T. dicoccum* and *A. squarrosa* and the protein pattern proved almost identical to that of *T. aestivum*.

Some other investigations have concerned the wild and cultivated species of the Solanaceae (Lester, 1979; Pearce and Lester, 1979; Edmonds and Glidewell, 1977); *Zea, Tripsacum* and other related genera (Smith and Lester, 1980); the evolution of cotton (*Gossypium*) species (Johnson and Thein, 1970); *Oryza* (Monod *et al.*, 1972); *Echinochloa* (Gasquez and Compoint, 1973); *Avena* (Ladizinsky and Johnson, 1972); *Hordeum* (McDaniel, 1971); Dioscoreaceae (Miège and Miège, 1971a and b); *Yucca* (Smith and Smith, 1970); *Juglans* (Clarkson *et al.*, 1974); Umbelliferae (Crowden *et al.*, 1969); *Coffea* (Centi-Grossi *et al.*, 1969; Payne *et al.*, 1973).

The difficulties encountered in comparing seed proteins in gel electrophoresis by general staining may be obviated to some extent by enzyme identification (Boulter *et al.*, 1966). This allows a more direct comparison between taxa. However, before proceeding with the technique, the taxonomist should carefully consider the background situation.

The number of enzymes present in seeds is quite considerable (Scandalios, 1974) and therefore the taxonomist must make some sort of choice. The subject of enzyme identification in taxonomy and population studies has recently been reviewed by Gottlieb (1977) and Hurka (1980), although many of the examples cited relate to leaf enzymes. It is suggested that the different molecular forms of an enzyme that catalyse the same reaction are called *isozymes* if their polypeptide constituents are coded by more than one gene locus and *allozymes* if their polypeptides are specified by different alleles at a single gene locus, the majority of enzymes routinely studied in natural populations having different allozymic forms. The advantages of enzyme studies in taxonomy are really the same as those accorded to general proteins, such as homology and that it is assumed that they are little influenced by the environment. Some disadvantages might also be stated. A small number of enzymes is often sampled. Allozymes that have identical mobilities do not necessarily have identical amino acid sequences. Consequently, Gottlieb (1977) suggests that more weight should be given to electrophoretic difference than similarity. He also suggests that genetic analysis should be carried out to distinguish between allozymes and isozymes and that highly polymorphic enzymes such as peroxidases and esterases might be avoided. Nevertheless Gottlieb feels that, under the correct conditions, electrophoretic analysis of enzymes in taxonomy is a powerful tool.

Enzymes have been utilized in studies of *Brassica* and *Sinapis* taxonomy at various levels of the hierarchy. Vaughan and Waite (1967) investigated the β-galactosidases, β-glucosidases and esterases in various species of *Brassica* and *Sinapis*. All the enzyme evidence supported the generic distinction between *Brassica* and *Sinapis*. Within *Brassica*, the esterases distinguished between the species *campestris*, *oleracea* and *nigra* but did not provide information on natural relationships. On the other hand, the β-galactosidases and β-glucosidases supported the relationship between the three species, as indicated by the serological work already described.

As an example of the application of enzyme studies to a different problem of *Brassica* taxonomy, the work of Denford and Vaughan (1977) on the 10 chromosome *B. campestris* and its allies might be quoted. This group contains the Western turnips and turnip rapes, the oil seed crops (*toria* and *sarson*) of the Indian sub-continent and the Eastern cabbages, such as petsai and pakchoi. Some taxonomists have included all these taxa under one species while others have recognized some of the taxa as distinct species. Denford and Vaughan investigated eleven enzyme systems (acid phosphatase, alkaline phosphatase, α-amylase, catalase, esterase, β-galactosidase, β-glucosidase, glutamic dehydrogenase, leucine aminopeptidase, myrosinase, peroxidase) by means of acrylamide gel electrophoresis. A consideration of all the enzyme systems suggested that the taxa might be classified into two groups—Western and Eastern.

The same enzyme systems were used by Phelan and Vaughan (1976) in an investigation of the 9 chromosome *B. oleracea* and *B. alboglabra*. The results obtained did not support the distinction of the taxa into two species but suggested varietal status for *alboglabra*. In the same approach with *B. juncea*, Vaughan and Gordon (1973) found that the enzymes investigated were not useful in supporting the hypothesis that there were two centres of evolution for the species, as suggested by glucosinolates and seed structure.

Bearing in mind the emphasis placed by Gottlieb (1977) on the necessity for genetic analysis in enzyme studies, reference might be made to Wills *et al.* (1979) concerning their work on determining sib frequencies in hybrid cultivars of *Brassica oleracea*. These workers investigated fourteen enzyme systems by means of acrylamide gel electrophoresis of seed extracts. Acid phosphatase seemed to be the most acceptable enzyme, for the purpose of the study. Wills *et al.* emphasized the desirability of using protein extracts of single seeds, as opposed to bulk samples, because products of less frequent alleles or gene combinations will be present in insufficient concentration in a bulk sample to allow easy identification.

Some other investigations have concerned the Leguminosae (Thurman *et al.*, 1967); Umbelliferae (Crowden *et al.*, 1969); *Nicotiana* (Smith *et al.*, 1970; Sheen, 1972); *Leucanthemum* (Villard, 1971); *Triticum* (Joudrier and Bourdet, 1972); *Juglans* (Clarkson *et al.*, 1974).

A technique involving electrophoresis which has appeared in relatively recent times is that of isoelectric focusing. This gives a very fine resolution of proteins based in a separation at their isoelectric points. Following isoelectric focusing, if gel electrophoresis is used in a second dimension then a far greater resolution of proteins may be obtained.

An excellent survey of the isoelectric focusing of seed proteins has been given by Wrigley (1977). Although there is already a considerable amount of published work concerning the application of the technique to the separation of seed proteins, relatively few projects have been concerned with plant relationships although it might be thought that the fine resolution afforded by the technique would be an asset to the taxonomist. If general seed proteins are separated by isoelectric focusing, particularly in two dimensions, then the very large number of constituents that become apparent are very difficult to interpret when a number of plant taxa are being considered. No doubt for this reason, isoelectric focusing of seed proteins in taxonomy has been concerned with enzymes. Esterases have been used in studies of *Triticum* and *Aegilops* (Bozzini *et al.*, 1973; Nakag and Tsunewaki, 1971), *Brassica* (Nakai, 1970) and *Heteranthelium* and *Taeniatherum* (Nakai and Sakamoto, 1977). In the latter study, there was good correlation with the genetic behaviour of the plants. In studies of *Triticum*, α-amylases have been utilized (Nishikawa, 1973; Nishikawa and Nobuhara, 1971). Postel *et al.* (1978) used isoelectric focusing for general proteins, peroxidases, phenolases, esterases and trypsin inhibitors in an investigation of distinctions between *Glycine* varieties. The work of Phelan and Vaughan (1980) on myrosinase in *Sinapis alba* suggests that isoelectric focusing of this enzyme in cruciferous plants might well have taxonomic value (Fig. 2). Isoelectric focusing would seem to have great potential in plant taxonomy although at present, the relatively high cost of the technique presents difficulties if the taxonomist is interested in analyzing large numbers of accessions.

V. Amino Acid Sequence Data

A highly original approach to the use of seed proteins in taxonomy and phylogeny has been the work of Boulter and his school (e.g. Boulter, 1973; Boulter, 1980; Ramshaw *et al.*, 1972). The amino acid sequence has been used to estimate the degree of homology of cytochrome *c* taken from the seeds of different species. It might be argued that this is a better test of

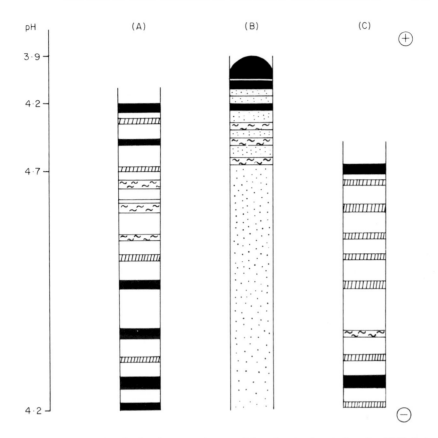

FIG. 2. Patterns of bands showing myrosinase activity after isoelectric focusing ▮ dark ⌇⌇⌇ medium ▨ pale. (A) Seeds, seedlings, very young leaf primordia. (B) Leaves, stems, flower parts and pods. (C) Roots. Shading of bands is a subjective assessment of density of sulphate precipitation (Phelan and Vaughan, 1980).

homology than the rate of mobility in a gel. Boulter analysed his results by computer with the so-called ancestral sequence method and presented his findings as a phylogenetic tree.

Although the tree was claimed to represent phylogeny, it might be best to examine it in the first instance as a representation of phenetic relationships. In this respect, most of the information presented is in accord with accepted taxonomy e.g. *Ginkgo* is distinct from the angiosperms, the monocotyledons form a group and the near relationship of *Guizotia* to *Helianthus* (Asteraceae), *Abutilon* to *Gossypium* (Malvaceae) and *Brassica oleracea* to *B. napus* (Cruciferae) is supported. As regards the phylogenetic implications of the tree, the early emergence of the one gymnosperm (*Ginkgo*) studied would be supported by most botanists.

However, some taxonomists (e.g. Cronquist, 1976) have been unhappy about the phylogeny presented for the relatively few angiosperms studied by Boulter, such as early appearance of the Asteraceae, often regarded as an advanced family. Nevertheless it must be pointed out that phylogenetic trees constructed by morphologists are highly subjective and there is very little fossil evidence to support their views. Boulter also used the cytochrome c information to estimate the time of origin of the angiosperms. In his opinion, this predated the Cretaceous period. Although the Cretaceous is usually described as the "Age of Angiosperms", the variety of angiosperm fossils present would suggest to all botanists an earlier origin.

The results presented by Boulter for cytochrome c have been an original contribution to the use of seed proteins in taxonomy and evolution. Criticisms have been levelled by taxonomists with regard to the use of just one protein (although Boulter has also studied plastocyanins and ferredoxins), the restricted number of taxa studied and difficulties such as back mutations. However, the amino acid sequence is a highly sophisticated means of determining protein homology and one would hope for technical advances so that the technique might be applied to a greater variety of proteins in a greater number of taxa.

VI. INDUSTRIAL AND AGRICULTURAL APPLICATIONS

In recent times, the exact identification of cultivars (sometimes erroneously referred to as varieties) has assumed importance. This is because of breeders rights, financial considerations in relation to the price of grain, enforcement of legislation, the correct identification of the seed before sowing and, in the case of barley, the distinction of cultivars suitable for feed from those suitable for malting. A review of the early work in this area has been presented by Grant (1973).

Most of the recent investigations have concerned wheat cultivars and considerable success has been achieved in identification using the seed gliadins. The electrophoretic methods are now well accepted in quality control although some identification may be carried out on seed morphology and chemical methods such as phenol tests. There is already a considerable amount of literature concerning wheat cultivar identification by the electrophoresis of seed gliadins (e.g. Andersson, 1980; Almgard and Clapham, 1977; Autran, 1975; Autran et al., 1979; Bushuk and Zillman, 1978; Coles and Wrigley, 1976; Du Cros and Wrigley, 1979; Ellis and Beminster, 1977; Konarev et al., 1979; Nierle, 1976; Shewry et al., 1978a; Zillman and Bushuk, 1979). In most of this work, starch gel electrophoresis has been suitable for distinguishing between cultivars but, in some difficult cases polyacrylamide gel electrophoresis (gradient and SDS) has been utilized.

Similar investigations have been carried out into barley cultivars (e.g. Shewry *et al.*, 1978b; Shewry *et al.*, 1979) although the range of patterns is very limited compared to wheat probably because the pattern is controlled only by two linked loci and genes conferring resistance to powdery mildew.

There is no doubt that other crop species will be investigated if legislation demands cultivar identification at the seed level. This type of work, particularly the results achieved for wheat, has shown conclusively that the seed protein spectrum can be an excellent reflection of the genome, a situation that should appeal to the taxonomist.

Seed protein analysis may also be useful for the identification of various food products. Wrigley (1977) has summarized the work carried out into the distinction of hexaploid wheat (*Triticum aestivum*) from macaroni wheat (*T. durum*) in pasta products.

Most of this type of work in food analysis has concerned the identification of soya (*Glycine max*) in meat products. Some analysts have used microscopy exclusively for this purpose (Flint, 1979). The method involving a stereological technique has been used for the quantitative determination of soya (Flint and Meech, 1978). Olsman and Hitchcock (1980) have reviewed in a very comprehensive manner the chemical methods investigated for soya identification and quantification. Using the soya proteins as markers, almost every type of electrophoretic and serological technique has been investigated and, in addition, peptide analysis by chromatography (Llewellyn *et al.*, 1978). A recent paper (Hitchcock *et al.*, 1981) has employed an enzyme-linked immunosorbent assay (ELISA) procedure. One advantage claimed for the procedure is that, as only very small volumes of antisera are required, antiserum from one animal can be fully characterized and used for inter-laboratory assessment and the analysis of thousands of samples. At the present stage of knowledge, Olsman and Hitchcock (1980) suggest that it would be wise to rely on more than one method of analysis.

VII. DISCUSSION

It is now clear that seed protein studies have contributed significantly to taxonomic investigations. However, real success will be achieved when seed proteins are studied in a routine manner by the practising taxonomist. Some of the approaches, such as amino acid sequencing and lectin investigations, may, for reasons of technique, be unsuitable at the present time for most botanists. However, this does not apply to the methods of gel electrophoresis and probably serology and there is no reason why these techniques should not be used more widely and with effect but a survey of the work already carried out on seed proteins and taxonomy indicates that there is variation in technique from the point of protein extraction to the analysis of protein patterns. This, coupled with the situation that a group of plants is rarely studied by more than one research school, means that it is somewhat difficult to compare results. To obviate this difficulty the seed protein biochemist would help the taxonomist greatly if he could state clearly the methods that lead to the maximum resolution of seed proteins.

REFERENCES

Almgard, G. and Clapham, D. (1977). *Swedish J. Agric. Res.* **7**, 137–142.
Altosar, I., Bohm, B. A. and Ornduff, R. (1974). *Biochem. Syst. Ecol.* **2**, 67–72.
Andersson, G. (1980). *Seed Sci. Technol.* **8**, 415–486.
Aniol, A. (1974). *Z. Pfanzenzuecht* **73**, 194–203.
Autran, J. C. (1975). *Ind. Aliment. Agric.* **9–10**, 1075–1094.
Autran, J. C., Bushuk, W., Wrigley, C. W. and Zillman, R. R. (1979). *Cereal Fds. World* **24**, 471–475.
Boulter, D. (1973). *In* "Chemistry in Botanical Classification" (G. Bendz and J. Santesson, eds), pp. 211–216. Academic Press, London and New York.
Boulter, D. (1980) *In* "Chemosystematics: Principles and Practice" (F. A. Bisby, J. G. Vaughan and C. A. Wright, eds.), pp. 235–240. Academic Press, London and New York.
Boulter, D., Thurman, D. A. and Turner, B. L. (1966). *Taxon* **15**, 135–143.
Boulter, D., Thurman, D. A. and Derbyshire, E. (1967). *New Phytol.* **66**, 27–36.
Boyden, A. and DeFalco, R. J. (1943). *Physiol. Zool.* **16**, 229–241.
Bozzini, A., Cubadda, T. and Quattruci, E. (1973). *In* "Proceedings of the Fourth International Wheat Genetics Symposium" (E. R. Sears and L. M. Sears, eds.) pp. 783–789. University of Missouri Press.
Bushuk, W. and Zillman, R. R. (1978). *Can J. Plant Sci.* **58**, 505–515.
Centi-Grossi, M., Tassi-Micco, C. and Silvano, V. (1969), *Phytochemistry* **8**, 1749–1751.
Chester, K. S. (1937). *Q. Rev. Biol.* **12**, 19–46, 165–190, 294–321.
Clarkson, R. B., Huang, F. H., Cech, F. C. and Gingerich, L. A. (1974) *Biochem. Syst. Ecol.* **2**, 59–66.
Coles, G. D. and Wrigley, C. W. (1976). *N.Z. J. Agric. Res.* **19**, 499–503.
Cristofilini, G. (1980). *In* "Chemosystematics: Principles and Practice" (F. A. Bisby, J. G. Vaughan and C. A. Wright, eds), pp. 269–288. Academic Press, London and New York.
Cristofilini, G. and Chiapella, L. (1970). *G. Bot. Ital.* **104**, 91–115.
Cronquist, A. (1968). "The Evolution and Classification of Flowering Plants". Houghton Mifflin Co., Boston.
Cronquist, A. (1976). *Brittonia.* **28**, 1–27.
Cronquist, A. (1980). *In* "Chemosystematics: Principles and Practice" (F. A. Bisby, J. G. Vaughan and C. A. Wright, eds), pp. 1–27. Academic Press, London and New York.
Crowden, R. K., Harborne, J. B. and Heywood, V. H. (1969). *Phytochemistry* **8**, 1964–1984.

Crowle, A. J. (1961). "Immunodiffusion". Academic Press, London and New York.
Daussant, J. (1975). *In* "The Chemistry and Biochemistry of Plant Proteins" (J. B. Harborne and C. F. Van Sumere, eds), pp 31–70. Academic Press, London and New York.
Denford, K. E. and Vaughan, J. G. (1977). *Ann. Bot (London)* **41**, 411–418.
Du Cros, D. L. and Wrigley, C. W. (1979). *J. Sci. Fd Agric.* **30**, 785–794.
Edmonds, J. M. and Glidewell, S. M. (1977). *Plant Syst. Evol.* **127**, 277–291.
El-Lakany, M. H., Samaan, L. G. and El-Rahim, M. A. (1977). *Aust. For. Res.* **7**, 219–224.
El Tinay, A. H. and Karamalla, K. A. (1979). *J. Exp. Bot.* **30**, 607–615.
Ellis, J. R. S. and Beminster, C. H. (1977). *J. Natl. Inst. Agric. Bot.* **14**, 221–231.
Fairbrothers, D. E. (1969). *Bull. Serol. Mus. New Brunsw.* **41**, 1–10.
Fairbrothers, D. E. (1977). *Ann. Mo. Bot. Gard.* **64**, 147–160.
Fairbrothers, D. E. (1980). *Taxon.* **29**, 412–416.
Fairbrothers, D. E., Mabry, T. J., Scogin, R. L. and Turner, B. L. (1975) *Ann. Mo. Bot. Gard.* **62**, 765–800.
Flint, F. O. (1979). *In* "Food Microscopy" (J. G. Vaughan, ed.), pp. 531–550. Academic Press, London and New York.
Flint, F. O. and Meech, M. V. (1978). *Analyst* **103**, 252–258.
Foriers, A., Baumann, C., Lieber, S., DeNeve, R. and Strosberg, A. D. (1980). *In* "Protides of the Biological Fluids" (H. Peeters, ed.), pp. 439–442. Pergamon Press, Oxford.
Gasquez, J. and Compoint, J. P. (1973). *C. R. Hebd. Seances Acad. Sci.* **277**, 837–840.
Gell, P. J. H., Hawkes, J. G. and Wright, S. T. C. (1960). *Proc. R. Soc. B* **151**, 364–383.
Gibbs, R. D. (1963). *In* "Chemical Plant Taxonomy" (T. Swain, ed.), pp. 41–88. Academic Press, London and New York.
Gottlieb, L. D. (1977). *Ann. Mo. Bot. Gard.* **64**, 161–180.
Grabar, P. and Burtin, P. (1964). "Immuno-electrophoretic Analysis". Elsevier, Amsterdam.
Grant, W. F. (1973). *In* "Chemistry in Botanical Classification" (G. Bendz and J. Santesson, eds), pp. 293–302. Academic Press, London and New York.
Hitchcock, C. H. S., Bailey, F. J., Crimes, A. A., Dean, D. A. G. and Davis, P. J. (1981). *J. Sci. Fd. Agric.* **32**, 157–165.
Houts, K. P. and Hillebrand, G. R. (1976). *Am. J. Bot.* **63**, 156–165.
Hurka, H. (1980) *In* "Chemosystematics: Principles and Practice" (F. A. Bisby, J. G. Vaughan and C. A. Wright, eds), pp. 103–121. Academic Press, London and New York.
Jackson, P., Milton, J. M. and Boulter, D. (1967). *New Phytol.* **66**, 47–56.
Jensen, U. (1973). *In* "Chemistry in Botanical Classification" (G. Bendz and J. Santesson, eds), pp. 217–227. Academic Press, London and New York.
Jensen, U. and Penner, R. (1980). *Biochem. Syst. Ecol.* **8**, 161–170.
Johnson, B. L. (1972). *Am. J. Bot.* **59**, 952–960.
Johnson, B. L. and Hall, O. (1965). *Am. J. Bot.* **52**, 506–513.
Johnson, B. L. and Thein, M. M. (1970). *Am. J. Bot.* **57**, 1081–1092.
Johnson, M. A. (1954). *Bull. Serol. Mus. Brunsw.* **13**, 1–5.
Joudrier, P. and Bourdet, A. (1972). *Ann. Amelior. Plant.* **22**, 263–279.
Kloz, J., Turkova, V. and Klozova, E. (1960). *Biologia (Torun, Pol).* **2**, 126–137.

Konarev, V. G., Gavrilyuk, I. P., Gubareva, N. K. and Peneva, I. I. (1979). *Cereal Chem.* **56**, 272–278.
Ladizinsky, G. and Hymowitz, T. (1979). *Theor. Appl. Genet.* **54**, 145–151.
Ladizinsky, G. and Johnson, B. L. (1972). *Can. J. Gen. Cytol.* **14**, 875–888.
Lee, Y. S. (1977). *Syst. Bot.* **2**, 169–179.
Lee, Y. S. and Fairbrothers, D. E. (1978). *Taxon.* **27**, 159–185.
Lee, D. W., Pin, Y. K. and Yew, L. F. (1975). *Bot. J. Linn. Soc.* **70**, 77–81.
Lester, R. N. (1979). *In* "The Biology and Taxonomy of the Solanaceae" (J. G. Hawkes, R. N. Lester and A. D. Skelding, eds), pp. 285–304. Academic Press, London and New York.
Llewellyn, J. W., Dean, A. C., Sawyer, R., Bailey, F. J. and Hitchcock, C. H. S. (1978). *J. Fd. Technol.* **13**, 249–252.
McDaniel, R. G. (1971). *J. Heredity* **61**, 143–247.
MacKenzie, S. L. and Blakely, J. A. (1972). *Can. J. Bot.* **50**, 1825–1834.
Mäkelä, O. (1957). *Ann. Med. Exp. Biol. Fenn. Suppl.* **35**, Supplement 11.
Mez, C. and Ziegenspeck, H. (1926). *Bot. Arch.* **13**, 483–485.
Miège, J. (1975). *In* "Les Proteines Des Graines" (J. Miège, ed,) pp. 305–365. Georg et Cie, Geneva.
Miège, M. N. and Miège, J. (1971a). *C. R. Hebd. Seances Acad. Sci.* **272**, 2536–2539.
Miège, M. N. and Miège, J. (1971b). *Arch. Sci.* **24**, 177–205.
Mies, D. W. and Hymowitz, T. (1973). *Bot. Gaz. (Chicago).* **134**, 121–125.
Monod, M., Marie, R. and Feillet, P. (1972). *C. R. Hebd. Seances Acad. Sci.* **274**, 1957–1960.
Nakai, Y. (1970). *Jpn. J. Breeding* **20**, 75–81.
Nakai, Y, and Sakamoto, S. (1977). *Bot. Mag.* **90**, 269–276.
Nakai, Y. and Tsunewaki, K. (1971). *Jpn. J. Genet.* **46**, 321–336.
Nierle, W. (1976). *Getriede Mehl Brot.* **30**, 207–210.
Nishikawa, K. (1973). *In* "Proceedings of the Fourth International Wheat Genetics Symposium" (E. R. Sears and L. M. Sears, eds), pp. 851–855. University of Missouri Press.
Nishikawa, K. and Nobuhara, M. (1971). *Jpn. J. Genet.* **46**, 345–353.
Nowacki, E. and Jaworski, A. (1978). *Genet. Pol.* **19**, 153–163.
Olsman, W. J. and Hitchcock, C, (1980). *In* "Developments in Food Analysis Techniques—2" (R. D. King, ed.), pp. 225–260. Applied Science Publishers, Barking, England.
Payne, R. C., Olneura, A. R. and Fairbrothers, D. E. (1973). *Rev. Roum. Biochim.* **10**, 55–59.
Pearce, K. and Lester, R. N. (1979). *In* "The Biology and Taxonomy of the Solanacea" (J. G. Hawkes, R. N. Lester and A. D. Skelding, eds), pp. 615–628. Academic Press, London and New York.
Phelan, J. R. and Vaughan, J. G. (1976). *Biochem. Syst. Ecol.* **4**, 173–176.
Phelan, J. R. and Vaughan, J. G. (1980). *J. Exp. Bot.* **31**, 1425–1433.
Postel, W., Goerg, A. and Westermeier, R. (1978). *Lebensm. Wiss. Technol.* **11**, 202–205.
Przybylska, J., Hurich, J. and Przybylska, Z. (1979). *Genet. Pol.* **20**, 518–528.
Ramshaw, J. A. M., Richardson, D. L., Meatyard, B. T., Brown, R. H., Richardson, M., Thompson, E. W. and Boulter, D. (1972). *New Phytol.* **71**, 773–779.
Scandalios, J. G. (1974). *Annu. Rev. Pl. Physiol.* **25**, 225–258.
Schechter, Y. and de Wet, J. M. J. (1975). *Am. J. Bot.* **62**, 254–261.

Schulz, O. E. (1919) "Cruciferae-Brassiceae" *In* Englers "Das Pflanzenreich" **1**.

Sheen, S. J. (1972). *Evolution* **26**, 143–154.

Shewry, P. R., Faulks, A. J., Pratt, H. M. and Miflin, B. J. (1978a). *J. Sci. Fd. Agric.* **29**, 847–849.

Shewry, P. R., Pratt, H. M. and Miflin, B. J. (1978b). *J. Sci. Fd. Agric.* **29**, 587–596.

Shewry, P. R., Pratt, H. M., Faulks, A. J., Parmar, S. and Miflin, B. J. (1979). *J. Natl. Inst. Agric Bot.* **15**, 34–50.

Simon, J. P. (1979). *Bot. Gaz.* **140**, 452–460.

Smith, C. M. and Smith, G. A. (1970). *Bot. Gaz.* **131**, 201–205.

Smith, H. H., Hamill, D. E., Weaver, E. A. and Thompson, K. H. (1970). *J. Heredity* **61**, 203–212.

Smith, J. S. C. and Lester, R. N. (1980). *Econ. Bot.* **34**, 201–218.

Smith, P. M. (1976). "The Chemotaxonomy of Plants". Edward Arnold, London.

Stace, C. A. (1980). "Plant Taxonomy and Systematics". Edward Arnold, London.

Stegemann, H. (1975). *In* "The Chemistry and Biochemistry of Plant Proteins" (J. B. Harborne and C. R. Van Sumere, eds), pp. 71–88. Academic Press, London and New York.

Takhtajan, A. (1969). "Flowering Plants, Origin and Dispersal". Oliver and Boyd, Edinburgh.

Thurman, D. A., Boulter, D., Derbyshire, E. and Turner, B. L. (1967). *New Phytol.* **66**, 37–45.

Toms, G. C. (1981). *In* "Advances in Legume Systematics" (R. M. Polhill and P. H. Raven eds), pp. 561–577. Royal Botanic Gardens, Kew.

Ungar, C. 'A. and Boucard, J. (1974). *Am. J. Bot.* **61**, 325–330.

Vaughan, J. G. (1975). *In* "The Chemistry and Biochemistry of Plant Proteins" (J. B. Harborne and C. F. Van Sumere, eds), pp. 281–298. Academic Press, London and New York.

Vaughan, J. G. and Denford, K. E. (1968). *J. Exp. Bot.* **19**, 724–732.

Vaughan, J. G. and Gordon, E. I. (1973). *Ann. Bot.* **37**, 167–184.

Vaughan, J. G. and Waite, A. (1967). *J. Exp. Bot.* **18**, 100–109.

Vaughan, J. G., Waite, A., Boulter, D. and Waiters, S. (1966). *J. Exp. Bot.* **17**, 332–343.

Villard, M. (1971). *Ber. Schweiz. Bot. Ges.* **80**, 96–188.

Wills, A. B., Fyfe, S. A. and Wiseman, E. M. (1979). *Ann. Appl. Biol.* **91**, 263–270.

Wrigley, C. W. (1977). *In* "Biological and Biomedical Applications of Isoelectric Focusing" (N. Catsimpoolas and J. Drysdale, eds), pp. 211–264. Plenum Press, New York and London.

Ziegenfus, T. T. and Clarkson, R. B. (1971). *Can. J. Bot.* **49**, 1951–1957.

Zillman, R. R. and Bushuk, W. (1979). *Can J. Plant Sci.* **59**, 287–298.

CHAPTER 7

Structure and Location of Legume and Cereal Seed Storage Proteins

J-C. PERNOLLET AND J. MOSSÉ

Laboratoire d'Etude des Protéines, Physiologie et Biochimie Végétales, Centre INRA. 78000, Versailles, France

I. INTRODUCTION

The storage protein topic has been surveyed in several recent reviews (Derbyshire *et al.*, 1976; Kasarda *et al.*, 1976; Wall and Paulis, 1978; Thomson and Doll, 1979; Mossé and Landry, 1980; Larkins, 1981). The subcellular location of these proteins has been reviewed by Briarty (1978), Pernollet (1978) and Larkins (1981).

Seed storage proteins can be characterized by several main features: (a) their main function which is providing amino acid or nitrogen to the young seedling; (b) the general absence of any other known function; (c) their abundance among seed nitrogen compounds; (d) their peculiar amino acid composition in the case of cereal and legume seeds; (e) their localization within storage organelles called protein bodies, at least during seed development; (f) their distribution in two major groups: legumin and vicilin in legumes, prolamine and glutelin in cereals.

Table I recalls the amino acid composition of some storage proteins.

TABLE I

Amino acid compositions (in residues per 1000 recovered residues) of various seed storage proteins

Amino acids	Legumin (pea)	Vicilin (pea)	Standard (egg)	Zein (maize)	ω-gliadin (wheat)	β-gliadin (wheat)
Gly	78	52	56	18	16	25
Ala	59	45	83	132	6	27
Val	58	57	79	41	5	48
Leu	83	103	88	189	34	72
Ile	45	55	55	39	38	42
Ser	63	78	55	64	34	54
Thr	37	30	54	30	12	15
Tyr	22	23	34	24	11	33
Phe	38	50	48	52	91	35
Trp	7	0·5	8	0	–	4
Pro	51	44	48	101	229	169
Met	8	1·3	28	8	1	6
Cys	7	0·5	28	10	4	23
Lys	55	72	65	1	4	5·5
His	21	16	20	9	10	13
Arg	77	57	46	10	7	16
Asx	117	134	101	49	5	24
Glx	176	182	108	213	501	389

They are characterized by a very high Glx amount that becomes very large in cereals which are also high in proline but very low in basic amino acids. Legume storage proteins are frequently close to standard egg composition, except for sulphur amino acids. Table II shows that the relative proportion of the major groups greatly differs from one species to another, in cereals as well as in legumes within a species. It can also vary from one genotype to another (Rao and Pernollet, 1981).

The preparation and isolation of the storage proteins have been reviewed elsewhere and will not be discussed here. Our purpose is to discuss the structure and the location of legume and cereal storage proteins. We

TABLE II
Relative proportions of the major storage protein groups

	Botanical groups	Ratio
Legumes (Legumin/vicilin ratio)	Pea	3
	French bean	0·1
Cereals (Prolamin/glutelin ratio)	Panicoideae	1·5
	Triticeae	1·15
	Oat	0·15
	Rice	0·1

intend to emphasize the structural homologies which result from recent data and lead to the maximal packing hypothesis.

II. LEGUME SEED STORAGE PROTEINS

It was at the end of the last century that Osborne and Campbell (1896) proposed the distinction of two kinds of globulins in pea: legumin and vicilin. This has been extended to numerous legume species by Danielsson (1949). In 1976, Derbyshire *et al.*, gathered data and arguments which make such a distinction practically unquestionable.

The first distinctive features of legume seed globulins of which legumins and vicilins are the major part, are summarized in Table III. Beside these

TABLE III
Distinctive characters of legume seed storage globulins

Characters	Legumins	Vicilins
Size	11S 300 to 400 kD	7S 200 ± 50 kD
Solubility in salt solution	less soluble	more soluble
Temperature of coagulation	higher more stable	lower less stable
N and S amount	higher	lower

two groups, which amount to some 80% of seed nitrogen, 2 S globulin fractions, lectins and some albumins may constitute other storage proteins which will not be discussed here.

A. LOCALIZATION

In the legume seed cotyledon, protein bodies are embedded between starch granules. They are membrane-bound organelles, a few microns in diameter, mainly filled with storage proteins and phytates. They are usually devoid of inclusions such as globoid or crystalloid except for *Arachis hypogaea* (Sharma and Dickert, 1975) and *Lupinus luteus* (Sobolev *et al.*, 1977). Fig. 1 represents a scanning electron microscopy view

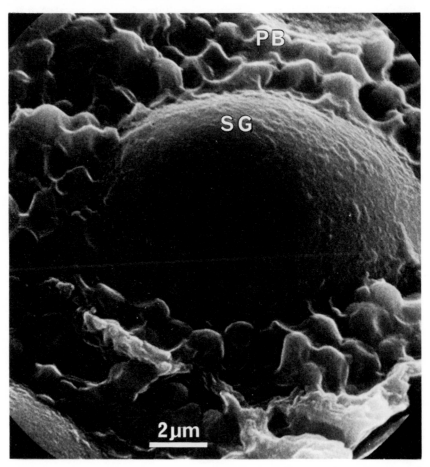

Fig. 1. Scanning electron microscope photograph of pea cotyledon. PB, protein body; SG, starch granule; ×5,400. Scale bar = 2 μm.

of protein bodies of pea cotyledon. A recent paper by Lott and Buttrose (1978) reviews the occurrence of globoids in protein bodies of legume seed cotyledons. These authors have shown that cotyledon protein bodies in some legumes have large and frequent globoids (*Arachis, Elianthus, Cassia*), whereas others have only small and rare inclusions (*Acacia, Glycine, Phaseolus, Pisum, Vicia*).

The protein content of legume protein bodies is about $80 \pm 10\%$. As far as we know, all storage protein polypeptide chains, either legumin or vicilin, have been restricted to the inside of protein bodies.

In *Pisum sativum*, vicilin and legumin are the only proteins which are mainly present within protein bodies and none or very few of these storage proteins are located outside (Varner and Schidlovsky, 1963; Alekseeva and Kovarskaya, 1978; Thomson et al., 1978; Rao and Pernollet, 1981). No phytohaemagglutinin can be located within pea protein bodies (Millerd et al., 1978, Rougé, 1977). In *Vicia faba*, Graham and Gunning (1970) have clearly shown that most protein bodies contain both legumin and vicilin but they could only detect vicilin in some storage organelles and in others neither vicilin nor legumin. This immunofluorescence study has been confirmed by biochemical analysis performed by Morris et al. (1970). The phytohaemagglutinins are found restricted to protein bodies in *Vicia faba*; they represent 1% of protein bodies extracted by Weber et al. (1978) who have used serological techniques. In the case of *Phaseolus*, phytohaemagglutinins (isolectins) are nearly all located within protein bodies (Barker et al., 1976; Pusztai et al., 1977 and 1978; Bollini and Chrispeels, 1978; Begbie, 1979), reaching 10% of the proteins which seem largely to be constituted by much vicilin and little legumin (Ericson and Chrispeels, 1973). In *Glycine max*, Catsimpoolas et al. (1968) have shown that protein bodies are filled with larger amounts of legumin, some vicilin and five other components. Tombs (1967) had previously given evidence that all glycinins (legumins) are stored within protein bodies. These results have been confirmed by Alekseeva and Kovarskaya (1978). Studies on lupin are less advanced but Sobolev et al. (1977) have shown that globulins are the main reserve proteins within protein bodies. We end this description with *Arachis hypogaea* protein bodies which have been the subject of earlier studies (Bagley et al., 1963; Dieckert et al., 1962; Sharma and Dieckert, 1975; Dieckert and Dieckert, 1976, for review). The presence of globulins (arachin and conarachin) within these organelles has been emphasized by Altschul et al. (1964) who have proposed a new classification for seed storage proteins, calling the proteins filling the protein bodies "aleurins".

Beside storage proteins, protein bodies also contain other proteins in lesser amounts such as enzymes or lectins. Sometimes lectins are almost all located within protein bodies. This is the case with broadbean and French bean lectins. Pusztai et al. (1979) have shown that French bean lectins reach 10% of protein body nitrogen.

Legume cotyledon protein bodies have been shown to be heterogeneous concerning their content. For instance, Graham and Gunning (1970) have shown that some broadbean protein bodies only contain vicilins while most of them are filled with both vicilins and legumins. As already discussed elsewhere (Pernollet 1978 and 1982; Mossé and Pernollet, 1982) such heterogeneity can be related to the ontogeny of protein bodies which can arise either directly from the endoplasmic reticulum or from vacuoles.

B. VICILIN STRUCTURE

The vicilin structure has been investigated in a small number of species, so that it is rather early and perhaps reckless to draw general conclusions. Concerning their quaternary structure, they frequently appear as homo- or heterotrimers, sometimes able to associate in hexamers.

For instance, in soybean seeds, the structure of β-conglycinin, the major of the three known conglycinins, has been partially elucidated by Thanh *et al.* (1975a and b), Thanh and Shibasaki (1976a and b, 1977, 1978a and b, and 1979) and also by Iibuchi and Imahori (1978a and b). Yamauchi *et al.* (1975 and 1976) and Yamauchi and Yamagishi (1979) have shown it is a glycoprotein. It consists of isotrimers of molecular weights ranging from 140 000 to 175 000. It is made of three main kinds of subunits called α, α' and β^*. Among the ten theoretically possible multiple forms, six (B_1 to B_6) have been demonstrated to occur (Thanh and Shibasaki, 1978a). They always contain either one α or one α' subunit associated with either two α or two β for one α plus one β subunits as follows: B_1, $\alpha'\beta_2$; B_2, $\alpha\beta_2$; B_3, $\alpha\alpha'\beta$; B_4, $\alpha_2\beta$; B_5, $\alpha_2\alpha'$; B_6, α_3. It can be seen that B_1 to B_5 are heterotrimers, B_3 being the only one to contain each kind of subunit, and B_6 is an homotrimer. Moreover, these six trimers are able to reversibly dimerize at low ionic strength or in the pH region 4·8–11·0 (Thanh and Shilbasaki, 1979). The resulting 9 S form is a superdimer (a hexamer) of two trimers facing each other. At extreme pH (2·0 and 12·0), dissociation into subunits (α, α' and β) are also reversible, so that the six molecular species B_1 to B_6 can be reconstituted by mixing the three subunits in urea solution and subsequently dialysing the solution against phosphate buffer.

In fact, the real quaternary structure is probably slightly more complex, for the possible number of subunit species is higher than three. Thanh and Shibasaki (1977) have isolated two other minor subunits (γ^* and δ) and evidenced the microheterogeneity of β subunits which is a mixture of four very similar components (β_1 to β_4).

All the main subunits α, α' and β are acid. The two subunits α and α' are very similar in amino acid composition with the same molecular weight

---

* In the literature, the letters β and γ indicate both two kinds of 7 S conglycinins and two kinds of subunits of β-conglycinin, which unfortunately may give rise to some confusion.

(54 000), and are devoid of cysteine like β in which methionine is lacking. The three subunits contain 4–5% of carbohydrate. Several glucidic sequences have been determined by Yamauchi and Yamagishi (1979). They are Asn $(GlcNAc)_2$–$(Man)_n$ with 7, 8 or 9 for n values.

Broadbean vicilin has been shown by Schlesier et al. (1978) to be associated into dimers and trimers. As regard to their sedimentation coefficient ($\sim 8\,S$) peanut conarachin (Dechary et al., 1961; Shetty and Rao, 1977), mung bean (Ericson, 1975) and cowpea (Carasco et al., 1978; Sefa-Dedeh and Stanley, 1979a and b) vicilins might have the same kind of quaternary structure.

On another hand, French bean phaseolin is made of at least three different subunits able to associate in heterotrimers or -tetramers. Pusztai (1966) and Pusztai and Watt (1970) have successively isolated two kinds of globulins linked with carbohydrates: glycoproteins I and II. Glycoprotein II was later suggested to be a vicilin-like protein by Derbyshire et al. (1976). Two proteins very similar to glycoprotein II were prepared, the first one by Racusen and Foote (1971) and the other one, named globulin G_1, by McLeester et al. (1973). McLeester et al. (1973) have shown that phaseolin, the G_1 globulin, is made of three different polypeptides of 43 000, 47 000 and 50 000. The same number of subunits, with practically the same molecular weights were then demonstrated by Barker et al. (1976), Bollini and Chrispeels (1978) and by Murray and Crump (1979). All these authors have first inclined towards a heterotrimeric structure for the first step of association between subunits. Sun et al. (1974), Stockman et al. (1976) and Bollini and Chrispeels (1978) suggested a tetrameric structure. Presently, the most plausible arrangement has been presented by Pusztai and Stewart (1980): phaseolin is made of four different subunits which, in a first step, are associated into different heterotrimers. These trimers can then be converted into more stable super-tetramers containing 12 subunits (i.e. 4 trimers). As for phaseolin primary structure, an important result has been demonstrated by Ma et al. (1980): peptide mapping of the subunits reveals a considerable sequence homology for about 60% of the length.

Phaseolin has recently been crystallized and submitted to X-ray diffraction by McPherson et al. (personal communication). They confirm that phaseolin is made of three subunits arranged in a trimeric trigonal antiprism of about 150 000 daltons. Each of the subunits has two very similar structural domains.

On the other hand, canavalin, the Jack Bean vicilin, is the only one legume storage protein the three dimensional structure of which has been published: McPherson (1980) has demonstrated it is composed of two trimers facing each other, arising in a hexameric pattern. Just like phaseolin, canavalin monomers are made of two domains which are virtually identical in structure. Concerning the secondary structure of canavalin monomer, it is almost entirely of β type.

Concerning pea viciln, Gatehouse et al. (1981) have shown that it is also a trimer mainly composed of about 50 000 dalton subunits. Croy et al. (1980c) have also distinguished another vicilin-like protein, they called "convicilin", which is made of 4 monomers of 71 000 daltons.

In lupin, γ conglutin is a homohexamer, the vicilin nature of which is uncertain (Blagrove et al., 1980). It is the only legume seed storage protein of which a moiety has been thoroughly sequenced by Elleman (1977).

Briefly speaking, our present knowledge of vicilin structure may be summarized by four main features: they are glycoproteins; they are devoid of disulphide bonds; they are frequently non covalently associated in trimers or even sometimes hexamers; they exhibit both a genetic and post-translational heterogeneity.

C. LEGUMIN STRUCTURE

Studies dealing with the quaternary structure of legumins are now well advanced and allow us to draw a general model. Neither tertiary nor secondary structures have been the subject of many investigations. So, we shall only mention N-terminal sequences which have permitted us to predict secondary structures.

1. Quarternary Structure of Legumins

Glycinin, the soybean legumin, was first demonstrated to be composed of twelve subunits (Catsimpoolas et al., 1967), corresponding to six different kinds of polypeptide chains (Catsimpoolas et al., 1971), with molecular weights of about 22 000 and 37 000. The subunits seem to be devoid of carbohydrate (Koshiyama and Fukushima, 1976a). They have been iso-lated by isoelectric focusing and their amino acid composition analysed, which enables a division into two groups: three acidic subunits have a higher content of glutamic acid or glutamine and proline, but a lower content of alanine valine, leucine, phenylalanine and tyrosine than the three basic subunits.

The quaternary structure of glycinin has been elucidated by Badley et al. (1975) and by Koshiyama and Fukushima (1976b) who have confirmed the results of Saio et al. (1970) and of Catsimpoolas et al. (1971). By means of electron microscopy, X-ray scattering measurements and by determining several physical parameters, Badley et al. (1975) have demonstrated that the twelve subunits are really packed in two identical hexagons placed one on the other in such a way that every acidic subunit is associated with three basic ones and *vice versa*: two in the same hexagon and the third in the other one. The structure involved is roughly similar to an oblate cylinder of 55 Å radius and 75 Å thick with a sedimentation coefficient $S^o_{20,w} = 12 \cdot 3$ and a molecular weight of 300 000. Some possible disulphide links between subunits have been suggested by Badley *et al.* (1975) and studied by Kitamura *et al.* (1976), Draper and Catsimpoolas (1978) and Mori *et al.* (1979). It appears that some interchain disulphide bonds could exist between some of the subunits. However Catsimpoolas and Wang (1971) have found a microheterogeneity for glycinin and Kitamura *et al.* (1976) have reported results consistent with the hypothesis that there are at least four different basic subunits and possibly four different acidic ones. At last, Staswick *et al.* (1981) have shown that the pairing between subunits is non random. This is consistent with the results obtained by Tumer *et al.* (1981) who have evidence that glycinin is synthesized as precursors which are post-translationally nicked into acidic and basic subunits.

Considering arachin, the peanut legumin, the existence of a quaternary structure with different kinds of subunits has been shown in a noteworthy study by Tombs (1965). The recent conclusions of Yamada *et al.* (1979a and b) summarize the knowledge on arachin structure. They have found that purified arachin can be separated into two components A_I and A_{II} through DEAE-Sephadex chromatography and that the ratio of these components depends on the variety. The two components have sedimentation coefficients of $9 \cdot 1$ and $14 \cdot 4$ S and molecular weight of 170–180 000 and 340–350 000 respectively. By changing either ionic strength or pH, A_I undergoes a reversible interconversion between 9 S and 14 S, A_{II} only a partially reversible interconversion. But A_I and A_{II} have identical amino acid composition and exhibit the same subunit pattern on SDS-gel electrophoresis, with six main bands, when prepared with 6–7 M urea in the presence of $0 \cdot 2$ M β-mercaptoethanol. The six kinds of subunits S_1 to S_6 have been isolated by isoelectric focusing in sucrose density gradient and characterized by their isoelectric points, molecular weights, N-terminal amino acids, molar ratios and by their amino acid composition. The first three are acidic subunits. They have different molecular weights, ranging from 35 500 to 40 500, two N-terminal amino acids (valine and isoleucine) and amino acid compositions which are significantly different in spite of rough similarities. The three others are basic subunits. They have the same molecular weight (around half of that of acidic subunits), the same N-terminal amino acid (glycine) and their amino acid compositions are rather similar, particularly concerning S_5 and S_6 which are both devoid of cysteine and methionine. Moreover, the three acidic subunits have higher content of glutamic acid or glutamine, but lower content of alanine, valine and leucine. By an estimation of the molar ratios of these subunits in the 9 S component (172 000 daltons), it was concluded that arachin consists of six different, non convalently bound, subunits in equimoles. In other word, arachin is a heterohexamer which is itself a kind of "supermonomer" partially dimerized, forming a heterododecamer (a "superdimer" of twelve subunits in the 14 S component). Such a quaternary structure is a full confirmation of that already suggested by Tombs (1965) and Tombs and Lowe (1967). A last important point is the ring-like structure revealed by electron microscopy of heterohexamer when investigated in *in vitro* salt-water systems (Tombs *et al.*, 1974): one cannot refrain from comparing this result with the structure of *Glycine max* globulins (glycinin) discussed below.

Pea legumin was purified by immunoaffinity chromatography (Casey, 1979a) and by isoelectrofocusing (Gatehouse *et al.*, 1980). These latter have suggested a glycinin-like quaternary structure, involving oligomerization of several different acidic and basic subunits (Casey 1979b; Croy *et al.*, 1979; Krishna *et al.*, 1979). In spite of these numerous elaborate studies, there is not any unquestionable demonstration of such a quaternary structure, for nobody has still completely isolated and characterized the involved subunits, but everybody agrees with the very plausible structure summarized by Gatehouse *et al.* (1980). Legumin of *Pisum sativum* is a heterooligomer of circa 350 000 daltons made up of six pairs of subunits (40 and 20 000 daltons) probably linked by one or more disulphide bridges. Both types of subunits are heterogenous in charge and size. Recently, Croy *et al.* (1980a and b), investigating *in vitro* proteosynthesis of legumin, have concluded that the disulphide bound subunits (40 000 and 20 000) are synthesized as a unique 60 000 dalton polypeptide chain. As for glycinin, it would mean that each legumin pair of (apparent) subunits consists of one dicatenar protein (like insulin) with two chains covalently linked by disulphide bridges. In our opinion, although split by reducing agents in electrophoretic studies the basic and acidic molecules are not native subunits but only consist of moieties of dicatenar proteins. Strictly speaking, *Pisum sativum* legumin would be composed of heterohexamer multiple forms.

Concerning broadbean legumins, Vaintraub *et al.* (1962) have proposed that they contain twelve polypeptide chains of an average molecular weight of 40 000. Wright and Boulter (1974) inclined towards a glycinin-like structure. They have indeed identified two kinds of acidic (α) chains of 37 000 daltons and three kinds of basic (β) chains with molecular weights of about 20 000, 21 000 and 24 000, with equimolar amounts of disulphide-bound α and β chains. It should be noticed that this occurrence of subunits in which one acidic chain and one basic chain are held together by disulphide bridges explains the equimolar amounts of these two kinds of chains. The 11 S legumin (circa 350 000 daltons) is therefore a heterohexamer. Mori and Utsumi (1979) have confirmed this model of quaternary structure. They have also detected three sizes of basic chains with molecular weights of 19 000, 20 500 and 23 000 very close to those found by Wright and Boulter (1974). But, as might be expected, they have found a number of acidic chains higher than those indicated by the latter: four instead of two, with molecular weights of 36 000, 36 000, 49 000 and 50 000. This fully supports the existence of multiple forms of legumin heterohexamers. Moreover, by the use of sucrose density centrifugation, they have demonstrated that *Vicia faba* legumin, characterized by sedimentation coefficient of 12·3 S and molecular weight of 319 000 is able either to associate into a superdimer (17·2 and 599 000 daltons) which is a heterododecamer, or to split into heterotrimers of 6·7 S.

The legumin fraction of mungbean (*Vigna radiata*) has been studied by Derbyshire and Boulter (1976). It is a 11 S component dissociated (by SDS with β-mercaptoethanol) into subunits or polypeptide chains with molecular weights of 37 000, 34 000 and 20 000.

From these data, we infer a general model of legumin quaternary structure which is illustrated in Fig. 2. The legumin molecule is a polymer formed by the association of six monomers. Each monomer is a bicatenar protein resulting from the post-translational nicking of a single polypeptide chain. The two chains (one acidic and one basic subunit) are frequently associated by one or more disulphide bridges, but not always: in mungbean, each acidic subunit is covalently bound by disulphide bridges to a basic subunit: in soybean glycinin only some of the subunits are disulphide-bonded, while peanut arachin does not possess any interchain disulphide bonds. In any case, acidic and basic chains are in equimole numbers. If the two chains of the monomomer are disulphide-linked, they can be separated by a reducing agent and give rise to the acidic and basic subunits. In the hexamer, the subunits are packed in two identical hexagons placed one on the other. Every acidic subunit is opposite three basic ones and vice versa.

2. Association-dissociation Properties of Legumins

Due to the *in vitro* ability of legumin subunits to reassociate spontaneously, several kinds of subunit interchanges have been investigated: reassociation between all the subunits of the same species, between part of these subunits, association of subunits isolated from one legume species to subunits extracted from another legume species and hybridization of subunits of legume species to subunits of globulins extracted from another family.

Yamada *et al.* (1981) have shown that *in vitro* reconstitution of arachin from the six isolated subunits gives rise to molecules strictly identical to native arachin as regards many physico-chemical features. The most striking result of these authors is that arachin "hexamers" have also been

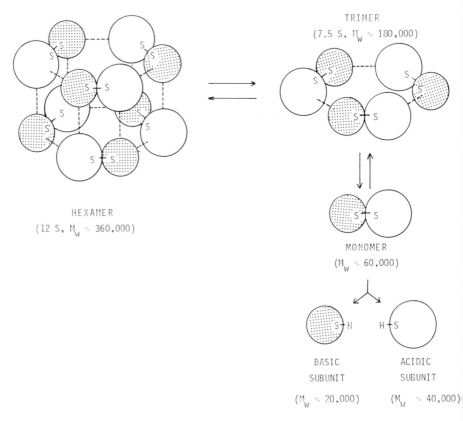

FIG. 2. Schematic model of legumin quaternary structure. Dashed lines, non covalent bonds; S–S, disulphide bridge.

obtained from the reassociation of only five or even four of the six different subunits. But these "hexamers" are more labile than intact arachin (or "hexamers" reconstituted from the six different subunits) against heating. These data strongly suggest that the subunits have a very similar but not identical folding. On the other hand, if the subunit pairing is non-random (as it can be inferred from the comparison with glycinin), the native monomers (one acidic and one basic subunit) have evolved in conserving (or in increasing) their ability to form stable trimers (constituted by six subunits).

Intergeneric molecular hybridizations have been successfully obtained between soybean acidic subunits and broadbean basic subunits and vice versa by Utsumi et al. (1980). The reassociated polymers are similar to native globulins. It is worth noticing that disulphide bridges between acidic and basic subunits are reconstituted.

At last, Hasegawa *et al.* (1981) have succeeded in preparing hybrid globulins from the combination of sesame globulin acidic subunits and soybean legumin basic subunits. Not only are acidic and basic subunits able to associate into a hybrid bicatenar monomer with reconstitution of disulphide linkage but moreover the monomers associate into hexamers containing 6 acidic and 6 basic subunits, as with reassociation of glycinin acidic and basic subunits. On the contrary, the reassociation of glycinin acidic subunits with sesame basic subunits only gives rise to monomer unable to polymerize.

We are led to conclude that, in spite of a general microheterogeneity inside each species and cultivar, that we have only briefly mentioned, legumin subunits as well as globulin subunits of other families exhibit some marked common features in their three-dimensional folding. This involves a severe conservatism for at least 3 parts of the molecule. The first one is the area of association between the acidic and the basic subunits originating from the same precursor polypeptide chain (area which is implicated in the formation of the disulphide linkage). The second and the third ones are the areas implicated in the non covalent association of each acidic (or basic) subunit with two basic (or acidic) subunits.

3. Sequence Comparison and Predicted Secondary Structure of Legumins

The similarity of legumins either extracted from the same species or even from different ones is emphasized by the comparison of the N terminal sequences.

When properly aligned, the N terminal sequences of soybean glycinin acidic subunits, in spite of deletions and replacements, exhibit conservative zones (Fig. 3). Moreira *et al.* (1981) have extended these homologies to internal cyanogen bromide-cleaved segments. Casey *et al.* (1981a) have sequenced one acidic subunit of pea legumin. It is quite comparable to the N terminal end of the A_2 acidic subunit sequenced by Moreira *et al.* (1979) except for Ala in position 8 (see Fig. 3 for numbering) which is replaced by Pro (analogous to A_1 glycinin acidic subunits) for the residues 15, 16, 17 replaced by Leu-Glu-Arg and for Lys in position 22 replaced by Glu (analogous to A_3 glycinin acidic subunit). Interspecific variations are not more frequent than differences between subunits isolated from one species.

The interspecific analogies are emphasized by the comparison of legumin basic subunits (Fig. 4). In this case, subunits can be classified in two groups depending on the amino acid residue in position 16. N terminal sequences of pea (Casey *et al.*, 1981b) and of broadbean (Gilroy *et al.*, 1979) have been obtained with unseparated legumin subunits; we have artificially reconstituted two different sequences corresponding to the

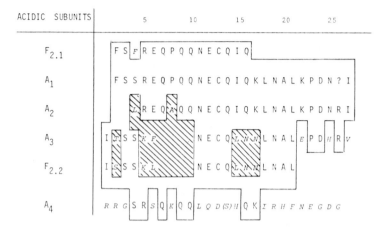

FIG. 3. N terminal amino acid sequences of glycinin acidic subunits (after Moreira *et al.*, 1979).

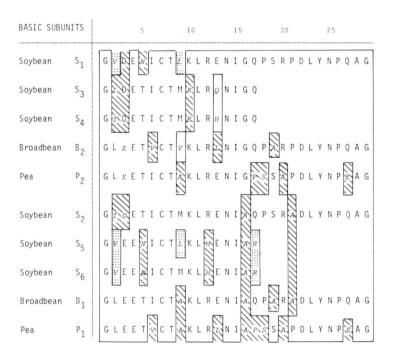

FIG. 4. N terminal amino acid sequences of legumin basic subunits. Broadbean sequences after Gilroy *et al.* (1979); Pea sequences after Casey *et al.* (1981b); Soybean sequences after Moreira *et al.* (1979).

uncertainties of residue determination. It is obvious that only a few amino acid replacements occur when comparing these sequences and that the residues at position 1, 4, 7, 8, 11, 14, 15, 22, 23, 24, 25, 26, 28 and 29 are perfectly constant from one polypeptide chain to another: roughly speaking about half of the residues are constant. It is striking to note that the short analysed N terminal sequences of pumpkin seed globulin, published by Ohmiya *et al.* (1980), are quite homologous with legumin basic subunit ones. This leads to the question of the importance of residue replacements on the structure of the molecule. One answer can be obtained with the prediction of secondary structure from the amino acid sequences (Pernollet, unpublished results).

We used the prediction algorithm of Garnier *et al.* (1978) which takes into account the influence of 8 residues preceding and following the residue, the secondary structure of which is computed. It allows not only the prediction of α-helix conformation (h), extended β-structure (e) and β-turns (t) as can be inferred from the Chou and Fasman rules (1978), but also to predict directly the aperiodic (or coil) structure (c): all the main structures which characterize the protein secondary structure can be predicted from amino acid sequences.

In Fig. 5, the predicted secondary structures of glycinin acidic subunits

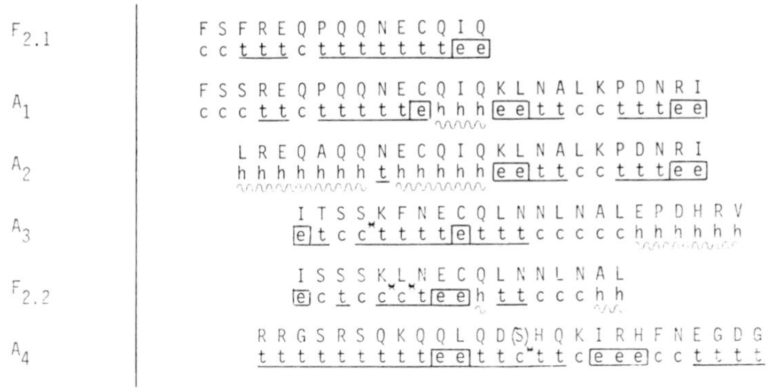

FIG. 5. Predicted secondary structures of glycinin acidic subunits. h, α-helix; e, extended β-structure; t, β-turn; c, aperiodic structure; c*, structure predicted as well as aperiodic as β-turn.

are presented, using a different alignment than for sequence comparison. Most of the point mutations alter the structure of the N terminal end but nevertheless conserve the main features of this structure. It can be characterized by successive β-turns and by a relatively low content of aperiodic structure.

The conservatism of the structure is more evident when comparing basic subunit predicted structures (Fig. 6). They all begin with an α-helix

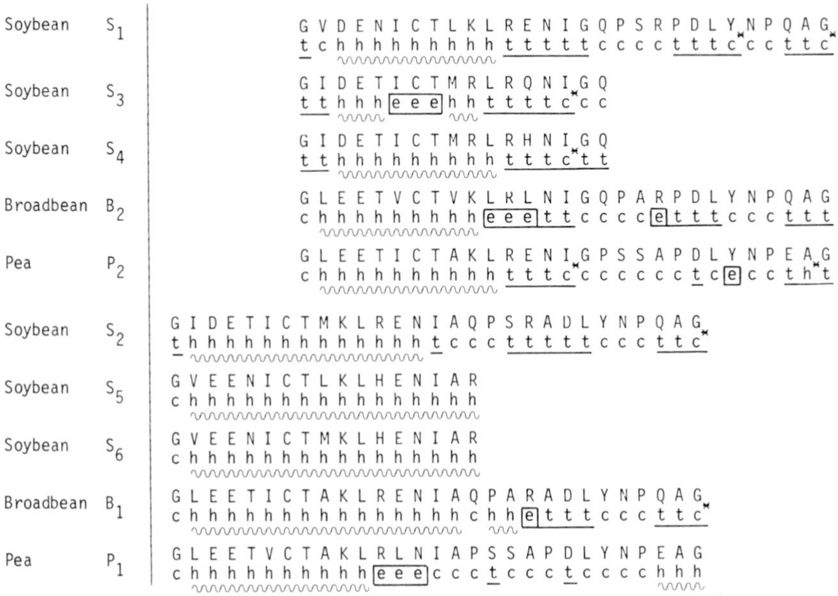

Soybean S_1

```
G V D E N I C T L K L R E N I G Q P S R P D L Y N P Q A G
t c h h h h h h h h h h t t t t t c c c c t t t c c c t t c
```

Soybean S_3

```
G I D E T I C T M R L R Q N I G Q
t t h h h h e e e h h t t t t t c c c
```

Soybean S_4

```
G I D E T I C T M R L R H N I G Q
t t h h h h h h h h h h t t t c t t
```

Broadbean B_2

```
G L E E T V C T V K L R L N I G Q P A R P D L Y N P Q A G
c h h h h h h h h h e e e t t c c c c e t t t c c c t t t
```

Pea P_2

```
G L E E T I C T A K L R E N I G P S S A P D L Y N P E A G
c h h h h h h h h h h t t t c c c c c c c t c e c c t h t
```

Soybean S_2

```
G I D E T I C T M K L R E N I A Q P S R A D L Y N P Q A G
t h h h h h h h h h h h h h t c c c t t t t t c c c t t c
```

Soybean S_5

```
G V E E N I C T L K L H E N I A R
c h h h h h h h h h h h h h h h h
```

Soybean S_6

```
G V E E N I C T M K L H E N I A R
c h h h h h h h h h h h h h h h h
```

Broadbean B_1

```
G L E E T I C T A K L R E N I A Q P A R A D L Y N P Q A G
c h h h h h h h h h h h h h h h h c h h e t t t c c c t t c
```

Pea P_1

```
G L E E T V C T A K L R L N I A P S S A P D L Y N P E A G
c h h h h h h h h h h e e e c c c t c c c t c c c c h h h
```

FIG. 6. Predicted secondary structures of legumin basic subunits. h, α-helix; e, extended β-structure; t, β-turn; c, aperiodic structure; c*, structure predicted as well as aperiodic as β-turn; h*, structure predicted as well as α-helix as β-turn.

followed by successive β-turns. Only the length of the α-helix varies. It corresponds to the mutation of glycine into alanine at position 16. All the other point mutations are roughly conservative ones.

As for the tertiary structure, primary and secondary structures exhibit large homologies which implies a noteworthy conservatism in the legumins.

4. General Features of Legumin Structure

Summarizing legumin structure we will underline the major points. They are hexamers composed of bicatenar monomers arising from post-translational modification of a single polypeptide chain. The two chains of the monomers are one acidic and one basic subunit, often but not always linked by disulphide bridges.

Within a cultivar one can see a polymorphism due to point mutations and to C terminal variations. It does not greatly affect the structure of the molecules. The homologies of structure are not restricted to the legumin polypeptide chains of one cultivar but extend not only to the different genera of legumes but also to globulins extracted from seeds of other families such as sesame and pumpkin.

Contrary to vicilins, legumins are not glycoproteins. Although discussed for a long time, the absence of covalently linked carbohydrates is now well established in the case of pea legumin (Gatehouse *et al.*, 1980; Hurkman and Beevers, 1980).

A complete survey of legume seed storage proteins ought to take into account some other minor proteins such as 2 S globulins, but in the frame of the present review it would have resulted in complicating the presentation of the common general features of these proteins. Besides, from a physiological point of view, storage proteins have a primordial trophic role, which is to supply amino acids and nitrogen compounds to the young seedlings for the short but critical period during which it is still heterotrophic. In this situation, every protein which is really degraded during germination is storage protein, as noticed for pea albumins by Murray (1979). But this trophic role may not be exclusive. Some of them have a lectin function, as was shown in *Phaseolus vulgaris* (Bollini and Chrispeels, 1978) and in *Pisum sativum* (Horisberger and Vonlanthen, 1980) and are the subject of another chapter. It is known that in drastic nitrogen deficiencies, living organisms utilize progressively structural or functional proteins for trophic purposes in order to supply amino acids for a proper turn over of the most vital proteins. Some others may have still undiscovered functions.

III. CEREAL SEED STORAGE PROTEINS

Located in the triploid starchy endosperm of cereal caryopses, the storage proteins considerably differ from the legume globulins; not only their amino acid composition but overall their solubility properties are peculiar characteristics of these proteins. Cereals may be classified in different groups on the basis of the relative proportion of the major storage protein groups: alcohol-soluble prolamins, glutelins and globulins (Mossé, 1968). As we shall see further the subcellular location of these proteins in the mature seed also depends on this classification which is quite comparable to the botanical classification. This is the reason why we have chosen to follow three main headings: the first one corresponds to the Panicoideae sub-family, the second one the Triticeae tribe and the last one to oat and rice storage proteins.

A. PANICOIDEAE

This sub-family comprises principally maize, sorghum and millet. The major storage proteins are prolamins (50 to 60% of seed proteins) and glutelins (35 to 40% of seed proteins). Everybody agrees now that prolamins are mainly and perhaps exclusively stored within protein bodies. Glutelins which are a heterogeneous group, are located both inside and outside these organelles. The ultrastructure of mature pearl millet starchy endosperm is presented in Fig. 7. Protein bodies are ovoid organelles, a

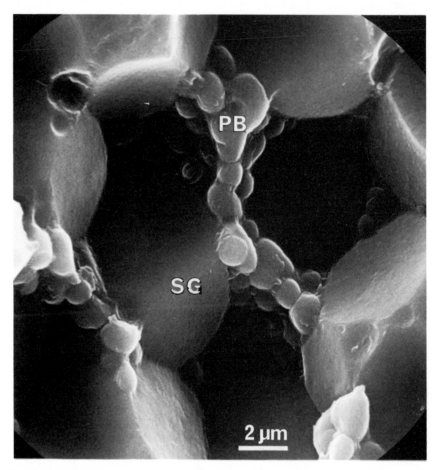

Fig. 7. Scanning electron microscope photograph of pearl millet starchy endosperm. PB, protein body; SG, starch granule; ×4,800. Scale bar = 2 μm.

few microns in diameter, embedded between starch granules. In maize and sorghum, they are quite similar in size and shape. Their internal ultrastructure is devoid of inclusions such as globoids or crystalloids.

In maize, it is now clearly established that protein bodies originate directly from the endoplasmic reticulum, the lumen of which is progressively filled with storage proteins in the course of seed development (Khoo and Wolf, 1970; Burr and Burr, 1976).

Concerning prolamin quaternary structure, zein, the maize prolamin, is composed of dimers (45 000 daltons) and of monomers (22 000 daltons). Under reducing conditions the dimers are dissociated (Mossé and Landry, 1980). Fig. 8 illustrates the occurrence of polymers of a greater degree of

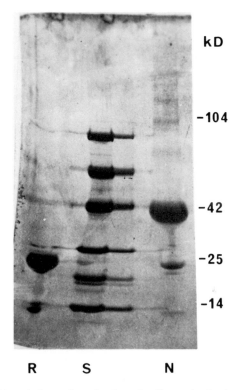

FIG. 8. Polyacrylamide gel electrophoresis of pearl millet prolamins in presence of sodium dodecyl sulfate. R, β-mercaptoethanol reducted prolamins; S, Pharmacia low molecular weight standard; N, Non-reduced prolamins. Molecular weights are indicated in kilodaltons (kD).

association in pennisetin, the pearl millet prolamin. It is too early to claim the existence of intercatenar disulphide bridges between subunits because under appropriate conditions one can dissociate pennisetin without reducing agent. Nevertheless, because of the insolubility of prolamins which requires the use of dissociating agents (SDS or urea) for their study, one cannot exclude the existence of a quaternary structure of these prolamins before their extraction.

Only zein has been submitted to sequence analysis. Because of zein polymorphism (Mossé and Landry, 1980), only 33 residues have been determined at the N terminal end (Bietz *et al.*, 1979). This result shows that like legume storage globulin the N terminal end of zein is conserved in spite of a wider varibility of the C terminal part of the molecule. The point mutations do not alter considerably the secondary structure we have predicted (Fig. 9). Zein secondary structure begins with one β-turn

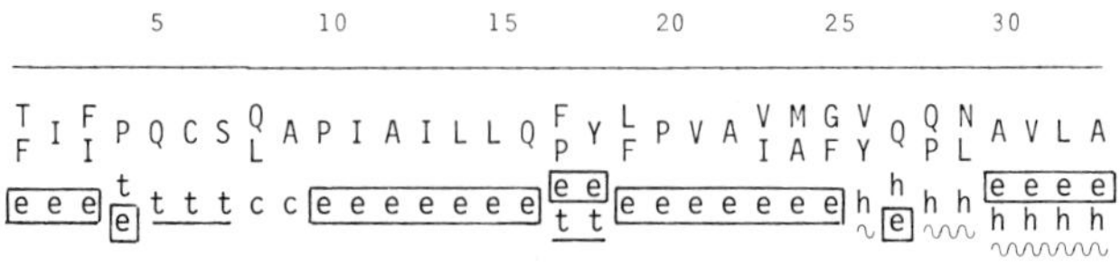

FIG. 9. N terminal amino acid sequence and predicted secondary structure of zein. Sequence after Bietz *et al.* (1979). h, α-helix; e, extended β-structure; t, β-turn; c, aperiodic structure.

followed by a long extended β-structure which is in good agreement with the well-known rod-shaped fibrillar structure of zein (Foster and Edsall, 1945).

By the use of cloned c-DNA sequencing Geraghty *et al.* (1981) have overcome the heterogeneity of zein and obtained the complete amino acid sequence of one molecule of zein. They have shown that conserved repetitive sequences of about 20 amino acids occur in this molecule. Besides they have sequenced the N terminal peptide signal, demonstrated by Burr *et al.* (1978) and Larkins and Hurkman (1978). Like legumins or vicilins, zein is synthesized as a preprotein, the N terminal signal peptide of which is cleaved during the protein body membrane crossing.

Generally speaking, glutelins are not a so well defined group as prolamins. Nevertheless maize glutelins have been much investigated. Some ten years ago, in our laboratory, Landry and Moureaux (1970) fractionated maize glutelins into three subgroups by using different extraction conditions. One of these subgroups appears to be a prolamin-like protein with practically the same amino acid composition as zein. It can now be admitted that it corresponds to polymerized zein subunits.

B. TRITICEAE

The Triticeae tribe which includes wheat, barley and rye, mainly differs from the Panicoideae in the storage protein localization and structure. In the starchy endosperm of the seeds belonging to this tribe no protein body is left at maturity (Pernollet and Mossé, 1980): at the end of the seed development the protein bodies vanish, clusters of proteins are then deposited between starch granules, but they are no more surrounded by a membrane. Nevertheless, protein bodies can be shown during the development of the endosperm (Lorenz *et al.*, 1978; Parker, 1980, 1981; Cameron-Mills and von Wettstein, 1980). Parker (1980) has shown that in wheat some protein bodies have internal inclusions.

In barley, Cameron-Mills and Ingversen (1978) and Brandt and Ingversen (1978) have shown that hordeins, the barley prolamins, are vectorially discharged into the lumen of the endoplasmic reticulum. Contrary to maize, Cameron-Mills and von Wettstein (1980) have not shown any continuity between protein bodies and the endoplasmic reticulum. They have observed small vacuoles implicated in the transport of storage proteins from the endoplasmic reticulum to the protein bodies.

In wheat, Parker (1981) has shown that protein bodies tend to associate one to another, forming clusters. She has also demonstrated the role of the Golgi apparatus in the transport of proteins from the endoplasmic reticulum to *de novo* formed vacuoles. By isolating protein bodies using isopycnic gradient density sedimentation (Fig. 10), we have shown that the wheat protein bodies associate one another to give rise to huge clusters of proteins. Miflin *et al.* (1981) have observed that, contrary to maize, the wheat protein body membrane is not continuous. These two facts are likely to be implicated in the protein body vanishing.

Concerning their content, we have recently shown that the protein bodies of wheat starchy endosperm are filled both with prolamins and glutelins. This is illustrated in Fig. 11.

In the Triticeae tribe, prolamins account for about 40–50% of seed proteins and glutelins for about 35–40%. In spite of several interesting works about wheat and barley glutelins, it is too early to draw any conclusion about their structure. They appear to be very complex and difficult to study. As for Panicoideae they are a very heterogeneous group of proteins and it is likely that one subgroup would consist of polymerized prolamin polypeptide chains. We will only discuss data on prolamin structure.

No prolamin quaternary structure has been shown in wheat or barley. This could be due to the necessary use of dissociating agents to solubilize prolamins, the association constant of which would be weaker than in maize zein.

The prolamins of wheat, barley and rye, respectively called gliadins, hordeins and secalins, are all composed of two different sets of polypeptides, one about 30 000 daltons in molecular weight (α-β or γ-gliadins, B-hordeins and B-secalins) and another about 60 000 daltons in molecular weight (ω-gliadins, C-hordeins and C-secalins). Physicochemical properties of α-gliadins have been reviewed by Kasarda (1980) and β-, γ and ω-gliadins by Charbonnier *et al.* (1980) who have emphasized the polymorphism of gliadins even within each sub-group.

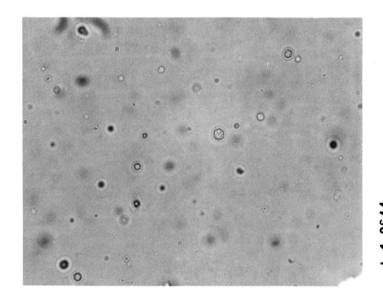

d = 1.2701

—— 25 μm

d = 1.2644

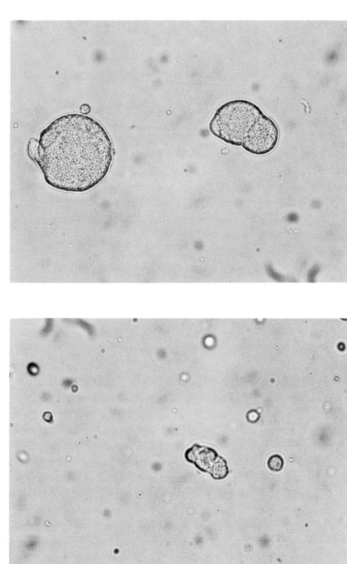

d=1.2736   d=1.2748

Fig. 10. Light photomicrograph of wheat developing endosperm isolated protein bodies. ×400. The 4 areas correspond to protein bodies differing in their apparent density determined by isopycnic ultracentrifugation in sucrose gradient. Densities are expressed in g/ml.

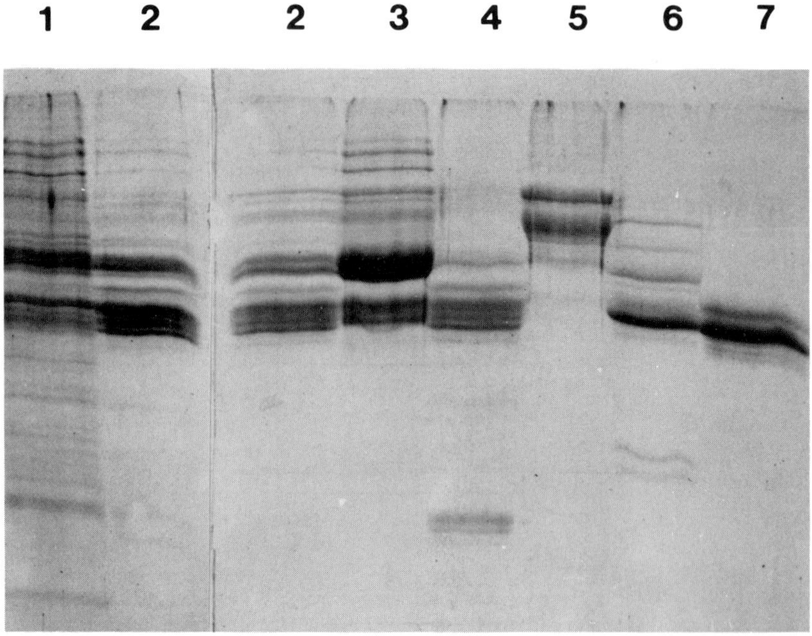

FIG. 11. Polyacrylamide gel electrophoresis in presence of sodium dodecyl sulfate of wheat storage proteins. 1. Proteins extracted from protein bodies isolated by sucrose gradient isopycnic centrifugation from developing endosperm. 2. Alcohol-soluble proteins of mature endosperm. 3. Alcohol-soluble glutelins and ω-gliadins. 4. α-, β- and γ-gliadins. 5. ω-gliadins. 6. γ-gliadins. 7. $β_2$- and $β_3$-gliadins.

Concerning their primary structure, the N terminal sequences of α-, β and γ-gliadins are presented in Fig. 12. They exhibit large homologies and have been postulated by Kasarda (1980) to arise from a common ancestral form consisting in the repetitive pentapeptide Pro—Tyr (or Phe)—Gln—Gln—Gln which is presented and aligned in Fig. 12. As for B-hordeins, the low molecular weight barley prolamins, their N-terminal end is blocked to Edman degradation (Schmitt and Svendsen, 1980a). Partial amino acid sequences of peptides cleaved with cyanogen bromide and proteases indicate internal sequence repetitions (Schmitt and Svendsen, 1980b).

Intraspecific homologies become obvious when comparing properly aligned N terminal sequences of high molecular weight prolamins (Fig. 13). Whereas the 9 first N terminal residues may undergo point mutations or deletions, the sequence tends to become unique and very close to the ancestral ω-gliadins form postulated by Kasarda (1980). This sequence is characterized by the repetitive pentapeptide: Pro—Gln—Gln—Pro—Tyr. The high molecular weight prolamins of rye called C-secalins, isolated by Charbonnier *et al.* (1981) and sequenced by Bietz (personal communica-

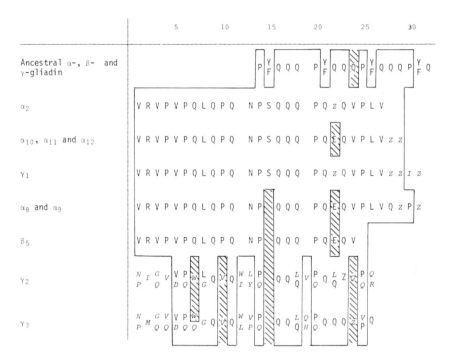

FIG. 12. N terminal amino acid sequences of low molecular weight gliadins. Ancestral sequence after Kasarda (1980); α_2-gliadin after Kasarda *et al.* (1974); other sequences after Bietz *et al.* (1977).

tion) have also a very similar N terminal end. Within a species, namely barley, Shewry *et al.* (1981) have isolated 3 C-hordein fractions different in molecular weight and in electrophoretic behaviour which exhibit the same 10 residue N terminal sequence. Moreover, they have identical sequences for the first five residues at the C terminal end.

By predicting conformations of low molecular weight gliadins we have shown (Fig. 14) that they are all characterized by a β-structure N terminal tail followed by three β-turns and another extended β-structure. The postulated ancestral form conformation is a striking succession of β-turns predicted with a very high information value. The present α-, β or γ-gliadins have only partially conserved this structure. It has been replaced by extended β-structure with large homology between the different polypeptide chains.

In the Fig. 15 is shown the same striking predicted conformation for the ω-gliadins ancestral form. Here, this succession of β-turns has been largely conserved in the present forms, even if the exact position of these turns has been modified.

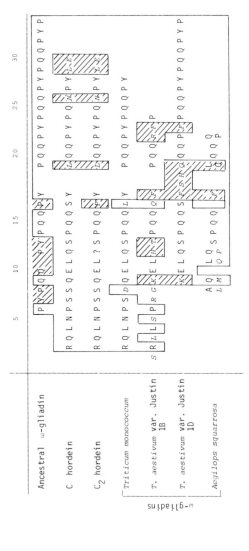

Fig. 13. N terminal amino acid sequences of high molecular weight prolamins of Triticeae. Ancestral sequence after Kasarda (1980); *Triticum monococcum* ω-gliadin and C-hordein after Shewry *et al.* (1980); *Triticum aestivum* var. Justin and *Aegilops squarrosa* ω-gliadins after Autran *et al.* (1979); C₂-hordein after Schmitt and Svendsen (1980a).

Ancestral α-, β- and γ-gliadin

P Y/F Q Q Q P Y/F Q Q Q P Y/F Q Q Q
t t t t c t t t t c t t t t c

α2

V R V P V P Q L Q P Q N P S Q Q Q P Q E Q V P L V
e e e e e e t e c t t c c t t c c c t t e e e e e e

α10, α11 and α12

V R V P V P Q L Q P Q N P S Q Q Q P Q E Q V P L V Q Q
e e e e e e t e c t t c c t t c c c t t e e e e e e e

γ1

V R V P V P Q L Q P Q N P S Q Q Q P Q Q Q V P L V Q Q I Q
e e e e e e e c t t c c c t t c c t t t e e e e e e e e e

α8 and α9

V R V P V P Q L Q P Q N P Q Q Q P Q E Q V P L V Q Q P Q
e e e e e e t c c t t c c t c c c t t e e e t e e c c t

β5

V R V P V P Q L Q P Q N P Q Q Q P Q E Q V
e e e e e e t c t t c c c t c c c t e e e

γ2

N I G V V P L Q V Q W L P Q Q L P Q L Q Q P Q
P Q V D Q W G Q V Q I Y Q Q Q V Q Q Q Q Q R
e e t c c t t c t c t c c t t t c c c t c c c c c t
e e e e e e e e e e e e e e c c t t t t t t

γ3

N M G V V P W G Q V Q W V P Q Q L Q P Q Q Q Q V Q
P M Q Q D Q Q Q V Q L P Q Q Q H Q Q Q Q P P Q
t t t c c t t c t c t e c t t t t c c t t e e e e
c c e e e e t t c c c t t c t t c c c t

FIG. 14. Predicted secondary structures of low molecular weight gliadins. h, α-helix; e, extended β-structure; t, β-turn; c, aperiodic structure; c*, structure predicted as well as aperiodic as β-turn.

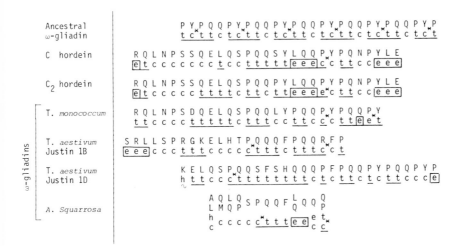

Ancestral ω-gliadin

P Y/P Q Q P Y/P Q Q P Y/P Q Q P Y/P Q Q P Y/P Q Q P Y/P
t c t t c t c t t c t c t t c t c t t c t c t t c t c t

C hordein

R Q L N P S S Q E L Q S P Q Q S Y L Q Q P Y/P P Q N P Y L E
e t c c c c c c c t c c t t t t t e e e c c t t c c e e e

C2 hordein

R Q L N P S S Q E L Q S P Q Q P Y L Q Q P Y/P P Q N P Y L E
e t c c c c c t t t t c t t c t t e e e e c t t c c e e e

T. monococcum

R Q L N P S D Q E L Q S P Q Q L Y P Q Q P Y/P P Q Q P Y
t t c c c c t t t t t c t t t c c t t c c c t t e e t

T. aestivum Justin 1B

S R L L S P R G K E L H T P Q Q Q F P Q Q R F P
e e e c c c t t t c c c c c t t t c t t t c c t

T. aestivum Justin 1D

K E L Q S P Q Q S F S H Q Q Q P F P Q Q P Y P Q Q P Y P
h t t c c c c t t t t t t t t t c t c t t c t c t t c c c e

A. Squarrosa

A Q L Q L Q Q Q
L M Q P S P Q Q F Q Q Q P
h c c c c c c t t t e e e c t
c c

ω-gliadins

FIG. 15. Predicted secondary structures of high molecular weight prolamins of Triticeae. h, α-helix; e, extended β-structure; t, β-turn; c, aperiodic structure; c*, structure predicted as well as aperiodic as β-turn; e*, structure predicted as well as extended β-structure as β-turn.

The last two species we shall discuss are oat and rice. Both exhibit a lower content of prolamins: rice storage proteins are composed of nearly 80% of glutelins, while prolamins only reach 8% of seed proteins. The same proportions have roughly been found in oat. Nevertheless, Kim (1978) has shown that a large part of oat glutelins can be considered as globulins, according to their extraction behaviour. In spite of numerous data on storage protein of oat and rice, no structural study has so far been published for these two species, so we shall only discuss their localization.

As for other cereals, storage proteins are partly confined to protein bodies which are still present in the mature seed. In oat, these organelles have been isolated and shown to store prolamins and part of the other storage protein groups (Pernollet *et al.*, 1982). Tanaka *et al.* (1980) have demonstrated two types of protein bodies in the starchy endosperm of rice. Different in density, the two kinds of protein bodies also differ in their protein content, some storing only prolamins while others are filled with glutelins. It is the only case of protein body polymorphism presently shown for cereal starchy endosperm protein bodies.

IV. MAXIMAL PACKING HYPOTHESIS

If the literature is still extremely short of results about secondary and tertiary structure, the different data on amino acid sequence and subunit arrangement already allow us to suggest a tentative interpretation: the maximal packing hypothesis.

If storage proteins were only copolymerized amino acids devoted to the young seedling nutrition, a very high mutability would be expected. Then how can we explain this severe conservatism, that we have pointed out? N terminal sequence and predicted secondary structure homologies strongly suggest some conservatism for the known moiety of polypeptide chains, in spite of their general microheterogeneity. Above all, subunit interchange and molecular hybridization of legumins and even with other dicot globulins are a definite proof of a common three-dimensional structure. So the question arises: what can be the function of such a conservative structure?

One answer would be that the storage protein folding has been maintained during evolution in order to fit peculiar endoproteases which hydrolyse seed proteins in the course of germination. In our opinion this reason is not very plausible, so we suggest another cause of this severe conservatism: that storage protein structure is mainly adapted to a maximal packing of storage proteins within protein bodies.

The maximal packing can be achieved in two degrees. On one hand, the compacting of proteins is increased by the formation of a quaternary structure and on the other hand, the folding of the polypeptide chain may be in favour of a maximal packing of amino acids within the protein molecule.

With regard to the first point, it is important to consider that for Panicoideae prolamins a quaternary structure is maintained even in denaturing conditions and that high degrees of polymerization can be observed in pearl millet pennisetin. Besides, in the Triticeae tribe, mainly in wheat, prolamins and glutelins are known to associate into aggregates arising in the formation of gluten (Kasarda, 1980; Bietz and Huebner, 1980). It is also striking to state the high frequency of legume globulin subunit association into either trimers or hexamers. Now one cannot refrain from noticing that these two kinds of quaternary structure are exactly those which allow the most compact packing: such a characteristic is possibly an important one for storing protein inside protein bodies. We think it is not by chance that reserve proteins are able to associate into these kinds of quaternary structure. Three other arguments are in favour of this explanation. Firstly, there is the sequence homology of basic polypeptide chains of legumin from soybean, broadbean, pea and globulins of pumpkin. It is indeed known that quaternary structure is mainly a consequence of primary structure. Secondly, the fact that globulin subunits of plant species as distant as *Glycine max* and *Sesamum indicum* can associate between themselves and undergo *in vitro* molecular hybridization, as shown by Mori *et al.* (1979). Thirdly, there is the conservatism of predicted secondary structures which are hardly altered by point mutations.

Concerning the second point, i.e. the maximal packing of amino acids within a minimal molecular volume, one must notice that secondary structure predictions have often resulted in successive β-turns. Fig. 16 illustrates a plausible pseudohelicoidal conformation of ancestral ω-gliadin repetitive sequence $(PYPQQ)_n$ taken as an example. Due to the fixed dihedral angle of proline and the formation of the β-turn hydrogen bond, a plausible structure would be successive planes joined by a glutamyl residue. This would allow a compact packing of amino acid residues. The X-ray diffraction analyses which are being undertaken in vicilins will bring new data to support this hypothesis.

V. Conclusion

Seed storage proteins are presently known to only play a role in the nutrition of the young seedling when it is still heterotrophic. One may

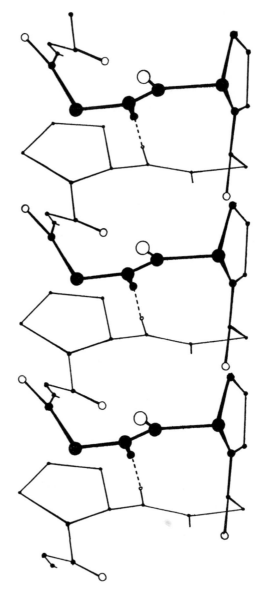

FIG. 16. Possible pseudo helicoidal conformation of ancestral ω-gliadin repetitive sequence (PYPQQ)$_n$. β-turns form parallel planes joined by a glutamyl residue. No side chain is shown, except proline cycle; ----, β-turn hydrogen bond; open circle, oxygen atom of the C = O bond.

wonder whether they derive from ancestors which had a functional role or whether they may have such a role in another physiological stage of the plant. The first point can be supported by the results of MacPherson and Smith (1980) and (personal communication) who have identified Jack bean canavalin as a modified form of α-D-mannosidase. This can be compared to mammary gland lactose synthetase which consists of a catalytic subunit (galactosyl transferase) and a modifier subunit (α-lactalbumin) changing the specificity of the catalytic subunit into lactose synthetase activity. The α-lactalbumin, devoid of any enzymatic function, is a molecule structurally very close to lysozyme.

On the other hand, it is important to note that some seed storage proteins have a lectin function, for instance in French bean and in pea. Moreover, it is interesting to mention the particular case of urease. It can act as an enzyme during seed development and as a storage protein during germination. It is indeed present in many legume seeds, and amounts to 1% of total seed protein of *Canavalia ensiformis* (Bailey and Boulter, 1971).

We have shown that seed storage proteins share in common the potentiality of associating into compact arrangements well fitted to increase the storage of amino acids in a smallest volume. In other words, we arrive at the conclusion that seed storage proteins exhibit a particular fitness to compacting.

Cytologically speaking, a large amount of storage proteins are accumulated and sequestred in protein bodies, such as starch in starch granules or oil in spherosomes and these reserve organelles are themselves located in reserve tissues. But some of these storage proteins, like some glutelins, are accumulated outside of the protein body.

It is worth emphasizing that seed storage proteins have to be put in the same category as secretory proteins: when crossing the membrane of the endoplasmic reticulum or of the protein body, the N terminal end of the nascent polypeptide (signal peptide) is cleaved. Other post-translational modifications are known to occur such as nicking of the legumin molecule into acidic and basic subunits or the vicilin glycosylation. The biological role of these modifications is not clear. Perhaps the nicked legumin monomers have an increased affinity one to another in order to polymerize but no evidence at present supports this hypothesis. With regard to the vicilin glycosylation, it has recently been shown by Badenoch-Jones *et al.* (1981) that it is not an essential step in the synthesis of vicilin nor in the formation of their quaternary structure.

Another conclusion to draw is the heterogeneity of the polymorphism of seed storage proteins. The different reasons of plant protein heterogeneity have already been analysed and reviewed elsewhere (Mossé, 1973). The occurrence of genetic variants which often explains the secondary minor bands of electrophoretic patterns is a first cause of heterogeneity, as well as the multiplicity of a few kinds of subunits. Other causes are post-translational, among which one can mention: the occurrence of several degrees of oligomerization (which results in a series of sedimentation coefficients); the possible interchange of some of the subunits, as shown for β-conglycinin which arises in multiple forms; the contingent variation of size or nature of bound carbohydrate groups (of which the exact function remains still unknown). All these reasons combine to produce a great heterogeneity both of size and charge. If their quaternary structure appears as resulting into very few models of association and packing, it also arises into a great multiplicity of these proteins due to the numerous multiple forms involved, which is a complementary source of heterogeneity, beside genetic variants.

Lastly, they appear as both stable and variable. Stable at the life scale, in the sense that their phenotypical expression almost does not depend on the environmental conditions. Very variable on the genetic level at the evolutionary scale, due to the fact that non lethal point mutations may reach a much higher level than in functional proteins. For this reason, they appear as noteworthy genetic markers.

As a last point, we want to insist that these proteins share in common numerous features with animal storage proteins such as caseins, for instance: they are all trophic secretory proteins, able to associate, exhibiting a relative multiplicity and polymorphism and their deposition is close to phosphate reserves.

A few months ago, Kemp in Hall's team in Madison (USA) succeeded in transferring a phaseolin gene into sunflower crown gall (Hall, personal communication). But, although present in tumoral tissues, the gene is not translated. At the same time a further step was made by Kurtz (1981) in obtaining mouse cells in which rat liver protein α_{2u}-globulin genes were introduced. The copies of the α_{2u}-globulin genes were not only detected but they were able to be induced by dexamethasone to produce α_{2u}-globulin mRNA and protein. In our opinion, one of the reasons for this success lies in strong homologies between mammal globulins. We think that the homology of structure and arrangement we have underlined, whatever it can be fit for, is particularly in favour of some successful breeding of seeds producing a cocktail of hybridized molecular species with an improved amino acid composition.

References

Alekseeva, M. V. and Kovarskaya, N. V. (1978). *Sov. Plant Physiol.* **25**, 365–370.
Altschul, A. M., Neucere, N. J., Woodman, A. A. and Dechary, J. M. (1964). *Nature* **203**, 501–504.

Autran, J. C., Nimmo, C. C., Lew, E. J. L. and Kasarda, D. D. (1979) *In* "Seed Improvements in Cereals and Grain Legumes" Vol. II, pp. 427–428. IAEA, Vienna.

Badenoch-Jones, J., Spencer, D., Higgins, T. J. V. and Millerd, A. (1981). *Planta* **153**, 201–209.

Badley, R. A., Atkinson, D., Hauser, H., Oldani, D., Green, J. P. and Stubbs, J. M. (1975). *Biochim. Biophys. Acta* **412**, 214–228.

Bagley, B. W., Cherry, J. H., Rollins, M. L. and Altschul, A. M. (1963). *Am. J. Bot.* **50**, 523–532.

Bailey, C. J. and Boulter, D. (1971). *In* "Chemotaxonomy of Leguminosae" (J. B. Harbone, D. Boulter and B. L. Turner, eds), pp. 485–502. Academic Press, London and New York.

Barker, R. D. J., Derbyshire, E., Yarwood, A. and Boulter, D. (1976). *Phytochemistry* **15**, 751–757.

Begbie, R. (1979). *Planta* **147**, 103–110.

Bietz, J. A. and Huebner, F. R. (1980). *Ann. Technol. Agric.* **29**, 249–277.

Bietz, J. A., Huebner, F. R., Sanderson, J. E. and Wall, J. S. (1977). *Cereal Chem.* **54**, 1070–1083.

Bietz, J. A., Paulis, J. W., and Wall, J. S. (1979). *Cereal Chem.* **56**, 327–332.

Blagrove, R. J., Gillespie, J. M., Lilley, G. G. and Woods, E. F. (1980). *Aust. J. Plant Physiol.* **7**, 1–13.

Bollini, R. and Chrispeels, M. J. (1978). *Planta* **142**, 291–298.

Brandt, A. and Ingversen, J. (1978). *Carlsberg Res. Commun.* **43**, 451–469.

Briarty, L. G. (1978). *In* "Plant Proteins" (G. Norton, ed.) pp. 81–106. Butterworths, London.

Burr, B. and Burr, F. A. (1976). *Proc. Natl. Acad. Sci. U.S.A.* **73**, 515–519.

Burr, B., Burr, F. A., Rubenstein, I. and Simon, M. N. (1978). *Proc. Natl. Acad. Sci. U.S.A.* **75**, 696–700.

Cameron-Mills, V. and Ingversen, J. (1978). *Carlsberg Res. Commun.* **43**, 471–489.

Cameron-Mills, V. and von Wettstein, D. (1980). *Carlsberg Res. Commun.* **45**, 577–594.

Carasco, J. F., Croy, R., Derbyshire, E. and Boulter, D. (1978). *J. Exp. Bot.* **29**, 309–323.

Casey, R. (1979a). *Biochem. J.* **177**, 509–520.

Casey, R. (1979b). *Heredity* **43**, 265–272.

Casey, R., March, J. F., Sharman, J. E. and Short, M. N. (1981a) *Biochim. Biophys. Acta* **670**, 428–432.

Casey, R., March, J. F. and Sanger, E. (1981b). *Phytochemistry* **20**, 161–163.

Catsimpoolas, N. and Wang, J. (1971). *Anal. Biochem.* **44**, 436–444.

Catsimpoolas, N., Rogers, D. A., Circle, J. and Meyer, E. W. (1967). *Cereal Chem.* **44**, 631–637.

Catsimpoolas, N., Campbell, T. G. and Meyer, E. W. (1968). *Plant Physiol.* **43**, 799–805.

Catsimpoolas, N., Keeney, J. A., Meyer, E. W. and Szubaj, B. F. (1971). *J. Sci. Food Agric.* **22**, 448–450.

Charbonnier, L., Tercé-Laforgue, T. and Mossé, J. (1980). *Ann. Technol. Agric.* **29**, 175–190.

Charbonnier, L., Tercé-Laforgue, T. and Mossé, J. (1981). *J. Agric. Food. Chem.* **29**, 968–973.

Chou, P. Y. and Fasman, G. D. (1978). *Annu. Rev. Biochem.* **47**, 251–276.

Croy, R. R. D., Derbyshire, E., Krishna, T. G. and Boulter, D. (1979). *New Phytol.* **83**, 29–35.

Croy, R. R. D., Gatehouse, J. A., Evans, I. M. and Boulter, D. (1980a). *Planta* **148**, 49–56.

Croy, R. R. D., Gatehouse, J. A., Evans, I. M. and Boulter, D. (1980b). *Planta* **148**, 57–63.

Croy, R. R. D., Gatehouse, J. A., Tyler, M. and Boulter, D. (1980c). *Biochem. J.* **191**, 509–516.

Danielsson, C. E. (1949). *Biochem. J.* **44**, 387–400.

Dechary, J. M., Talluto, K. F., Evans, W. J., Carney, W. B. and Altschul, A. M. (1961). *Nature* **190**, 1125–1126.

Derbyshire, E. and Boulter, D. (1976). *Phytochemistry* **15**, 411–414.

Derbyshire, E., Wright, D. S. and Boulter, D. (1976). *Phytochemistry* **15**, 3–24.

Dieckert, J. W. and Dieckert, M. C. (1976). *J. Food Sci.* **41**, 475–482.

Dieckert, J. W., Snowden, J. E., Moore, A. T., Heinzelman, D. C. and Altschul, A. M. (1962), *J. Food. Sci.* **27**, 321–325.

Draper, M. and Catsimpoolas, N. (1978). *Cereal Chem.* **55**, 16–23.

Elleman, T. C. (1977). *Aust. J. Biol. Sci.* **30**, 33–45.

Ericson, M. C. (1975). *Diss. Abstr.* **35**, 4767-B.

Ericson, M. C. and Chrispeels, M. J. (1973) *Plant Physiol.* **52**, 98–104.

Foster, J. F. and Edsall, J. T. (1945). *J. Am. Chem. Soc.* **67**, 617–625.

Garnier, J., Osguthorpe, D. J. and Robson, B. (1978). *J. Mol. Biol.* **120**, 97–120.

Gatehouse, J. A., Croy, R. R. D. and Boulter, D. (1980). *Biochem. J.* **185**, 497–503.

Gatehouse, J. A., Croy, R. R. D., Morton, H., Tyler, M. and Boulter, D. (1981). *Eur. J. Biochem.* **118**, 627–633.

Geraghty, D., Peifer, M. A. and Rubenstein, I. (1981). *Nucleic Acids Res.*, **9**, 5163–5174.

Gilroy, J., Wright, D. J. and Boulter, D. (1979). *Phytochemistry* **18**, 315–316.

Graham, T. A. and Gunning, B. E. S. (1970). *Nature* **228**, 81–82.

Hasegawa, K., Tanaka, K. and Tamai, S. (1981). *Agric. Biol. Chem.* **45**, 809–814.

Horisberger, M. and Vonlanthen, M. (1980). *Histochemistry* **65**, 181–186.

Hurkman, W. J. and Beevers, L. (1980). *Planta* **150**, 82–88.

Iibuchi, C. and Imahori, K. (1978a). *Agric. Biol. Chem.* **42**, 25–30.

Iibuchi, C. and Imahori, K. (1978b). *Agric. Biol. Chem.* **42**, 31–36.

Kasarda, D. D. (1980). *Ann. Technol. Agric.* **29**, 151–173.

Kasarda, D. D., Da Rosa, D. A. and Ohms, J. I. (1974). *Biochim. Biophys. Acta* **351**, 290–294.

Kasarda, D. D., Bernardin, J. E. and Nimmo, C. C. (1976). *In* "Advances in Cereal Science and Technology" (Y. Pomeranz, ed) Vol. 1, pp. 158–236. American Association of Cereal Chemists, St. Paul, Minnesota, USA.

Khoo, U. and Wolf, M. J. (1970). *Am. J. Bot.* **57**, 1042–1050.

Kim, S. I. (1978). Thèse d'Etat. Université de Paris VI.

Kitamura, K., Takagi, T. and Shibasaki, K. (1976). *Agric. Biol. Chem.* **40**, 1837–1844.

Koshiyama, I. and Fukushima, D. (1976a). *Cereal Chem.* **53**, 768–769.

Koshiyama, I. and Fukushima, D. (1976b). *Int. J. Peptide Protein Res.* **8**, 283–289.

Krishna, T. G., Croy, R. R. D. and Boulter, D. (1979). *Phytochemistry* **18**, 1879–1880.

Kurtz, D. T. (1981). *Nature* **291**, 629–631.

Landry, J. and Moureaux, T. (1970). *Bull. Soc. Chim. Biol.* **52**, 1021–1037.

Larkins, B. A. (1981). *Biochemistry of Plants* **6**, 449–489.
Larkins, B. A. and Hurkman, W. J. (1978). *Plant Physiol.* **62**, 256–263.
Lorenz, K., Yetter, M. and Saunders, R. M. (1978). *Cereal Chem.* **55**, 66–76.
Lott, J. N. A. and Buttrose, M. S. (1978). *Aust. J. Plant Physiol.* **5**, 89–111.
Ma, Y., Bliss, F. A. and Hall, T. C. (1980). *Plant Physiol.* **66**, 897–902.
McLeester, R. C., Hall, T. C., Sun, S. M. and Bliss, F. A. (1973). *Phytochemistry* **2**, 85–93.
McPherson, A. (1980). *J. Biol. Chem.* **255**, 10472–10480.
McPherson, A. and Smith, S. C. (1980). *Phytochemistry* **19**, 957–959.
Miflin, B. J., Burgess, S. R. and Shewry, P. R. (1981). *J. Exp. Bot.* **32**, 199–219.
Millerd, A., Thomson, J. A. and Schroeder, H. E. (1978). *Aust. J. Plant Physiol.* **5**, 519–534.
Moreira, M. A., Hermodson, M. A., Larkins, B. A. and Nielsen, N. C. (1979). *J. Biol. Chem.* **254**, 9921–9926.
Moreira, M. A., Hermodson, M. A., Larkins, B. A. and Nielsen, N. C. (1981). *Arch. Biochem. Biophys.* **210**, 633–642.
Mori, T. and Utsumi, S. (1979). *Agric. Biol. Chem.* **43**, 577–583.
Mori, T., Utsumi, S. and Inaba, H. (1979). *Agric. Biol. Chem.* **43**, 2317–2322.
Morris, G. F. I., Thurman, D. A. and Boulter, D. (1970). *Phytochemistry* **9**, 1707–1714.
Mossé, J. (1968). *In* "Progrès en Chimie Agricole et Alimentaire", pp. 47–81. Herman, Paris.
Mossé, J. (1973). *Physiol. Veg.* **11**, 361–384.
Mossé, J. and Landry, J. (1980). *In* "Cereals for Food and Beverages. Recent Progress in Cereal Chemistry" (G. E. Inglett and L. Munck, eds), pp. 255–273. Academic Press, London and New York.
Mossé, J. and Pernollet, J. C. (1982). *In* "Chemistry and Biochemistry of Legumes" (S. K. Arora, ed). Oxford and IBH Publishing Co., New Dehli (in press).
Murray, D. R. (1979). *Plant Cell Environ.* **2**, 221–226.
Murray, D. R. and Crump, J. A. (1979). *Z. Pflanzenphysiol.* **94**, 339–350.
Ohmiya, M., Hara, I. and Matsubara, H. (1980). *Plant Cell Physiol.* **21**, 157–167.
Osborne, T. B. and Campbell, G. F. (1896). *J. Am. Chem. Soc.* **18**, 583–609.
Parker, M. L. (1980). *Ann. Bot.* **46**, 29–36.
Parker, M. L. (1981). *Micron* **12**, 187–188.
Pernollet, J. C. (1978). *Phytochemistry* **17**, 1473–1480.
Pernollet, J. C. (1982). *Physiol. Veg.* **20**, 259–276.
Pernollett, J. C. and Mossé, J. (1980). *C. R. Acad. Sci.*, Ser. D, *Sci. Nat.*, **290**, 267–271.
Pernollet, J. C., Kim, S. I. and Mossé, J. (1982). *J. Agric. Food Chem.* **30**, 32–36.
Pusztai, A. (1966). *Biochem. J.* **101**, 379–384.
Pusztai, A. and Stewart, J. C. (1980). *Biochim. Biophys. Acta* **623**, 418–428.
Pusztai, A. and Watt, W. B. (1970). *Biochim. Biophys. Acta* **207**, 413–431.
Pusztai, A., Croy, R. R. D., Grant, G. and Watt, W. B. (1977). *New Phytol.* **79**, 61–71.
Pusztai, A., Stewart, J. C. and Watt, W. B. (1978). *Plant Sci. Lett.* **12**, 9–15.
Pusztai, A., Croy, R. R. D., Stewart, J. S. and Watt, W. B. (1979). *New Phytol.* **83**, 371–378.
Racusen, D. and Foote, M. (1971). *Can. J. Bot.* **49**, 2107–2111.
Rao, R. and Pernollet, J. C. (1981). *Agronomie* 1, 909–916.

Rougé, P. (1977). *Thèse de Doctorat d'Etat* n° 746, Université Paul Sabatier, Toulouse.

Saio, K., Matsuo, T. and Watanabe, T. (1970). *Agric. Biol. Chem.* **34**, 1851–1854.

Schlesier, B., Manteuffel, R. and Scholz, G. (1978). *Biochem. Physiol. Pflanzen* **172**, 285–290.

Schmitt, J. and Svendsen, I. (1980a). *Carlsberg Res. Commun.* **45**, 143–148.

Schmitt, J. and Svendsen, I. (1980b). *Carlsberg Res. Commun.* **45**, 549–556.

Sefa-Dedeh, S. and Stanley D. (1979a). *J. Agric. Food Chem.* **27**, 1238–1243.

Sefa-Dedeh S. and Stanley, D. (1979b). *J. Agric. Food Chem.* **27**, 1244–1247.

Sharma, C. B. and Dieckert, J. W. (1975). *Physiol. Plant.* **33**, 1–7.

Shetty, K. J. and Rao, M. S. N. (1977). *Int. J. Peptide Protein Res.* **9**, 11–17.

Shewry, P. R., Autran, J. C., Nimmo, C. C., Lew, E. J. L. and Kasarda, D. D. (1980). *Nature* **286**, 520–522.

Shewry, P. R., Lew, E. J. L. and Kasarda, D. D. (1981). *Planta* **153**, 246–253.

Sobolev, A. M., Buzulukova, N. P., Dmitrieva, M. I. and Barbashova, A. K. (1977). *Sov. Plant. Physiol.* **23**, 621–628.

Staswick, P. E., Hermodson, M. A. and Nielsen, N. C. (1981). *J. Biol. Chem.* **256**, 8752–8755.

Stockman, D. R., Hall, T. C. and Ryan, D. S. (1976). *Plant Physiol.* **58**, 272–275.

Sun, S. M., McLeester, R. C., Bliss, F. A. and Hall, T. C. (1974). *J. Biol. Chem.* **249**, 2118–2121.

Tanaka, K., Sugimoto, T., Ogawa, M. and Kasai, Z. (1980). *Agric. Biol. Chem.* **44**, 1633–1639.

Thanh, V. H. and Shibasaki, K. (1976a). *Biochim. Biophys. Acta* **439**, 326–338.

Thanh, V. H. and Shibasaki, K. (1976b). *J. Agric. Food Chem.* **24**, 1117–1121.

Thanh, V. H. and Shibasaki, K. (1977). *Biochim. Biophys. Acta* **490**, 370–384.

Thanh, V. H. and Shibasaki, K. (1978a). *J. Agric. Food Chem.* **26**, 692–695.

Thanh, V. H. and Shibasaki, K. (1978b). *J. Agric. Food Chem.* **26**, 695–698.

Thanh, V. H. and Shibasaki, K. (1979). *J. Agric. Food Chem.* **27**, 805–809.

Thanh, V. H., Okubo, K. and Shibasaki, K. (1975a). *Plant Physiol.* **56**, 19–22.

Thanh, V. H., Okubo, K. and Shibasaki, K. (1975b). *Agric. Biol. Chem.* **39**, 1501–1503.

Thomson, J. A. and Doll, H. (1979). *In* "Seed Protein Improvement of Cereals and Grain Legumes" Vol. I, pp. 109–123. IAEF, Vienna.

Thomson, J. A., Schroeder, H. E. and Dudman, W. F. (1978). *Aust. J. Plant Physiol.* **5**, 263–279.

Tombs, M. P. (1965). *Biochem. J.* **96**, 119–133.

Tombs, M. P. (1967). *Plant Physiol.* **42**, 797–813.

Tombs, M. P. and Lowe, M. (1967). *Biochem. J.* **105**, 181–187.

Tombs, M. P., Newson, B. F. and Wilding, P. (1974). *Int. J. Peptide Protein Res.* **6**, 253–277.

Tumer, N. E., Thanh, V. H. and Nielsen, N. C. (1981). *J. Biol. Chem.* **256**, 8756–8760.

Utsumi, S., Inaba, H. and Mori, T. (1980). *Agric. Biol. Chem.* **44**, 1891–1896.

Vaintraub, I. A., Shutov, A. D. and Klimenko, V. G. (1962). *Biokhimiya* **27**, 349–358.

Varner, J. E. and Schidlovsky, G. (1963). *Plant Physiol.* **38**, 139–144.

Wall, J. S. and Paulis, J. W. (1978). *In* "Advances in Cereal Science and Technology" (Y. Pomeranz, ed), pp. 135–219. American Association of Cereal Chemists, St. Paul, Minnesota, U.S.A.

Weber, E., Manteuffel, R. and Neumann, D. (1978). *Biochem. Physiol. Pflanzen* **172,** 597–614.

Wright, D. J. and Boulter, D. (1974). *Biochem. J.* **141,** 413–418.

Yamada, T., Aibara, S. and Morita, Y. (1979a). *Agric. Biol. Chem.* **43,** 2549–2556.

Yamada, T., Aibara, S. and Morita, Y. (1979b). *Agric. Biol. Chem.* **43,** 2563–2568.

Yamada, T., Aibara, S. and Morita, Y. (1981). *Agric. Biol. Chem.* **45,** 1243–1250.

Yamauchi, F. and Yamagishi, T. (1979). *Agric. Biol. Chem.* **43,** 505–510.

Yamauchi, F., Kawase, M., Kanbe, M. and Shibazaki, K. (1975). *Agric. Biol. Chem.* **39,** 873–878.

Yamauchi, F., Thanh, V. H., Kawase, M. and Shibasaki, K. (1976). *Agric. Biol. Chem.* **40,** 691–696.

CHAPTER 8

The Molecular Biology of Storage Protein Synthesis in Maize and Barley Endosperm

J. INGVERSEN

Department of Physiology, Carlsberg Laboratory, Gamle Carlsberg Vej 10, DK–2500 Copenhagen Valby, Denmark

I. INTRODUCTION

Cereal grains are the major source of the protein that is consumed by man and thus the need for better nutritionally balanced cereal proteins is urgent. The amino acid composition of the cereal endosperm proteins is characterized by a high content of proline and glutamine while the amount of the essential amino acid lysine in particular is limiting for the efficient metabolization of the grain when used as animal or human food. After the first high lysine maize mutant was reported by Mertz *et al.* in 1964, screening programmes were initiated to find similar mutants in other cereal species. So far high lysine barley, maize and sorghum mutants have been identified (Mertz *et al.*, 1964; Munck *et al.*, 1970; Ingversen *et al.*, 1973; Singh and Axtell, 1973; Doll *et al.*, 1974). The biochemical explanation for the overall increase in the lysine content of the endosperm proteins from these mutants is a decrease in the amount of the lysine poor storage protein fraction or by an increase in lysine rich proteins (Ingversen et al., 1973; Doll, 1980; Jonassen, 1980a and b; Shewry et al., 1980a). A major obstacle for the commercial exploitation of these barley, maize and sorghum high lysine lines, however, is that the starch synthesis of the mutant endosperms is also partly reduced, resulting in decreased grain yield. Despite intensive cross breeding no normal yielding high lysine cereal variety has yet appeared on the market, and it is clear that a better understanding of the cellular and biochemical processes involved in the synthesis and deposition of cereal endosperm protein and starch is required to permit genetic improvement of the nutritional quality of the cereal grain. Studies in pursuance of this goal have been carried out in several laboratories during the last 6 years. High lysine mutants of maize and barley and their parent varieties have been the most extensively studied, particularly with regard to biosynthesis of the endosperm storage proteins.

Endosperm storage proteins are deposited in protein bodies during the grain filling period and preferentially synthesized in response to increasing levels of nitrogen fertilizer. The major storage protein fraction of maize, barley, wheat and sorghum is the prolamin fraction. Prolamin is soluble in 55% isopropanol containing 2% β-mercaptoethanol and contributes 30–50% of the endosperm protein (Shewry et al., 1978a, Miflin and Shewry, 1979); it is deposited in protein bodies (Ingversen, 1975; Burr and Burr, 1976; Cameron-Mills and von Wettstein, 1980; Miflin et al., 1981) and contains a high proportion of proline and glutamine, whereas the content of lysine is very low. This review is limited to a summary of the present knowledge on the molecular composition, amino acid sequence, biosynthesis and deposition of prolamin from the two diploid species maize and barley.

II. Description and Analysis of the Prolamin Fraction from Maize and Barley

Hordein—the prolamin isolated from barley endosperm—fractionates by electrophoresis on polyacrylamide gels containing SDS into 8 to 10 polypeptide bands with apparent molecular weights from 70 000 to 27 000, whereas zein—the prolamin from maize endosperm—only yields two major polypeptide bands with molecular weights of 22 000 and 19 000 (Burr and Burr, 1976; Holder and Ingversen, 1978). When zein and hordein are fractionated on isoelectric focusing gels, however, a considerable charge heterogeneity is revealed (Righetti et al., 1977; Shewry et al., 1977). By combination of SDS-polyacrylamide gel electrophoresis and isoelectric focusing in a two dimensional protein separation system, 15 zein and more than 15 hordein components, depending on the variety, can be identified (Righetti *et al.*, 1977; Shewry *et al.*, 1977; Hagen and Ruben-stein, 1980). The prolamin fraction of maize and barley is thus composed of a number of polypeptides differing in apparent molecular weight and/or charge. Righetti *et al.* (1977) demonstrated that the extensive charge heterogeneity of zein reflects the actual complexity of the primary transla-tion product. The isoelectric focusing patterns were independent of the developmental stage of the endosperm, not a result of differential ampho-line-polypeptide interaction and inherited in a Mendelian way in agree-ment with the triploid constitution of the maize endosperm. Patterns due to different conformational states of a single polypeptide chain cannot be distinguished from patterns due to different polypeptide chains by the employed techniques. Faulks *et al.* (1981) isolated individual hordein polypeptides from the major spots, as they emerge on a polyacrylamide gel slab after two-dimensional separation with electrophoresis in the first dimension and isoelectric focusing in the second. The extracted hordein polypeptides were cleaved with cyanogen bromide and the resulting peptides analysed on a gradient polyacrylamide gel containing SDS. The maps from different hordein polypeptides showed some similarities, but distinct differences were also recognized.

The most convincing evidence for the existence of multigene families for barley and maize prolamin is based on results obtained from *in vitro* translation of messenger RNA and polysomes coding for prolamin. Brandt and Ingversen (1976) isolated initially membrane bound polysomes from immature barley endosperm, translated the polysome bound messen-ger RNA *in vitro*, and fractionated the products on an SDS-polyacryl-amide gel. The *in vitro* products were found to have the same apparent molecular weights as those of hordein synthesized *in vivo* (Brandt and Ingversen, 1976). Matthews and Miflin (1980) subjected hordein synthe-sized *in vitro* from polyadenylated barley endosperm RNA to isoelectric focusing. Again the *in vitro* translation products had similar isoelectric focusing patterns as authentic hordein. A similar study (Viotti *et al.*, 1978) reveals that *in vitro* synthesized zein has the same spectrum of net charges as authentic zein.

In order to compare the molecular complexity of the individual *in vitro* synthesized hordein polypeptides Holder and Ingversen (1978) excised 8 *in vitro* synthesized ^{35}S-labelled hordein polypeptides from a polyacrylamide gel, treated them with chymotrypsin, and analysed the resultant peptides by two dimensional thin layer chromatography and autoradiography. They concluded that the primary structure of the 8 polypeptides was polymor-phic but similar, suggesting a common origin of the genetic material coding for them. In summary, it appears that the prolamin fraction of barley and maize is composed of 15 or more polypeptides each specified for by a separate structural gene. The hordein polypeptides are structurally related, suggesting that in barley at least the prolamins are coded for by a group of closely related genes.

Total hordein and zein mixtures have been subjected to amino acid sequence analysis by the automated Edman degradation technique (Schmitt, 1979; Bietz, 1981). One major sequence was obtained for the 31 N-terminal residues (Schmitt and Svendsen, 1980a,b) suggesting extensive sequence homology between the separate hordein polypeptides. Amino acid polymorphisms occurred at 6 positions in this sequence. Using gel filtration and ion exchange chromatography (Schmitt, 1979; Shewry et al., 1980c) hordein could be fractionated into aC and aB fraction, the former comprising the polypeptides with apparent molecular weights ranging from 70 000 to 45 000 and the latter from 40 000 to 25 000, respectively. The B-hordeins were blocked at the N-terminus (Schmitt and Svendsen, 1980a; Shewry et al., 1980c), whereas the N terminal sequence of the C-hordein mixture was identical to the sequence previously found for a mixture of B- and C-hordeins (Schmitt and Svendsen, 1980b; Bietz, 1981). Recently four individual C-hordein polypeptides have been isolated in sufficient amounts to permit a determination of their N termini (Schmitt, 1979; Shewry et al., in press). The four preparations had identical amino acid sequences for the first 10 N terminal residues and for the first three C terminal residues, although their apparent molecular weights were different. Schmitt (1979) isolated one hordein polypeptide belonging to the N terminal blocked B-group with an apparent molecular weight of 27 000. Fragments corresponding to one quarter of the molecule were isolated after proteolytic cleavage and sequenced (Schmitt and Svendsen 1980b). No extensive sequence homologies between the B-polypeptide and C-hordein were detected. N terminal sequence analysis of the whole zein mixture isolated from corn endosperm indicates a high degree of sequence homology among the individual zein polypeptides (Bietz, 1979). No similarities exist, however, between the presently known zein and hordein sequences.

Genetic studies (for a review see Nelson, 1980) show that the electrophoretic pattern observed for hordein on an SDS-polyacrylamide gel is controlled by two linked loci on chromosome 5, the B and C-hordeins being controlled by the Hor 2 and Hor 1 locus respectively (Shewry et al., 1978b; Doll and Brown, 1979). No recombination has been observed within the two hordein loci, and it has been suggested that each locus consists of a series of tandemly linked genes, derived by duplication and divergence of a single gene (Holder and Ingversen, 1978; Miflin and Shewry, 1979). The high degree of N terminal amino acid sequence homology among the C-hordeins (Hor 1) supports this theory. In contrast to the hordein genes, the genes coding for the zein polypeptides are located in at least three positions in the maize genome and are not linked with each other (Soave et al., 1978).

The cDNA copies of zein mRNA have been used to study the molecular complexity of zein mRNA and the frequency and location of zein genes in the maize genome (Park *et al.*, 1980; Pedersen *et al.*, 1980; Wienand and Felix, 1980; Hagen and Rubenstein, 1981). Park *et al.* (1980) suggest on the basis of hybridization experiments that the *in vitro* synthesized zein polypeptides belong to three families of mRNA's. By hybridizing zein cDNA to fragments of maize chromosomal DNA produced by restriction enzyme cleavage, evidence was obtained for a complex multigene system coding for zein polypeptides (Wienand and Feix, 1980; Hagen and Rubenstein, 1981). Pedersen *et al.* (1980) hybridized zein cDNA to sheared maize chromosomal DNA. The results suggest the presence of 1–5 copies of zein genes per haploid genome.

Hybridization of maize chromosomal DNA with ^{125}I-labelled zein mRNA or its ^{3}H-labelled cDNA copy *in situ* indicates the presence of zein structural genes on chromosomes 4, 5, 7, and 10 (Viotti *et al.*, 1980).

The final answer as to the complexity of the prolamin genes depends on the successful isolation of the structural genes coding for prolamin. Maize and barley genomic libraries have been constructed (Brandt *et al.*, 1981; Lewis *et al.*, 1981). Screening of the maize library with cDNA clones of zein mRNA resulted in 19 hybridizing λ-clones, 3 of which were analysed in detail (Lewis *et al.*, 1981). They all contained DNA inserts, which specifically hybridized to the zein cDNA probe, and showed sequence homology to zein mRNA's.

III. Biosynthesis of Prolamin

A. isolation and *in vitro* translation of mRNA isolated from maize and barley endosperm

Extraction and quantitative estimation of maize zein and barley hordein at different times during the grain filling period shows that zein as well as hordein is intensively synthesized from 15 to 35 days after fertilization (Brandt, 1976; Shewry *et al.*, 1979; Tsai, 1979). Ultrastructural studies of the developing endosperm from barley and maize reveal an extensive development of the endoplasmic reticulum during storage protein synthesis (Munck and von Wettstein, 1976; Larkins and Hurkman, 1978; von Wettstein, 1979; Miflin *et al.*, 1981). The occurrence of numerous polyribosomes attached to the endoplasmic reticulum suggested it to be the major site for storage protein synthesis. In order to investigate this point membrane bound and free polyribosomes from developing endosperms of barley and maize have been isolated and translated *in vitro*. Brandt and Ingversen (1976) showed that only the initially membrane bound polyribosomes isolated from barley endosperm catalysed the synthesis of hordein polypeptides *in vitro*. Burr and Burr (1976) isolated protein bodies from young maize endosperm, stripped off the attached polyribosomes, and then translated them *in vitro*. Analysis of the translation products soluble in 70% alcohol by SDS polyacrylamide gel electrophoresis revealed that most if not all polypeptides synthesized *in vitro* were zein. Larkins *et al.* (1976a) found that membrane bound polyribosomes coded for zein *in vitro*. The *in vitro* translation products of polyribosomes initially associated with protein bodies were compared with those derived from the endoplasmic reticulum of the maize endosperm by Larkins and Hurkman (1978). Both contained zein. Whether the protein body bound polyribosomes constitute a specific pool with high template activity for zein is not certain, since the distribution of polyribosomes within distinctive subcellular fractions might be a result of artifacts created during the homogenization of the endosperm. Initially membrane bound polyribosomes from maize and barley endosperm were used as starting material for the isolation of zein (Larkins *et al.*, 1976a,b; Burr *et al.*, 1978; Larkins and Hurkman, 1978; Wienand and Feix, 1978; Melcher, 1979) and hordein (Brandt and Ingversen, 1978; Matthews and Miflin, 1980) mRNA.

Both Brandt and Ingversen (1978) and Mathews and Miflin (1980) isolated polyadenylated hordein mRNA. Although their size estimations of the isolated mRNA molecules differ, both preparations catalyse the *in vitro* synthesis of hordein precursor polypeptides soluble in 55% isopropanol which are 2–3 000 dalton larger in apparent molecular weight than the native hordein components. It was not possible to separate mRNA species for the individual hordein polypeptides by either sucrose gradient sedimentation or preparative agarose gel electrophoresis. Zein mRNA directs the synthesis of zein precursor molecules with apparent molecular weights of 21 000 and 24 000, compared with 19 000 and 22 000 for authentic zein (Burr *et al.*, 1978; Larkins and Hurkman, 1978; Melcher, 1979). Both sucrose gradient sedimentation (Larkins and Hurkman, 1978) and preparative polyacrylamide gel electrophoresis (Wienand and Feix, 1978) of zein mRNA preparations demonstrate that two groups of mRNA molecules exist for zein, one coding for the 22 000 and one of the 19 000 molecular weight group of zein polypeptides. The data summarized above show that hordein and zein are synthesized on polyribosomes attached to the endoplasmic reticulum of the endosperm cell. The primary translation products are precursor polypeptides with a 2–3 000 larger molecular weight than mature polypeptides.

B. INTRACELLULAR TRANSPORT AND DEPOSITION OF PROLAMIN

In analogy with the mechanism deduced for the synthesis and deposition of certain animal secretory proteins (Blobel and Dobberstein, 1975) it is conceivable that the prolamin precursor polypeptides contain an N terminal extension, a signal peptide which interacts with some specific site of the endoplasmic reticulum and triggers ribosome membrane binding, ensuring the subsequent passage of the nascent polypeptide chain across the membrane. To test this hypothesis Cameron-Mills et. al. (1978) isolated ribosome studded microsome vesicles from young barley endosperms, and used them as a template for in vitro protein synthesis. The hordein polypeptides synthesized co-migrated with native hordein polypeptides on SDS-polyacrylamide gels and were protected from proteolysis, in contrast to hordein synthesized in vitro with polysomes as the template. It was concluded that hordeins are synthesized on the rough endoplasmic reticulum and discarded vectorially into its lumen.

Recently Burr and Burr (1981) translated zein mRNA in the presence of an endoplasmic reticulum enriched fraction isolated from maize endosperm. The reaction mixture was posttranslationally treated with proteinase K to remove translation products not protected by the endoplasmic reticulum membrane. SDS-polyacrylamide gel electrophoresis of the proteinase resistant polypeptides revealed that a fraction of the zein precursor molecules were processed to their mature size. Both hordein synthesized in vitro on initially membrane bound polyribosomes and zein synthesized in vitro on polyadenylated mRNA from maize endosperm are thus partially protected against proteolytic degradation by the endoplasmic reticulum membrane, even when the membranes are supplied post-translationally (Cameron-Mills and Ingversen, 1978; Burr and Burr, 1981). The post-translational protection observed in vitro could be an experimental artifact, caused by nonspecific association of the hydrophobic prolamin molecules with the endoplasmic reticulum membrane. The route whereby storage proteins are transported from the rough endoplasmic reticulum to the final deposit, the protein body, has not yet been determined by biochemical methods. The information available is deduced from ultrastructural studies of the developing cereal endosperm (Khoo and Wolf, 1970; Munck and Wettstein, 1976; Wettstein, 1979, Cameron-Mills and Wettstein, 1980; Miflin et al., 1981). Khoo and Wolf (1970) postulated that maize prolamin was deposited inside the rough endoplasmic reticulum. This interpretation is supported by Larkins and Hurkman (1978) who present evidence for the physical continuity between the rough endoplasmic reticulum and the developing protein bodies of maize endosperm. In contrast to the monophasic small protein bodies present in maize endosperm, developing barley protein bodies are more complex and consist of clusters of homogeneous components embedded in a fibrillar matrix and associated with electron-dense spheres (Munck and Wettstein, 1976; Wettstein, 1979; Cameron-Mills and Wettstein, 1980). Miflin et al. (1981) suggest, in analogy with the proposal in the maize endosperm, that hordein aggregates into clumps inside the endoplasmic reticulum and is subsequently released into the cytoplasm. Cameron-Mills and von Wettstein (1980) have recently confirmed that developing protein bodies in barley are deposited inside vacuoles (Munck and Wettstein, 1976; Cameron-Mills and Wettstein, 1980). Numerous small vesicles aggregate around the growing protein body inside the vacuole, which might imply involvment of small vesicles in the transport of hordein from the rough endoplasmic reticulum to the vacuoles.

IV. CONSTRUCTION OF HORDEIN AND ZEIN cDNA CLONES

The isolation of polyadenylated mRNA coding for hordein and zein prompted the synthesis of double stranded DNA probes complementary to hordein and zein mRNA (dscDNA) (Brandt, 1979; Wienand et al., 1979; Park et al., 1980; Pedersen et al., 1980). The dscDNA was inserted into bacterial plasmids and used for transformation of E. coli cells. The cloned plasmid can be isolated in amounts sufficient to allow biochemical characterization of the inserted cDNA sequence. Wienand et al. (1979) obtained 25 recombinant colonies with DNA sequences which hybridize to a ^{125}I labelled zein mRNA fraction. Zein mRNA was hybridized to plasmid DNA from two of the recombinants, and the mRNA molecules which hybridized to the cDNA probe were subsequently released and translated in vitro. Analysis of the translation products by SDS-polyacrylamide gel electrophoresis and fluorography, indicated that the two plasmids harboured cDNA sequences for the 19 000 and the 22 000 molecular weight zein polypeptides respectively.

Using an almost identical strategy Brandt (1979) cloned a 850 base pair fragment, which upon hybridization arrested the in vitro synthesis of two hordein polypeptides of the B group.

Recently Geragthy et al. (1981) determined the nucleic acid sequence of a zein cDNA clone. The sequence contains one reading frame only, from which the sequence of 225 amino acids for the corresponding zein precursor polypeptide could be deduced. The first 12 amino acid residues account for the signal peptide, and from residue 13 the amino acid sequence parallels the N terminal amino acid sequence obtained by sequencing a mixture of zein proteins (Bietz, 1979). The total molecular weight of the unprocessed polypeptide amounts to 24 416 and that of the processed to 23 329. The central part of the zein molecule from residue 105 to residue 166 consists of two direct repeats each 28 amino acids long and separated by 5 residues.

V. Storage Protein Mutants in Barley and Maize

High lysine barley and maize mutants are characterized by an overall increase in the lysine content of the mature endosperm, compared with the parent varieties. In most cases the high lysine character involves a reduction in the synthesis of the lysine poor prolamin fraction while the mutant Risø 1508 (*lys* 3A) (Brandt, 1976; Shewry *et al.*, 1979) synthesizes decreased amounts of all the prolamin polypeptides, the mutation Risø 56 (*hor* 2ca) reduces specifically the synthesis of the B hordein polypeptides (Doll, 1980). Similarly the maize mutation opaque 2 (o_2) depresses the synthesis of the 22 000 molecular weight zein components (Jones *et al.*; 1976, Jones *et al.*, 1977a and b). One high lysine barley mutant, Hiproly (*lys*), displays none, or only a small reduction in prolamin synthesis, the elevated lysine content resulting from a change in the composition of the albumin and globulin fraction. The increased deposition of one polypeptide, SP II albumin, accounts alone for 37% of the overall lysine increase in the Hiproly endosperm when compared with a normal variety (Jonassen, 1980a and b; Svendsen *et al.*, 1980). The o_2 mutant in maize and the *hor* 2ca mutant in barley are of particular interest because both mutant genes have been mapped at or near the loci coding for prolamin polypeptides. Soave *et al.* (1978) reported that the genes coding for three distinct zein components differing in net charge were closely linked and positioned on chromosome 7, 5·5 cM from the o_2 locus. Two additional genes are located on other chromosomes. The recessive gene causing the high lysine character of the barley mutant Risø 56 (*hor* 2ca) is positioned at or near the *hor* 2 locus coding for the hordein–2 polypeptides (Doll, 1980).

Jones *et al.* (1976, 1977a and b) isolated initially membrane bound polyribosomes from developing endosperm of the o_2 maize mutant, and found that large polyribosomes (more than 5 ribosomes) were less abundant in the mutant endosperm compared with the parental line (Jones *et al.*, 1977a,b). The *in vitro* translation products of the mutant polyribosomes lacked especially the 22 000 molecular weight zein component. Although the sucrose density gradient profiles of membrane bound polyribosomes isolated from 20 days old endosperms of the barley mutant *lys* 3A and its parent variety Bomi were identical, *in vitro* translatable mRNA coding for hordein polypeptides were virtually absent from the *lys* 3A polyribosomes (Brandt and Ingversen, 1976). Hybridization experiments involving mRNA from mutant endosperms and the hordein and zein cDNA probes now available could settle the question of the presence of untranslatable mRNA in the mutants. The observation that functional microsomes cannot be reconstituted *in vitro* from stripped microsomes isolated from *lys* 3A endosperms, and initially membrane bound polyribosomes isolated from normal barley endosperms, suggests that the transport mechanism for hordein polypeptides is defective in the *lys* 3A endosperm (Cameron-Mills and Ingversen, 1978).

The condensation of the storage proteins inside the vacuole is markedly altered in the barley mutants *hor* 2ca and *lys* 3A (Munck and Wettstein, 1976; Cameron-Mills, 1980; Cameron-Mills and Wettstein, 1980). Twenty days after anthesis the protein bodies of normal barley comprise clusters of homogeneous components, embedded in a fibrillar matrix associated with electron-dense spheres. The fibrillar matrix appears to be a transient stage in the condensation of storage proteins into a homogeneous structure. Reduction in hordein synthesis increases the proportion of the fibrillar matrix, apparently retarding the condensation process.

VI. Concluding Remarks

The cereal endosperm is ideal experimental material for the molecular biologist, being highly specialized tissue primarily engaged in synthesis and accumulation of starch and storage protein. The existence of cereal mutants with drastic alterations in seed protein synthesis has provided an incentive for research on storage protein synthesis. Our knowledge on the molecular composition of storage proteins, their primary structure, biosynthesis and deposition has expanded considerably as has our knowledge on the genetic control of prolamin synthesis. Our information, however, on the primary genetic changes causing altered protein accumulation in the mutant endosperms of barley and maize remains as yet fragmentary.

References

Bietz, J. A. (1979). *Cereal Chem.* **56,** 327–332.
Bietz, J. A. (1981). *Cereal Chem.* **58,** 83–85.
Blobel, G. and Dobberstein, B. (1975). *J. Cell Biol.* **67,** 852–862.
Brandt, A. (1976). *Cereal Chem.* **53,** 890–901.
Brandt, A. (1979). *Carlsberg Res. Commun.* **44,** 255–267.
Brandt, A. and Ingversen, J. (1976). *Carlsberg Res. Commun.* **41,** 312–320.
Brandt, A. and Ingversen, J. (1978). *Carlsberg Res. Commun.* **43,** 451–469.
Brandt, A., Ingversen, J., Cameron-Mills, V., Schmitt, J. M. and Rasmussen, S. K. (1981). *Barley Genetics IV* (in press).
Burr, B. and Burr, F. A. (1976). *Proc. Natl. Acad. Sci. USA.* **73,** 515–519.

Burr, F. A. and Burr, B. (1981). *J. Cell Biol.* **90**, 427–434.
Burr, B., Burr, F. A., Rubenstein, I. and Simon, M. N. (1978). *Proc. Natl. Acad. Sci. USA.* **75**, 696–700.
Cameron-Mills, V. (1980). *Carlsberg Res. Commun.* **45**, 557–576.
Cameron-Mills, V. and Ingversen, J. (1978). *Carlsberg Res. Commun.* **43**, 471–489.
Cameron-Mills, V. and Wettstein, D. von (1980). *Carlsberg Res. Commun.* **45**, 577–594.
Cameron Mills, V., Ingversen, J. and Brandt, A. (1978). *Carlsberg Res. Commun.* **43**, 91–102.
Doll, H. (1980). *Hereditas* **93**, 217–222.
Doll, H. and Brown, A. H. D. (1979). *Can. J. Genet. Cytol.* **21**, 391–404.
Doll, H., Køie, B. and Eggum, B. O. (1974). *Radiat Bot.* **14**, 73–80.
Faulks, A. J., Shewry, P. R. and Miflin, B. J. (1981) (in press).
Geraghty, D., Peifer, M. A. and Rubenstein, I. (1981). XIII International Botanical Congress, Sydney, Australia, 21–28 August, 1981 (abstract p. 272).
Hagen, G. and Rubenstein, I. (1980). *Plant Science Lett.* **19**, 217–223.
Hagen, G. and Rubenstein, I. (1981). *Gene* **13**, 239–249,
Holder, A. A. and Ingversen, J. (1978). *Carlsberg Res. Commun.* **43**, 177–184.
Ingversen, J. (1975). *Hereditas* **81**, 69–76.
Ingversen, J., Køie, B. and Doll, H. (1973). *Experientia* **29**, 1151–1152.
Jonassen, I. (1980a). *Carlsberg Res. Commun.* **45**, 47–58.
Jonassen, I. (1980b). *Carlsberg Res. Commun.* **45**, 59–68.
Jones, R. A., Larkins, B. A. and Tsai, C. Y. (1976). *Biochem. Biophys. Res. Commun.* **69**, 404–410.
Jones, R. A., Larkins, B. A. and Tsai, C. Y. (1977a). *Plant Physiol.* **59**, 733–737.
Jones, R. A., Larkins, B. A. and Tsai, C. Y. (1977b). *Plant Physiol.* **59**, 525–529.
Khoo, U. and Wolf, M. J. (1970). *Am. J. Bot.* **57**, 1042–1050.
Larkins, B. A. and Hurkman, W. J. (1978). *Plant Physiol.* **62**, 256–263.
Larkins, B. A., Bracker, C. E. and Tsai, C. Y. (1976a). *Plant Physiol.* **57**, 740–745.
Larkins, B. A., Jones, R. A. and Tsai, C. Y. (1976b) *Biochemistry* **15**, 5506–5511.
Lewis, D. E., Hagen, G., Mullins, J. I., Mascia, P. N., Park, W. D., Bento, W. D. and Rubenstein, I. (1981). *Gene* **14**, 205–215.
Matthews, J. A. and Miflin, B. J. (1980). *Planta* **149**, 262–268.
Melcher, U. (1979). *Plant Physiol.* **63**, 354–358.
Mertz, E. T., Bates, L. S. and Nelson, O. E. (1964). *Science* **176**, 1425–1427.
Miflin, B. J. and Shewry, P. R. (1979). *In* "Recent Advances in the Biochemistry of Cereals" (D. L. Laidman and R. G. Wyn Jones eds), pp. 239–273 Academic Press, London and New York.
Miflin, B. J., Burgess, S. R. and Shewry, P. R. (1981). *J. Exp. Bot.* **32**, 199–219.
Munck, L. and Wettstein, D. von (1976). *In* "Genetic Improvement of Seed Proteins." Proc. of a workshop, March 1974, National Academy of Sciences, Washington D. C., USA, pp. 71–82.
Munck, L., Karlsson, K. E., Hagberg, A. and Eggum, B. O. (1970). *Science* **168**, 985–987.
Nelson, O. E. (1980). *In* "Advances in cereal science and technology" vol III (Y. Pomeranz ed.). pp. 41–71. Am. Assoc. of Cereal Chemists Inc., St. Paul Minnesota, USA.
Park, W. D., Lewis, E. D. and Rubenstein, I. (1980). *Plant Physiol.* **65**, 98–106.
Pedersen, K., Bloom, K. S., Anderson, J. N., Glover, D. V. and Larkins, B. A. (1980). *Biochemistry* **19**, 1644–1650.

Righetti, P. G., Gianazza, E., Viotti, A. and Soave, C. (1977). *Planta* **136**, 115–123.

Schmitt, J. M. (1979). *Carlsberg Res. Commun.* **44**, 431–438.

Schmitt, J. M. and Svendsen, I. (1980a). *Carlsberg Res. Commun.* **45**, 143–148.

Schmitt, J. M. and Svendsen, I. (1980b). *Carlsberg Res. Commun.* **45**, 549–555.

Shewry, P. R., Pratt, H. M., Charlton, M. J. and Miflin, B. J. (1977). *J. Exp. Bot.* **28**, 597–606.

Shewry, P. R., Ellis, J. K. S., Pratt, H. M., Finch, R. A. and Miflin, B. J. (1978a). *J. Sci. Food Agric.* **29**, 433–441.

Shewry, P. R., Pratt, H. M., Finch, R. A. and Miflin, B. J. (1978b). *Heredity* **40**, 463–466.

Shewry, P. R., Pratt, H. M., Leggat, M. M.and Miflin, B. J. (1979). *Cereal Chem.* **56**, 110–117.

Shewry, P. R., Faulks, A. J. and Miflin, B. J. (1980a). *Biochem. Genet.* **18**, 133–151. 133–151.

Shewry, P. R., Field, J. M., Kirkman, A., Faulks, A. J. and Miflin, B. J. (1980b). *J. Exp. Bot.* **31**, 393–407.

Shewry, P. R., March, J. F. and Miflin, B. J. (1980c). *Phytochemistry* **19**, 2113–2115.

Shewry, P. R., Lew, E. J.–L. and Kasarda, D. D., (in press).

Singh, R. and Axtell, J. D. (1973). *Crop Sci.* **13**, 535–539.

Soave, C., Righetti, P. G., Lorenzoni, C., Gentinetta, E. Salamini, F. (1976). *Maydica* **21**, 61–75.

Soave, C., Suman, N., Viotti, A. and Salamini, F. (1978). *Theor. Appl. Genet.* **52**, 263–267.

Svendsen, I., Martin B. and Jonassen, I. (1980). *Carlsberg Res. Commun.* **45**, 79–85.

Tsai, C. Y. (1979). *Maydica* **24**, 129–140.

Viotti, A., Sala, E., Alberi, P. and Soave, C. (1978). *Plant. Sci. Lett.* **13**, 365–375.

Viotti, A., Pogna, N. E., Balducci, C. and Durante, M. (1980). *Mol. Gen. Genet.* **178**, 35–41.

Wettstein, D. von (1979). *In* "Proc. Eur. Brewery Conv. Congr.", Berlin 1979, pp. 587–629.

Wienand, U. and Feix, G. (1978). *Eur. J. Biochem.* **92**, 605–611.

Wienand, U. and Feix, G. (1980). *FEBS Lett.* **116**, 14–16.

Wienand, U., Brüschke, C. and Feix, G. (1979). *Nucleic Acid Res.* **6**, 2707–2715.

CHAPTER 9

Genetic Organization and Regulation of Maize Storage Proteins

C. SOAVE[*] AND F. SALAMINI[†]

*Istituto Biosintesi Vegetali C.N.R, Milano, Italy
†Istituto Sperimentale per la Cerealicoltura, Sez. Mais coltura, Bergamo, Italy

1. INTRODUCTION

Gene regulation in procaryotes has been extensively investigated and some general features of gene control are taking shape. Regulation may occur at the very first step of the process, the transcriptional level, where the interaction of RNA polymerase with the promoter may be influenced by various activators or repressors, and at the translational level by changing the frequency of reading of internal cistrons in polycistronic messages or by autogenous translational regulation of specific mRNA molecules (see for a review Goldberger, 1980).

Despite the considerable increase in our knowledge of gene organization in recent years in eucaryotes, our understanding of regulation of gene expression remains primitive. Some examples of operon-like organization in eucaryotic genomes have been reported (Arst and McDonald, 1975) together with the description of positive regulation of specific metabolic pathways (Perlman and Hooper, 1979; Metzenberg, 1979). Furthermore in eucaryotes, mRNAs have longer life span than in procaryotes (discussed in Meyuhas and Perry, 1979) and stability or control of the translational activity of mRNAs are expected to be important in regulating gene expression. In this respect, the protein–protein interaction described in the control of the mouse reticulum-bound β-glucuronidase (Paigen, 1979), the hemin-controlled synthesis of the HCI inhibitor of globin synthesis in reticulocytes (Ochoa and de Haro, 1979) and the translational mutant resistant to diphtheria toxin in CHO cells (Moehring and Moehring, 1977), are some of the notable cases that can be found in the literature.

A model system suitable for studying gene regulation in higher plants is represented by the synthesis of zeins the major endosperm proteins of maize. The main features of the zein system are the following:

1. Zeins consist of several polypeptides, similar but not identical to each other, which can be partially fractionated by bidimensional electrophoresis (isoelectrofocusing-IEF-followed by SDS electrophoresis) in about 25 components (Fig. 1) (Vitale et al., 1980). By molecular weight, polypeptides with molecular weight of 23,22,20,19,14 and 10 K (kilodalton) can be identified; they are grouped for simplicity in four classes of 22,20,14 and 10 K (Righetti et al., 1977).

2. During endosperm development, zeins begin to be accumulated around 15 days after pollination (DAP) and all the polypeptides are synchronously synthesized by membrane-bound polysomes and deposited in protein bodies (Burr and Burr, 1976).

3. The structural genes encoding the zein polypeptides represent a multigene family whose members are scattered in the genome (Viotti et al., 1979; Soave et al., 1978a; Valentini et al., 1979). Two main sites were, up to now, identified: one on the short arm of chromosome 7, the other on chromosome 4. In each site the zein genes are not strictly clustered but slightly dispersed (Soave et al., 1981).

In principle, at least two kinds of regulation should be active in the zein system: one regulating the onset of zein synthesis, that is a temporal control, the other coordinating the rate of expression of the zein genes and leading to the synchronous accumulation of all the zein polypeptides. As a matter of fact, loci controlling either the onset of zein synthesis and the rate of accumulation of different groups of zein polypeptides have been described (Manzocchi et al., 1980; Mertz et al., 1964; Nelson et al., 1965; McWhirter, 1971; Ma and Nelson, 1975; Salamini et al., 1979). Here we summarize our present knowledge on the genetic organization of zein structural genes and on the control of their expression.

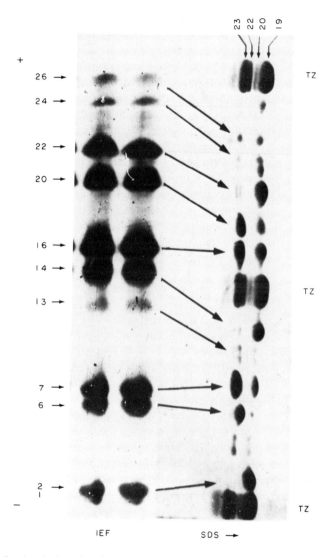

FIG. 1. Isoelectric focusing (IEF) and sodium dodecylsulphate (SDS) polyacrylamide gel electrophoresis of zein from the inbred OH43. Zeins were first fractionated by IEF and then the individual bands were excised, equilibrated in SDS sample buffer and run on 15% polyacrylamide SDS gel. TZ, OH43 total zein. Numbers on the left indicate the code position of the OH43 zein IEF components while those on the upper right the MW of the zein subunits in K.

II. GENETIC ORGANIZATION OF ZEIN GENES

There is now a reasonable agreement on the multiplicity of the zein genes in the maize genome; independent evidence suggests approximately 100 copies of zein genes per haploid genome (Viotti et al., 1979; Hagen and Rubenstein, 1981). We do not know how many of these genes represent active zein genes, since silent sequences, such as those present in the β-globin system of some mammalian organisms, could be present (Jahn et al., 1980; Lacy and Maniatis, 1980). We can only infer, from the number of different zein polypeptides identifiable after bidimensional gel elec-trophoresis in those inbred lines showing the highest resolution, a mini-mum estimate of about 30 expressed zein genes (Valentini et al., 1979). The variability in the isoelectric focusing (IEF) pattern of zein among the inbreds of maize permitted the investigation of the linkage relationship and of the chromosomal location of the structural genes coding for some zein IEF components. The genetic analysis was performed only on those IEF zein components which, after bidimensional gel electrophoresis appeared to be constituted by single polypeptide and showed 3 : 1 segregation ratios in F2.

Recently we gave evidence on the clustering of seven genes encoding different zein polypeptides belonging to 20 K family in a region of about 30 crossover units in the short arm of chromosome 7 (Fig. 2) (Soave et al., 1981). This genetic organization is of particular interest from several points of view: firstly, the scattering of seven 20 K zein genes encoding seven similar but not identical polypetides in this short chromosomal region suggests the idea of duplications of a short chromosomal region as a mechanism for generating a clustered geography of multiple genes. Secondly, at least three of these genes are stricly clustered: they are $Zp1$, $Zp2$, $Zp3$ encoding for zein IEF components always expressed or silent altogether. Although our analysis had a mean sensitivity in the range of one map unit, the possibility that $Zp1$, $Zp2$, $Zp3$, are coded by a multicistronic message cannot be, at least in principle, ruled out. Thirdly, the observation that in the same chromosomal region even loci controlling zein accumulation are present (i.e. the case of 02 and De^*B30) and duplicated, may be more than a coincidence (Salamini et al., 1979). We could speculate that the structural and regulatory elements derived from a common ancestor endowed with both functions which, after duplications, differentiated in elements retaining only the regulatory function and in other elements strengthening their structural role by redundancy.

The second chromosomal region where structural zein genes were found is on both arms of chromosome 4. Up to now we located six elements in a region around the $Fl2$ and $Su1$ loci on the short arm of chromosome 4 and one element on the long arm. All these elements encode for 22 K zein

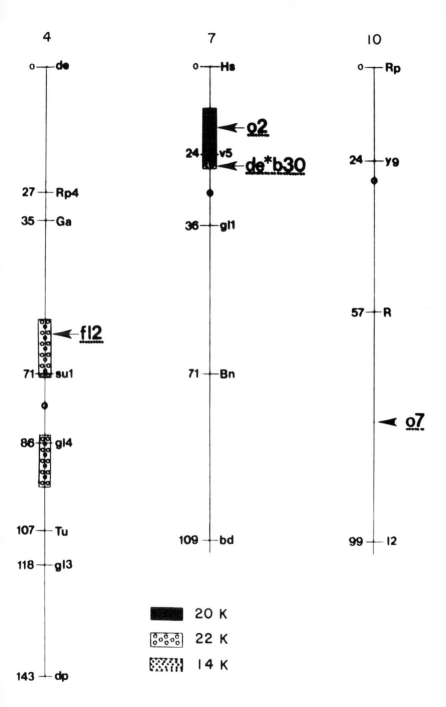

FIG. 2. Schematic diagram of maize chromosome 4, 7 and 10 showing the position of some structural zein genes and of the "regulatory" loci. 22, 20 and 14 K refer to genes encoding for zein polypeptides of 22 000, 20 000 and 14 000 daltons.

polypeptides and their distribution repeats to some extent that observed in chromosome 7: a family of zein genes scattered in a relatively short chromosomal region where also maps a zein regulatory element (i.e. the *Fl2* locus).

Very few data are available on the genetic relationship of the two lower molecular weight zein families (10 and 14 K) to the two major zein polypeptide families. Very recently, however, we found two allelic variants of the 14 K zein polypeptide slightly differring in their molecular weight. By studying the inheritance of this gene in connection with that of some 20 K zein genes we mapped the 14 K zein gene near the zein cluster of chromosome 7.

III. Control of Zein Expression

A. TEMPORAL CONTROL

Maize endosperm is a tissue which follows a very defined sequence of events during development. At least three phases can be identified: cell division, from fertilization to about 15 days after pollination (DAP), expansion and deposition of storage products from 15 to 45 DAP, and maturation. Zeins begin to accumulate in the endosperm around 15 DAP and the IEF pattern of this "early zein" is similar to that observed at later stages (Fig. 3). This result indicates that the active zein genes are simultaneously switched on at 15 DAP and thereafter they are synchronously transcribed and translated.

Developmental mutants altering the timing of the expression of zein polypeptides have been reported (Manzocchi *et al.*, 1980). These mutants can be grouped into two classes: those delaying the onset of the deposition of the zein polypeptides and, more interestingly, those delaying only the synthesis of some zein components (in the cases observed the 21–22 K zeins). The specificity of these last mutants recalls the effect of other zein regulatory loci, such as *02* and *De*B30* (see below). However here the peptide specificity is matched with time specificity suggesting that the simultaneous switching on of the members of the multigene zein family is under the control of several genetic elements.

B. RATE CONTROL

Several gene loci control the level of zein accumulation during endosperm development (see for a review Soave and Salamini, 1979). The mutant alleles at these loci reduce zein accumulation to a different extent (Table

ZEIN IEF PATTERN

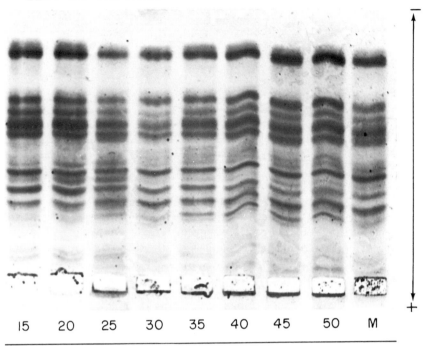

Fig. 3. Isoelectric focusing patterns of zeins from Illinois High Protein endosperms collected at different days after pollination (DAP). M, maturity.

I). The degree of repression brought about by 5 of these mutants follows the order: *o6 o7 o2 fl2 De* B30*. Furthermore, the accumulation of the two zein major MW classes(the 20 and 22 K classes) appears to be specifically affected by the mutations. Two of them, *o7* and *o2*, repress both the zein classes but exhibit a preferential action respectively on the 20 and 22 K classes (Di Fonzo *et al.*, 1979; Soave *et al.*, 1976); *fl2* and *o6* seem to control the accumulation of both classes to the same extent (Soave *et al.*, 1978b; and our unpublished observatious for *o6*). *De*B30* behaves in a particular way inhibiting only the 22 K class without affecting the other one (Salamini *et al.*, 1979). It should be remembered, however, that in the presence of the mutant alleles of these loci all the zein constituents which can be identified by the various electrophoretic separations are still present, even in a lower amount, throughout the zein synthesis time. This feature, together with the specific action of each mutant on the zein polypeptides and with their genomic location non coincident with those of

TABLE I

Effect (% of the normal values) of the mutant alleles *o2, fl2, o7, o6* and *De* B30* on total zein accumulation and on the 20 and 22 K zein components

Genotype	Total zein	20 K zein	22 K zein
Normal	100	100	100
Opaque–6	11·5	11·0	12·2
Opaque–7	27·5	41·4	20·4
Opaque–2	53·2	41·4	54·5
Floury–2	64·5	65·4	64·1
De*B30	89·4	65·4	138·5

The normal alleles, toward which the mutant alleles have been compared, were: mean values of 4 normal inbreds for *o6*, W22 for *o7*, mean values of 36 normal inbreds for *o2*, mean values of 40 normal inbreds for *fl2* and B37 for *De*B30*.

zein structural genes, suggest a "regulatory" role of the opaque mutants on zein synthesis. Our present knowledge of the biochemical level of the genes controlling zein deposition is however very incomplete and the term "regulatory" role should be intended in a broad sense. It could be that these loci interfere with the machinery of zein protein synthesis or with the stability of zein mRNAs, or they monitor the cellular environment and adjust the rate of zein synthesis accordingly. The only conclusion which can be derived up to now is that the simultaneous switching on and the synchronous expression of the zein genes are under the control of the balanced action of several "regulatory" elements whose role is, at least in part, component specific.

A deeper insight in this integrated control was obtained by looking at the interrelationships among these loci in double mutants (Di Fonzo et al., 1980). We investigated the interactions between mutants *o2* and *fl2* and *o2* and *o7* on the levels of the zein 20 and 21–22 K classes (Fig. 4). Our results indicated that in the presence of three doses (the maize endosperm is triploid) of the recessive *o2* allele, the dose-dependent *fl2* action is completely overshadowed and only the *o2* zein phenotype is maintained. Thus *o2* is apparently epistatic over *fl2* in controlling zein synthesis. A different situation was observed for the interaction between *o2* and *o7*. The mutant alleles at these loci are both recessive and, when homozygous, they repress mainly the higher and lower molecular weight zein peptides respectively. In the *o2o7* double mutant (Fig. 4H last column) both the alleles are active and additively reduce zein synthesis: thus the two loci seem to act independently on zein accumulation. These results lead us to propose the following conclusion. In zein synthesis, multiple regulatory pathways are active: at least one is related to zein 21–22 K and the other to zein 20 K accumulation with *o2* and *o7* being involved respectively in the first and second pathway. The data available moreover indicate that *Fl2* is active on the first pathway operating downstream or upstream to the *o2*

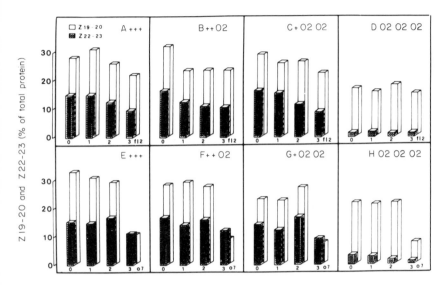

FIG. 4. Distribution of total zein into 22–23 K polypeptides (Z 22–23) and into 19–20 K polypeptides (Z 19–20) in endosperms containing different doses of the recessive alleles *o2*, *fl2*, *o7*. Endosperms were collected at maturity and the extracted zeins were fractionated by SDS gel electrophoresis. The gels were then scanned densitometrically and the amount of the Z 19–20 and Z 22–23 zein polypeptides was expressed in percent of total endosperm proteins.

locus. Experiments are now in progress to ascertain if *Fl2* is active also on the second pathway in conjunction with *o7*.

IV. MOLECULAR BASIS OF CONTROL OF ZEIN SYNTHESIS : SOME ADVANCEMENTS

Among the loci controlling the level of zeins in the endosperm, the *Opaque-2* locus (*O2*) has been particularly well studied because the recessive alleles at this locus confer a superior nutritive value to maize meal (Mertz *et al.*, 1964). At the molecular level, however, the mechanism of action of *O2* (as of the other regulatory loci) on zein synthesis has not been elucidated.

Studies on *O2* demonstrated that the locus positively controls the accumulation of the 20 and 22 K zein polypeptides. Recently we have analysed seven independently originated, recessive alleles at the *O2* locus confirming the preferential repression of the 22 K zein family, even if, at least in one case, both classes were affected to the same extent (Soave *et al.*, unpublished results). Furthermore in the mutant endosperm the cytoplasmic levels of zein mRNAs are lower than that observed in wild types and the reduction is more pronounced for those mRNAs encoding the 21–22 K zein peptides (Pedersen *et al.*, 1980). It is not clear, however, if the zein mRNA concentration is the primary effect of the mutant *o2* alleles (acting for example at the level of mRNA transcription or processing) or a consequence of a reduced efficiency of the translation machinery followed by degradation of the mRNAs.

Whatever the mechanism may be, it should be stressed that the *O2* locus must act by producing a diffusible factor which interferes with zein production since it controls the expression of several zein genes not linked to it and because the known mutant alleles at the locus are completely recessive. This was the basis of a search for proteins associated with the wildtype allele and modified or absent in *o2* endosperms. We have analysed the SDS electrophoretic patterns of water soluble proteins (S–30 supernatants) from 30 DAP endosperms of wildtype and *o2* alleles (Fig. 5).

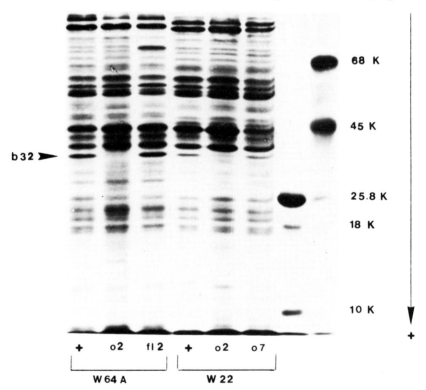

Fig. 5. Sodium dodecyl sulphate polyacrylamide gel electrophoresis of S-30 proteins from wildtype (+), *opaque-2(o2)*, *floury-2* (*fl2*) and *opaque-7* (*o7*) endosperms. Seeds were collected at 30 DAP, and the endosperms homogenized in 60 mM Tris-HCl pH 6·8, 1 mM phenylmethanesulphonylfluoride (PMSF). The extracts were centrifugated at 30 000 xg for 10 minutes and the supernatants equilibrated in SDS sample buffer. After denaturation at 90°C for 5 minutes, approximately 100 μg of protein were loaded on 18% SDS polyacrylamide gels. Numbers on the right refer to molecular weight markers.

The major difference was the presence in the wild type of a protein with a molecular weight of 32 000 daltons (designed as b–32 protein) absent or drastically reduced in the mutant. More interestingly, this protein was present in other mutants which control zein accumulation such as *fl2* and *o7*.

To eliminate the possibity that the absence of b–32 protein in *o2* endosperms could be related to some background variability, we tested the presence of b–32 protein in seven, independently originated, *o2* recessive alleles, all introduced by several backcrosses in the W64A genotype. b–32 protein was absent in all the *o2* alleles studied. Furthermore, by im-munoelectrophoresis of S–30 extracts against an anti b–32 rabbit antibody, we demonstrated that all the *o2* endosperms were CRM⁻ for b–32 protein. These data indicate that the *O2* locus, beside controlling the accumulation of the zein peptides, also acts on a water soluble protein, b–32, which from many points of view, is absolutely different, from the alcohol soluble zeins.

Furthermore, by analysing the time course of b–32 and zein accumula-tion during endosperm development together with their location in the maize plant and their response to selection, it appeared that: (a) zeins and b–32 are temporally coordinated during development, (b) both are confined only to endosperm (Fig. 6) and (c) long term selection

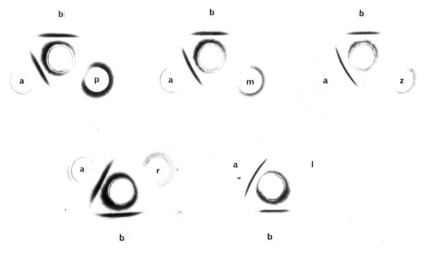

Fig. 6. Location of b-32 protein in maize tissue and in subcellular endosperm fractions by double immunodiffusion test. Root and leaf (from 7 days old seedlings), embryo and endosperm (from 30 DAP seeds) extracts were prepared homogenizing the tissue in 60 mM Tris-HCl pH 6·8, 1 mM PMSF, 1% (v/v) Triton X-100. Protein bodies membranes were obtained by treating zein protein bodies with Triton X-100. Immunodiffusion was in 1% Agarose, 1% Triton X-100. Center well, 8 μl of anti b-32 rabbit antiserum. a, 5 μl of a solution of pure b-32 protein (1 mg/ml). b, 8 μl of S-30 endosperm extgract. p, embryo extract. m, protein bodies membrane. z, zein granules. r, root extract. l, leaf extract.

TABLE II
Levels of zeins, albumins and globulins and b–32 protein in Illinois High Protein
(IHP) and Illinois Low Protein (ILP) strains of maize

Genotype	DAP	Zein mg/end.	Albumin-globulin* mg/end.	b–32[b] μg/end.	%on Albumin-globul.
IHP	31	12·5	3·2	219	6·8
ILP	31	2·5	2·6	48	1·8

[a] Albumins and globulins were extracted by shaking 2h at 4°C with 0·5 MNaCl lyophilized endosperms. The samples were centrifuged, the supernatants dialyzed against water and the protein content assayed by micro Kieldahl.
[b] The amount of b–32 protein was estimated by densitometric scan of SDS polyacrylamide gels of albumin and globulin proteins.

experiments in favour or against zein accumulation induced parallel enhancement or reduction of b–32 protein (Table II). These results support the indication that b–32 plays a role in zein accumulation (Soave et al., unpublished results). Using, however, two allelic electrophoretic variants of b–32 protein, we demonstrated that the O2 gene does not by itself encode protein b–32; instead the protein seems to be encoded by another locus, the O6 locus. What is important to point out is that the recessive o6 endosperms are not only devoid of b–32 protein but they are also almost deprived of zeins (see Table I). Our findings suggest that, out of the two mutants devoid of b–32 protein, O6 is apparently responsible for the synthesis of the polypeptide while O2 interacts with O6 in promoting b–32 synthesis; b–32 in turn sustains positively accumulation of zeins. This introduces the concept that there is, in controlling the rate of zein accumulation, a hierarchy between at least two "regulatory" genes (Fig. 7). Previous results already suggested that two genes involved in zein control, namely O2 and Fl2, were interacting in an epistatic fashion (Di Fonzo et al., 1980). Here, however, we are faced with a battery of zein

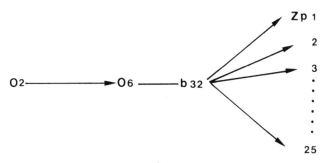

FIG. 7. Schematic diagram of the model proposed for the interaction between the loci o2 and o6 in controlling zein accumulation.

structural genes that seem positively controlled by the *O6* locus, in turn under the influence of *O2*. This situation is reminiscent of some of the regulatory models proposed in eukaryotic systems, where a cascade of events leading to structural gene activation is ascribable to multiple regulatory genes. Among these systems, the control of globin synthesis in the reticulocyte (Ochoa and de Haro, 1979), the regulation of phosphate absorption in *Neurospora* (Metzenberg, 1979) and the *gal4–gal80* interaction in regulating the galactose pathway in *Saccharomyces* (Perlman and Hooper, 1979) show some formal similarities with the hypothesis the *Opaque–2* controls zein synthesis through an intermediate step conditioned by *O6*.

References

Arst, N. H. and Mac Donald, D. W. (1975). *Nature* **254,** 26–31.
Burr, B. and Burr, F. A. (1976). *Proc. Natl. Acad. Sci. USA.* **73,** 515–519.
Di Fonzo, N., Gentinetta, E., Salamini, F. and Soave, C. (1979). *Plant Sci. Lett.* **14,** 345–354.
Di Fonzo, N., Fornasari, E., Salamini, F., Reggiani, R., and Soave, C.(1980). *J. Heredity* **71,** 397–402.
Goldberger, R. F. (1980). *In* "Biological Regulation and Development" (R. F. Goldberger, ed), Vol. 1 pp. 1–18. Plenum Press, London and NewYork.
Hagen, G. and Rubenstein, I. (1981). *Gene* **13,** 239–249.
Jahn, C. L., Hutchinson, III, C. A., Phyllips, S. J., Weaver, S., Haigwood, N. L., Voliva, C. F. and Edgell, M. H. (1980). *Cell* **21,** 159–168.
Lacy, E. and Maniatis, T. (1980). *Cell* **21,** 545–553.
Ma, Y., Nelson, O. E. (1975). *Cereal Chem.* **52,** 412–419.
Manzocchi, L. A., Daminati, M. G. and Gentinetta, E. (1980). *Maydica* **25,** 199–210.
McWhirter, K. S. (1971). *Maize Genet. Coop. News Lett.* **45,** 184.
Mertz, E. T., Bates, L. S., Nelson, O. E. (1964). *Science* **145,** 279–280.
Metzenberg, R. L. (1979). *Microb. Rev.* **43,** 361–383.
Meyuhas, O. and Perry, R. P. (1979). *Cell* **16,** 139–148.
Moehring, T. J. and Moehring, J. M. (1977). *Cell* **11,** 447–454.
Nelson, O. E., Mertz, E. T. and Bates, L. S. (1965). *Science* **150,** 1469–1470.
Ochoa S. and De Haro, C. (1979). *Annu. Rev. Biochem.* **48,** 549–580.
Paigen, K. (1979). *Annu. Rev. Genet.* **13,** 417–466.
Pedersen, K., Bloom, K. S., Anderson, J. N., Glover, U. V. and Larkins, B. A. (1980). *Biochemistry* **19,** 1644–1650.
Perlman, D. and Hooper, J. E. (1979). *Cell* **16,** 89–95.
Righetti, P. G., Gianazza, E., Viotti, A. and Soave, C. (1977). *Planta* **136,** 115–123.
Salamini, F., Di Fonzo, N., Gentinetta, E. and Soave, C. (1979). *In,* "Seed Protein Improvement in Cereals and Grain Legumes" pp. 97–108 FAO–IAEA, Neuherberg.
Soave, C. and Salamini, F. (1979). Monogr. IV, *Genet. Agrar.,* pp. 107–140.

Soave, C., Righetti, P. G., Lorenzoni, C., Gentinetta, E. and Salamini, F. (1976). *Maydica* **21,** 61–75.
Soave, C., Suman, N., Viotti, A. and Salamini, F. (1978a). *Theor. Appl. Genet.* **52,** 263–267.
Soave, C., Dassena, S., Lorenzoni. C., Di Fonzo, N. and Salamini, F. (1978b). *Maydica* **23,** 145–152.
Soave, C., Reggiani, R., Di Fonzo, N. and Salamini, F. (1981). *Genetics* (in press).
Valentini, G., Soave C. Ottaviano, E. (1979). *Heredity* **42,** 33–40.
Viotti, A., Sala, E., Marotta, R., Alberi, P., Balducci, C. and Soave, C. (1979). *Eur J. Biochem.* **102,** 211–222.
Vitale, A., Soave, C., Galante, E. (1980). *Plant Sci. Lett.* **18,** 57–64.

CHAPTER 10

Regulation of Storage Protein Synthesis and Deposition in Developing Legume Seeds

D. BOULTER

Department of Botany, University of Durham, Durham, England

I. Introduction

This paper describes results obtained mainly at Durham on the regulation of storage protein synthesis and deposition in developing pea seeds. However, several other legumes including *Phaseolus vulgaris*, *Vicia faba*, *Glycine max*, *Arachis hypogaea*, have been investigated sufficiently well to suggest that similar mechanisms operate there also; there is circumstantial evidence that the basic description and proposed underlying mechanisms of storage protein synthesis and deposition outlined in the paper apply to other grain legumes of the Papilionaceae.

II. General Description of the Biochemistry of Protein Deposition in Developing Cotyledon Cells

Legume seeds contain several different storage proteins. In peas these are of three types, legumin (Croy *et al.*, 1979; Krishna *et al.*, 1979; Casey, 1979; Matta *et al.*, 1981), vicilin (Gatehouse *et al.*, 1981) and convicilin (Croy *et al.*, 1980c). Sensitive, radioactive immunological techniques have shown these to be present in very small amounts during the very early stages of pea seed development (Domoney *et al.*, 1980).

About a third of the way through the course of development of the pea seed, these proteins increase dramatically in amount and continue to be synthesized over several days such that in the mature seed they constitute as much as 80% (varies in different varieties) of the total protein of the seed. At or just prior to the onset of increased storage protein synthesis, a remarkable change in the fine structure occurs in the cotyledon cells of the seed. Much rough endoplasmic reticulum (RER) (bearing polysomes) is laid down and soon afterwards storage protein is deposited in membrane bound protein bodies (Pernollet, 1978). At the same time, the highly vacuolate cells undergo a process of vacuole sub-division and protein is deposited within them. The origin of protein bodies has, however, not been clearly established. The most generally held opinion is that they are of dual origin, partly formed from sub-divided vacuoles and partly by coalescence of protein containing vesicles derived either from the ER or from the Golgi. In any event, it has been established that the storage proteins are synthesized on ribosomes attached to the endoplasmic reticulum (Evans *et al.*, unpublished observations). It is also clear that considerable post-translational modifications of the storage proteins occur subsequent to, and to some extent concomitant with, their synthesis on the polysomes (Croy *et al.*, 1980a and 1980b; Gatehouse *et al.*, 1981; Higgins and Spencer, 1981). Proposed mechanics for the regulation of protein synthesis must therefore take into account the mechanism of storage protein synthesis itself, the increased provision and change in the protein synthesizing machinery (RER), the switching on and off of the increased rates of storage protein synthesis, differences in the rates of storage protein accumulation and the duration of synthesis, co- and post-translational changes and the control of protein transport to, and deposition in, protein bodies; information is required not only on the amounts of the different proteins synthesized during seed development, but also on the amount of mRNAs and on the number of genes involved.

III. REGULATION

A comparison of the translation products of poly-A containing RNA and microsomes in *in vitro* protein synthesizing systems with isolated storage proteins shows that considerable co- and post-translational modifications of the storage protein polypeptides takes place. Vicilin, convicilin and probably legumin have additional sequences of about 1000 molecular weight when synthesized, which are subsequently removed by enzymes in the RER (Croy *et al.*, 1980a and 1980b; Gatehouse *et al.*, 1981; Higgins and Spencer, 1981; and unpublished observations). Furthermore, legumin is made as a subunit of about M_r 60 000 which is subsequently cleaved to the M_r 40 and 20 000 subunits (Croy et al., 1980a). Vicilin polypeptides are synthesized initially as M_r 50 and 47 000 subunits which are in part subsequently cleaved to the M_r 33 000, 19 000 13 000 and 12 500 subunits found in isolated vicilins (Croy et al., 1980b; Gatehouse et al., 1981; Higgins and Spencer, 1981). Other post-translational changes include the glycosylation of some of the vicilin polypeptides (Gatehouse et al., 1981), but not the convicilin (Croy et al., 1980c) or legumin subunits (Casey, 1979). The precise intracellular locations of the enzymes responsible have not yet been established; both the RER and Golgi have been implicated. Proteolytic cleavage of both vicilin and legumin takes place several hours after protein synthesis, as is evident from pulse/chase labelling experiments (Gatehouse et al., 1981 and unpublished observations) and could occur therefore in the protein bodies themselves since Bailey et al. (1970) showed that the half-life of transport of storage protein from the RER to the protein bodies in Vicia faba is about 30 minutes.

Vicilin accumulates initially faster than legumin, but the synthesis of the latter persists longer (till 14 and 20 days after flowering respectively, the total seed development period being 24 days). Since the proteins do not turnover significantly, it appears likely that the changed proteins levels would be reflected in a changed mRNA population. This has now been shown to be the case, by probing the amounts of particular messages for vicilin and legumin using cloned cDNA of the respective messages. Results from these experiments show the presence of increasing amounts of vicilin and legumin mRNAs in cotyledon cells up to 14 days; an α-amanitin sensitive "burst" of transcriptional activity in isolated nuclei, from cotyledons 8–10 days old, accounted for these increases (Boulter, unpublished observations). Data are not available for convicilin mRNA, but by analogy it is assumed to build-up also. The increased transcription which was observed could have resulted from the activation of one or a few genes, or alternatively by gene amplification. Investigation of pea genomic DNA with cDNA probes using Southern blotting shows that legumin genes are present in a few single copy genes, whereas vicilin genes occur in more copies but are not highly reiterated (Croy et al., 1982).

IV. Conclusions

The results presented in this paper demonstrate the involvement of transcription level control as a major regulatory mechanism for storage protein synthesis in the developing legume seed. Other controls, e.g. post-transcriptional and translational may also exist, although so far not demonstrated, and these possibilities are now under investigation in our laboratory.

REFERENCES

Bailey, C. J., Cobb, A. and Boulter, D. (1970). *Planta* **95,** 103–118.
Casey, R. (1979). *Biochem. J.* **177,** 509–520.
Croy, R. R. D., Derbyshire, E., Krishna, T. G. and Boulter, D. (1979). *New Phytol.* **82,** 29–35.
Croy, R. R. D., Gatehouse, J. A., Evans, I. M. and Boulter, D. (1980a). *Planta* **148,** 49–56.
Croy, R. R. D., Gatehouse, J. A., Evans, I. M. and Boulter, D. (1980b). *Planta* **148,** 57–63.
Croy, R. R. D., Gatehouse, J. A., Tyler, M. L. and Boulter, D. (1980c). *Biochem. J.* **191,** 509–516.
Croy, R. R. D., Lycett, G. W., Gatehouse, J. A., Yarwood, J. N. and Boulter, D. (1982) *Nature* **295,** 76–79.
Domoney, C., Davies, D. R. and Casey, R. (1980). *Planta* **149,** 454–460.
Gatehouse, J. A., Croy, R. R. D., Morton, H., Tyler, M. and Boulter, D. (1981). *Eur. J. Biochem.* **118,** 627–633.
Higgins, T. J. V. and Spencer, D. (1981). *Plant Physiol.* **67,** 205–211.
Krishna, T. G., Croy, R. R. D. and Boulter, D. (1979). *Phytochem.* **18,** 1879–1880.
Matta, N., Gatehouse, J. A. and Boulter, D. (1981). *J. Exp. Bot.* **32,** 1295–1307.
Pernollet, J. C. (1978). *Phytochem.* **17,** 1473–1480.

CHAPTER 11

Breeding for Protein Quantity and Protein Quality in Seed Crops

P. I. PAYNE

Plant Breeding Institute, Maris Lane, Trumpington, Cambridge, England

I. INTRODUCTION

The term "protein quality" has several different meanings, depending on which crop is being bred and for what purpose. Perhaps the most common meaning is in relation to the nutritional value of the crop for humans or for farm animals. Here the breeder selects for protein amount and for the balance of the essential amino acids in the protein. But there are other meanings. In breeding for protein quality in oil-seed rape the proteins themselves are not improved nutritionally but rather the toxic glucosino-lates which occur in the protein rich, oil-free meal are being removed. In food processed from wheat, protein quality is in terms of viscoelasticity for making leavened bread and chapatis, extensibility for biscuit-making and hardness for the manufacture of the pasta products, spaghetti, macaroni and noodles. In legumes it can mean the ability of the protein to be spun and textured for meat substitutes. Paradoxically, in breeding for protein quality in malting barley, the breeder aims for minimal protein content.

In this review, rather than giving a comprehensive cover of all the major crops of the world, I will dicuss in some detail a few examples of breeding for different types of quality in a selection of crops.

II. The World's Seed Crops: Their Protein Content and Amino acid Composition

The world's major crops in terms of dry matter production are shown in Fig. 1. They are dominated by the cereals, the cultivated members of the grass family, the Gramineae. The cereals provide the bulk of the protein and carbohydrate for man and his farm animals (Fig. 2). Of the remaining plant protein, nearly all comes from dictoyledonous seeds, especially the legumes. The only non-seed crop other than grass which produces signifi-cant quantities of protein is the potato (Fig. 2). The sweet potato, cassava and the yam produce very little protein.

Wheat, a plant of temperate climates, is widely grown and is the staple food of much of Europe, the nations of European descent and temperate regions of India and China. Rice is grown in the tropics where rain and sunshine are abundant and forms the staple diet of one half of the world's population. In tropical regions that have a limited rainfall, sorghum and various types of millet replace rice as the staple food. The remaining major cereal, maize, is grown extensively in tropical and subtropical areas and forms the staple food of populations in parts of South America, Eastern Europe and East and South Africa.

Cereals are also of prime importance in agriculture for feeding farm animals. In temperate regions, barely, oats and rye are grown for this purpose. They are grown in conditions where wheat is less profitable; for instance, rye is cultivated on light soils where winters are very cold and oats are grown in cool, wet climates. In tropical and subtropical areas, millet and particularly maize are prime sources of animal feeds.

The major seed crop of dictoyledonous plants is the soybean (Fig. 1), grown for its oil and protein. A legume, it is a native of China but is now cultivated in the USA (the principal producer), Korea, Japan and

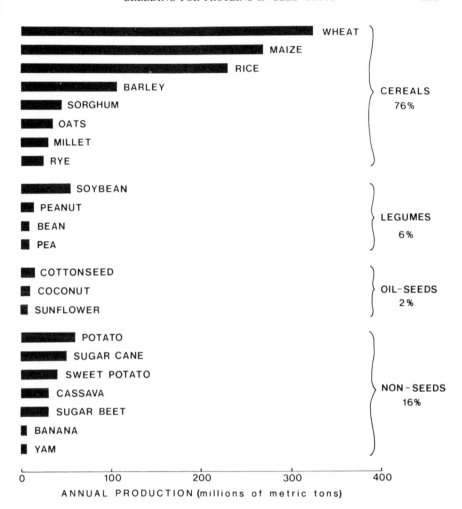

FIG. 1. Estimated dry matter production of the world's 22 major crops (excluding grass). Drawn from the data of Harlan and Starks (1980) with permission. The values for rice are estimated after de-hulling and for cassava after peeling.

Malaysia, countries which have very warm summers and frost-free nights. The oil is mainly for human consumption but most of the protein meal is used for feeding farm animals. The other leguminous oil-seed widely grown is the peanut. A native of South America, it is widely grown in the tropics and subtropics. The chief producers are India, the USA, China and Nigeria. Numerous grain legumes, storing starch and protein rather than oil and protein, are grown in most climates for protein foods. India in particular grows many different kinds but their total production compared with cereals is very low.

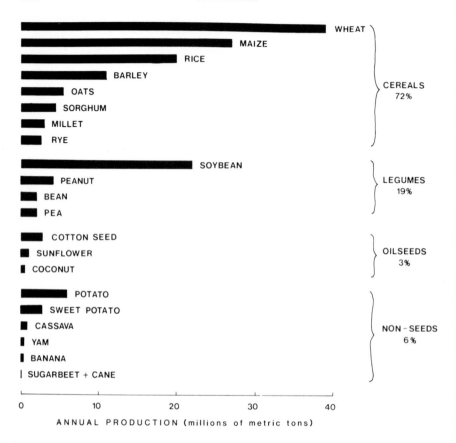

FIG. 2. Estimated protein production by the world's 22 major crops (excluding grass). Drawn from the data of Harlan and Starks (1980) with permission.

The protein content of the major seed crops is shown in Table I. Cereals usually do not exceed 15% in protein content. The grains of wheat and oats are the richest in protein and those of maize and rice are the poorest. Legumes and oilseeds contain much more protein than cereals, rarely less than 20% and often more, notably the soybean at over 30%.

In all grains the predominant protein type is storage protein. Synthesized on the rough endoplasmic reticulum (Cameron-Mills *et al.*, 1978) and stored in protein bodies (Pernollet, 1978) the proteins are usually of one of three types:

1. Globulins; soluble in salt solutions but insoluble in water.
2. Prolamins; soluble in aqueous alcohols.
3. Glutelins; soluble in dilute acid and alkalis.

The storage protein of dicotyledonous seeds is invariably globulin whereas in cereals it is usually prolamin (maize, barley), glutelin (rice) or

TABLE I
The protein content of crop seeds

Crop	Crude Protein content, %
Cereals	
Barley	8·2–11·6
Maize	7·2– 9·4
Oats	12·1–14·2
Rice	7·5– 9·0
Wheat	11–14
Oilseeds	
Cotton	17–21
Oil-seed rape	20–25
Peanut	25–28
Sesame	25
Soybean	32–42
Sunflower	27
Legumes	
Chickpea	20–28
Lentil	23–29
Lima bean	19–21
Pea	21–28
Field bean	20–30

The crude protein is calculated as crude protein (nitrogen × 6·25) per 100 g dry seed. Some of the data is taken with permission from Altschul (1965).

TABLE II
Essential amino acid composition of seed protein

	FAO pattern	Wheat	Rice	Maize	Soybean	Peanut	Broad bean
Isoleucine	278	253	290	225	319	224	333
Leucine	305	409	501	717	483	407	438
Lysine	279	174	239	169	429	218	476
Methionine + cystine	275	265	316	200	197	173	112
Phenylalanine + tyrosine	360	457	629	496	557	571	567
Threonine	180	192	235	225	269	171	284
Tryptophan	90	67	78	33	80	64	69
Valine	270	272	398	263	336	274	373

Data expressed as mg amino acid per g nitrogen and taken with permission from Altschul (1965). Those amino acids present at less than 90% of the FAO pattern have been underlined. Cystine is interconvertible with methionine so the two amino acids are usually considered together. The FAO pattern is that considered by the FAO (1957) as providing the ideal food protein for humans.

228   P. I. PAYNE

both (wheat) (Whitehouse, 1973). All these proteins are rich in either glutamine or asparagine and this is consistent with them acting as nitrogen stores for the rapidly-growing embryo during early germination. Barley prolamin for example has a glutamine content of 33·1 moles % (Laurière *et al.*, 1976) and so comprises 1 in every 3 amino acid residues. Storage proteins are usually insoluble in the cell sap for they have appreciable numbers of the hydrophobic amino acids proline, leucine, isoleucine, and, phenylalanine. The high levels of amidated and hydrophobic amino acids results in a balance of amino acids which is not ideal for the human diet. Table II lists the contents of the essential amino acids (those which cannot be synthesized by man *in vivo*) in the major seed crops.

The first limiting essential amino acid for humans in all cereals is lysine, although in maize, tryptophan is co-limiting. In contrast in legumes the limiting amino acid is methionine. Seeds are also often deficient in isoleucine and particularly trypotophan (Table II).

III. Breeding for Nutritional Quality

A. PROTEIN AND AMINO ACID REQUIREMENTS

Man obtains protein from plants, either directly or from animals fed on plants. In a world which has about 500 million people or about one tenth of the total population, who are seriously undernourished, there are pressures on the plant breeder to increase the content of protein in a seed and to improve its balance of essential amino acids. In attempting this the breeder has various approaches open to him but before these can be evaluated the protein requirements of man and his farm animals must be considered.

1. Humans

During the last 25 years or so, two extreme forms of malnutrition in infancy have been recognized: kwashiorkor and marasmus, the first being due, it is claimed, to a major deficiency in protein and the latter to a lack of sufficient calories. The incidence of kwashiorkor led to the concept of a protein gap (for review see McLaren, 1974) and the Protein Advisory Group of The United Nations was set up in the mid 1950's to advise on how the gap might be closed.

More recently, the concept of protein deficiencies has been contested (McLaren, 1974; Payne, 1978). It has been pointed out that man is a very slow growing animal even during the early years of life, compared to laboratory and farm animals (Carpenter, 1975). Generally, he also conserves the nitrogen in his body by re-using the amino acids liberated during protein turnover. Thus, only small amounts of nitrogen are lost in the urine and faeces, in growing hair, nails and the sloughing-off of cells. The exceptions are when body protein has to be metabolized to make up for an insufficiency of carbohydrate and when excess protein is eaten. Payne (1978) and Coward and Lunn (1981) have also contested whether kwashiorkor is in fact primarily a protein-deficiency disease.

Over the years, estimates of the minimum safe net dietary requirement for protein have steadily lowered (Payne, 1978). The latest estimate by the World Health Organization (FAO/WHO, 1973) was 36 g per day for a 65 kg man. Now it is generally agreed that a similar-sized man requires 3000 Kcal of energy per day (FAO/WHO, 1973). If placed on an all-wheat diet and assuming a protein content of 12% (Table I) then he would consume 105 g protein per day to satisfy his calorie requirements. However, although wheat protein is highly digestible (McCance and Widdowson, 1947) a correction factor is required in this calculation because wheat protein is low in lysine (Table II) and only about 40% of the ideal for the human diet. Even taking this into account the conclusion is that if adults in wheat-eating (and in practice grain-eating) communities have resources to satisfy their food energy needs they will also satisfy their protein requirements with a fair margin of safety, a conclusion also reached by others (Carpenter, 1975; Payne, 1978). For legumes, Blixt (1979) has concluded that only 30–115 g of high-protein peas and beans or 20–80 g of soybeans are required to satisfy human dietary requirements.

Adult man simply requires protein in the diet for the maintenance of tissues. However babies, children and pregnant women additionally require protein for growth and lactating women for the production of milk protein. A baby requires 5 times more protein on a body weight basis than an adult (FAO/WHO, 1973). However, caloric requirements are also increased and Payne (1978) has calculated that the protein/energy ratio only increases by 36%. A baby transferred from a diet of human milk to wheat porridge would probably marginally satisfy its protein requirements though it is doubtful whether it could take the sheer volume of food.

In practice, most communities do not have a diet which is exclusively from cereals but is mixed to varying extents with legumes, vegetables and meat. Mixing different foods can greatly improve the nutritional quality of the total protein. A good example is when whole wheat is consumed with *Phaseolus* bean meal (Fig. 3). The low lysine of the cereal and the low methionine of the bean are compensated so that the nutritional value of mixtures (expressed as chemical score in Fig. 3) is higher than for either food alone. The optimal mixture where lysine and methionine are co-limiting is approximately at a point where wheat is fortified with 15–20% bean meal (assuming a protein content of 25% in the latter). Other good combinations are wheat flour mixed with cottonseed meal or with milk.

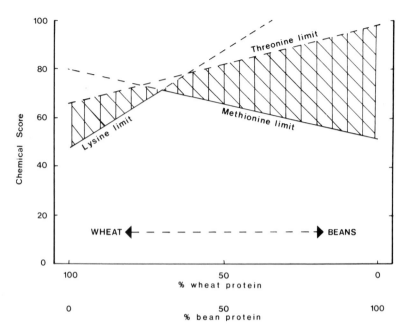

FIG. 3. Changes in the balance of the essential amino acids in different mixtures of wheat and bean protein. Modified from Carpenter (1975) with permission.

2. Animals Farmed for Human Consumption

During the last 30 years, there has been a rapid increase in the production of animal meat in Western Europe, particularly that of pigs and poultry. This has been made possible by the intensive rearing of animals; bringing them indoors and changing their basic diet from grass (or household swill in the case of pigs) to food concentrates often consisting of cereals, fish meal, soyabean meal and other high-protein legumes. Pigs, beef and poultry fed and treated in this way can reach the same weight as conventionally grown animals in only one third the time.

Thus, intensively reared animals grow very quickly and their growth rate is far greater than that at any stage in the life cycle of man. Consequently the protein requirements of single-stomached (monogastric) farm animals (mainly pigs and poultry) are correspondingly higher and more than can be met by cereal grains alone. The situation is less severe with ruminants (cattle and sheep) for they obtain their amino acids from micro-organisms living in the alimentary system. Here the requirement from food is nitrogen, not amino acids, so that plant storage protein is adequate in quality particularly because of its high glutamine content (Derbyshire *et al.*, 1976; Kasarda *et al.*, 1976).

1. Yield or protein content?

In cereals there is in general an inverse relationship between the yield of the grain crop and its protein content (Grant and McCalla, 1949; Pushman and Bingham, 1976; Mesdag, 1979). Such a relationship is shown in Fig. 4 with ten varieties of winter wheat grown at three levels of nitrogen fertilizer treatment. The reasons for the protein-yield relationship have been discussed (Austin *et al.*, 1977). In broad terms the breeder can opt either for high yield with low protein content, a high protein content with a yield penalty or a compromise. In relation to breeding for human foods especially in the developing countries where there is principally a calorie gap between supply and demand, the emphasis must be on yield. Even with the feeding of intensively-farmed monogastric animals, where there is a much greater need for protein, yield in cereals at least will be the most important factor. For instance in the feed barley breeding programme at the Plant Breeding Institute (PBI), Cambridge, yield has a much higher priority than protein quantity (PBI Annual Report, 1979). Although cereals will mainly contribute calories to animal feed, the quality of its low-lysine protein is greatly increased by blending with low-methionine starchy legumes such as the field bean (25% protein) or the protein-rich oil-free meal of oil-seeds such as soyabean (50% protein) (see Fig. 3).

Breeding improved varieties of legume crops is beset with problems. Temperate grain legumes are very susceptible to variable yields, unlike cereals, so yield stabilization is the major objective. In the tropics, legumes also suffer from numerous diseases so breeding disease resistance into varieties is the principal goal.

Thus, from the above discussions it is concluded that the yield of the crop is the most important objective for the plant breeder to improve. But this is not to say that protein content should be ignored. As seen in Fig. 4, there are some varieties which deviate favourably from the protein-yield relationships. Mesdag (1979) showed that the wheat variety Orca and the breeding line 6225–19–4 yield the same but the latter had 1·9% more protein in the grain. These exceptions occur also in other crops and they are currently being exploited. Some will be discussed in later sections.

2. Protein Quality

Breeding for protein quantity in cereals has the disadvantage that as protein increases in amount in seeds the overall nutritional quality decreases. For instance in wheat (Fig. 5), as the protein content increases from 7% to 15% there is a steady drop in lysine, expressed as a percentage of total protein from 4% to 3%. This is because increasing protein up to

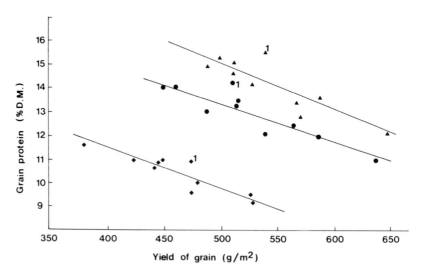

Fig. 4. The relationship between grain protein percentage and grain yield in 10 varieties/lines of winter wheat grown at three levels of nitrogen application. Taken from Pushman and Bingham (1976) with permission. Protein values for the breeding line TJB 54/224 marked "1" are significantly above all three regression lines.

15% increases the ratio of storage protein to metabolic and structural proteins in the grain the former being lysine deficient and the latter lysine rich. At levels of protein above 15%, lysine remains constant at 3%. Increasing the protein content does, however, raise the overall lysine content per seed (Fig. 6). Similarly, in legumes the percentage of methionine drops with increasing protein content (Adams, 1973).

The alternative strategy to breeding for protein quantity is to breed for protein quality. Here, there may be no inverse relationship with yield and the requirement for expensive nitrogen fertilizers may be less excessive. The most successful approach so far has been to change the proportions of the different proteins in the seed either by conventional crossing involving a rare line which has unusual proportions of protein types or by mutation programmes. Those protein groups richest in lysine (or methionine for legumes) would be increased at the expense of more lysine (or methionine)—poor groups. As will be seen in later sections, some success has been obtained with this approach in barley and maize.

The ideal way of improving quality would be to radically alter the nucleotide sequence of the structural genes which code for the storage proteins by mutation programmes so that a storage protein with an improved balance of essential amino acids is obtained. As far as is known, this has not so far been achieved. As pointed out by Boulter (1976) although storage protein genes are probably more open to mutational change than say multi-subunit enzymes, there must still be fairly stringent

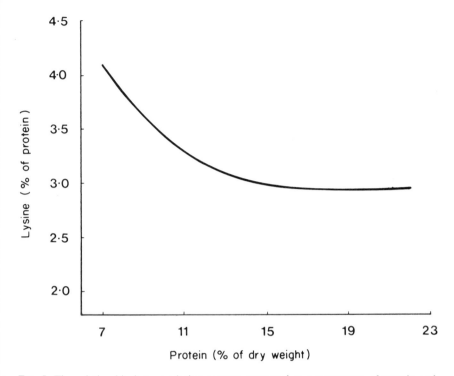

FIG. 5. The relationship between lysine content expressed as a percentage of protein and protein content expressed as a percentage of grain dry matter in the world collection of wheats. Redrawn from Anon (1980) with the permission of Dr. V. A. Johnson.

constraints on changes in amino-acid sequences. This is because storage proteins have evolved to be synthesized on membrane-bound polyribosomes, to penetrate the lumen of the endoplasmic reticulum, to be assembled in protein bodies and to be specifically broken down during germination by proteolytic enzymes adapted for "normal" storage protein. The chances of obtaining a mutant which only has alterations in its storage protein genes is also somewhat remote. This approach may not therefore be feasible or at least not until genetic manipulation methodology becomes available.

Another possible approach to protein improvement is to change the morphology of the seed. The aleurone layer in cereals is rich in basic amino acids and is consequently of superior nutritional quality but is usually only one cell thick. Multiple aleurone layers have been reported in barley (Sawicki, 1952), maize (Wolfe et al., 1972) and oats and rice (Hoshikawa, 1967). The embryo also is richer in protein and in lysine than the starchy endosperm and there may be scope here for increasing its size relative to the endosperm.

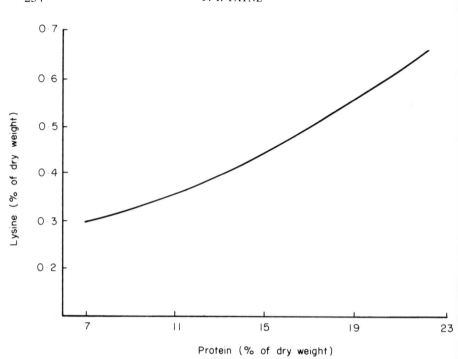

Fig. 6. The relationship between lysine and protein content, both expressed as a percentage of grain dry matter in the world collection of wheats. Redrawn from Anon (1980) with the permission of Dr. V. A. Johnson.

C. WHEAT

A thirteen-year project to study the possibilities for genetical improvement in protein levels and lysine content of wheat grains has recently been completed by Johnson and Mattern and colleagues at the University of Nebraska, USA (Anon, 1980). Initially, they examined over 20 000 entries of the World Wheat Collection maintained by the United States Department of Agriculture (Vogel *et al.*, 1973). The protein content varied from 7% to 25% and of this 18% variation, 5% was at the genetical level. One of the high-protein varieties was Atlas 66, a soft-milling spring wheat. It was crossed with the good bread-making, hard-milling wheat Comanche and the F_1 progeny were crossed to another good-quality wheat, Lancer. After several generations of inbreeding a final selection was made, the variety Lancota. It had elevated grain protein content, good milling and bread-making properties, good disease resistance and winter hardiness. In field trials over three years it had similar or slightly higher yields to the popular variety Centurk but contained about 1·5% more protein. The extra protein was in the starchy endosperm and so was also present in white flour milled from the grains.

Another high-protein variety, Nap Hal from India, has also been crossed with Atlas 66 (Johnson *et al.*, 1973). Some of the progenies had higher protein levels than either parent (transgressive segregation), indicating that the parent varieties carry different genes for elevated protein content. The progenies had poor agronomic characteristics and were low yielding with small grains. They were therefore crossed to superior varieties and high protein lines from them were sent to various breeding institutes in developing countries to be used as parents in breeding programmes.

In their survey of the World Wheat Collection, Vogel *et al.* (1973) calculated that the genetic component of total lysine varied by only 0·5%. This is only one third of the amount required to bring lysine in reasonable balance with the other essential amino acids in wheat proteins (Johnson and Mattern, 1978). Thus, large genetic effects on lysine content, as occurs for instance with the *opaque-2* gene of maize (section 3E), were not found. Nevertheless the two most promising wheat lines, Nap Hal (also high in protein) and CI 13559 were crossed and some progeny were higher in lysine than either parent. In some lines, this was combined with the high protein content of Nap Hal and high grain yield. Further breeding programmes will be necessary however for the elevated lysine trait to be incorporated into agronomically acceptable lines.

Although the upper limit for grain protein content is around 25% (Vogel *et al.* 1973), it is much higher still in several species which are related to bread wheat. In *Triticum dicoccoides* for example it can be as high as 42% (Avivi, 1978). As this species readily crosses with bread wheat and produces fertile offspring it should be possible to transfer the high-protein traits. This is currently being examined at the PBI and elsewhere, including Holland, Israel and Australia.

In some sophisticated cytogenetical experiments, entire homoeologous chromosomes from several different *Triticum* species have been substituted into three bread wheat varieties in the place of one of their group 2 chromosomes (Law *et al.*, 1978). It was found that when the 2M chromosome of *Aegilops comosa* replaced chromosome 2D, grains were produced which had an increased protein content and this occurred without any reduction in yield. A programme is now being set up to substitute chromosome 2M into bread wheat varieties currently grown in the UK.

D. BARLEY

In a comparable study to that by Vogel *et al.* (1973) with wheat, Munck *et al.* (1969) screened the entire USDA world collection of barleys and found

Hiproly, a primitive variety high in both lysine and in protein. The increased lysine is due mainly to increased amounts of several water-soluble polypeptides (Rhodes and Gill, 1980). Hiproly yields only one third that of the best modern varieties but nevertheless it has been used extensively in breeding programmes. Persson (1975) found that one cross to a modern commercial variety raised the yield in selected lines to about 60% of the standard. Continued backcrossing to the modern variety with selection for high lysine raised the yield further, though less dramatically, and at the sixth backcross it was about 85–95% of the recurrent commercial variety. The slightly lower yield was associated with small seed size and slightly reduced fertility. The lines also perform less well in poor weather. At the Swedish Seed Association, Svalöv, where Hiproly features strongly in their breeding programme, they are aiming to produce with some optimism a high lysine variety with normal protein content, a slightly smaller than average seed size but a grain yield as high as the best of the commercial varieties (Hagberg et al., 1979).

The alternative approach to screening for natural variation has been to create high-lysine mutants by chemical methods or by irradiation with gamma or X-rays. There are the Risø mutants from Denmark (Køie and Doll, 1979), Notch-1 and Notch-2 from India (Balaravi et al., 1976) and from Italy, Lys 95 and Lys 449 (Di Fonzo and Stanca, 1977). In these mutants, in contrast to Hiproly, high lysine content is caused by a reduced accumulation of hordein (Køie and Doll, 1979), the prolamin of barley. Unfortunately, starch accumulation is also reduced, producing shrivelled grains and low crop yields. Several of these mutants have been used in breeding programmes in attempts to produce high yielding, high lysine barleys. Results with the mutant highest in lysine, Risø 1508, have been particularly disappointing at various Institutes including PBI (Payne and Rhodes, 1979) because it has not been possible to separate the high lysine character from shrunken seed and thus low grain yield. In a study of carbohydrate and protein accumulation in the Risø mutants, Doll and Kreis (1979) speculated that hordein plays an essential role in seed development by removing excess nitrogen from the developing endo-sperm. When hordein synthesis is suppressed, as in Risø 1508, it is suggested that the excess nitrogen compounds interfere with carbohydrate metabolism. If this is the case then the pleiotropic complex, high lysine and low yield, may never be broken in mutants which have radical shifts to high lysine. Breeding work with Risø 1508 has practically ceased, though it still continues with other mutants such as Risø 7 which has a modest increase in lysine but plump grains.

At Svalöv, the possibilities of combining the high lysine traits of Hiproly and the Risø mutants to form high-lysine double recessives is being investigated. Use is being made of the finding that the high lysine gene *lys* (from Hiproly) and *lys 3* from the mutants are both located on chromo-some 7 and are linked to gene *S* (short rachilla hairs) (Karlsson, 1976). A breeding line, carrying the Hiproly high-lysine gene in the genetic back-ground of a modern, commercial variety (Hiproly × Mona[5]) was crossed to Risø mutant 7 (Hagberg *et al.*, 1979). High-lysine double recessives were selected firstly by the presence of short rachilla hairs and secondly by a dye binding procedure which estimates lysine content. Some of them have well filled grains without shrunkeness and these are being tested for practical importance by further breeding.

E. MAIZE

The endosperms of normal maize grains have translucent, vitreous areas which are concentrated at the embryo-end of the grain. The rest of the endosperm is whitish and starchy in appearance. Several spontaneous mutants of maize lack the translucent areas and Mertz *et al.* (1964) showed that one of them which contained the recessive gene, *opaque-2*, had a lysine content of 4% of the total protein, almost twice that found in typical hybrid maize. Soon afterwards a second mutant gene, *floury-2*, was discovered (Nelson *et al.*, 1965) which also had elevated lysine though not as high as *opaque-2* maize. However, it had a higher grain protein content and a much higher percentage of methionine. Both mutants had elevated tryprophan which in maize is co-limiting with lysine.

The improved amino-acid composition of these mutants was shown to be due to a 50% reduction in lysine-deficient zein (the prolamin of maize) and a concomitant increase in albumins, globulins and particularly glutelins. The *opaque-2* allele causes an overall suppression of the synthesis of most zeins during endosperm development but those with alkaline isoelectric points appear to be suppressed most (Soave *et al.*, 1976).

The expected improvement in protein nutritional quality of the mutants was confirmed by feeding *opaque-2* maize to pigs (Pickett, 1966). Bressani (1966) concluded that for children the nutritive value of *opaque-2* protein was about 90% that of skim milk and Clark (1966) found that adult man would receive 93% of his protein requirements when fed 300 g of *opaque-2* maize each day.

It was with great enthusiasm therefore that maize breeders all round the world started to transfer these mutant genes into their promising inbreds and open-pollinated varieties by backcrossing procedures. Because of the lack of biochemical facilities at many of the centres, the high-lysine trait was usually screened by selecting for the soft, chalky endosperm pheno-type. Commercial varieties with this endosperm type were produced in some countries but they suffered from strong disadvantages to standard varieties. In particular, grain yield was lower by some 10% and mechanical damage of the soft endosperm types during harvesting was almost double that of standard forms.

With these problems confronting opaques, breeding programmes in many centres were either abandoned or drastically reduced. Work, however, continues at certain better-equipped centres, such as The International Maize and Wheat Improvement Center at Mexico (CIMMYT) and important advances have been made. Work at CIMMYT is centred on the exploitation of genetic modifiers which can change the *opaque-2* trait from soft to hard endosperm types (Vasal *et al.*, 1979). These have been successfully obtained and chemical analyses at each generation have enabled protein quality to be maintained. A number of hard-endosperm *opaque-2* lines are being further selected for improved yield and a few selections actually had yields which approached those of standard varieties (Vasal *et al.*, 1979).

The outlook for developing maize varieties with 9–10% grain protein which have similar yields and agronomic characteristics to normals, but have high lysine and high tryptophan levels as a bonus, once more looks promising.

F. OIL SEED RAPE

Rape is the major oil-seed crop of temperate zones of the earth and is principally grown for its oil content. The seed is also rich in protein containing 20–25% by weight (Table I), and the meal after oil extraction consists of 40–45% protein. In addition the protein has a high nutritional value by plant standards, being somewhat similar to the protein of soyabean. The exploitation of the oil-free meal as a protein food is limited by the presence of glucosinolates. These compounds have the general formula:

$$R - C \underset{N - O - SO_3}{\overset{S - C_6H_{11}O_5}{<}}$$

There are many types of side chain, R, ranging from simple alkyl chains to heterocyclic side chains. The seeds of *Brassica napus*, the predominant rape in Western Europe, primarily stores progoitrin (Robbelen and Thies, 1980) which has the side chain:

$$CH_2 = CH. CH. CH_2 -$$
$$/$$
$$OH$$

Glucosinolates break down in the stomachs of animals by the action of glucosinolase (= myrosinase e.c. 3.2.3.1), which is also present in the meal giving bitter-flavoured toxic, isothiocyanates (the mustard oils), goitrin and cyanides. Monogastric animals which have been fed with appreciable amounts of rape meal rapidly lose weight, their thyroid glands become swollen and they frequently die. Several attempts have been made to detoxify rape meal, including destruction by fungi, heavy metals, hydrogen peroxide and heat (El Nockrashy, 1976). None of these methods is entirely satisfactory as they result in the loss of biological value of the protein. However, protein purification by isoelectric precipitation is showing promising results (El Nockrashy, 1976).

The objective in the breeding of oil-seed rape is not to increase the protein content as with wheat, barley and maize as this would inevitably reduce the yield of oil, or to improve the amino acid composition for it is already rich in most of the essential amino acids, but rather to reduce the amount of glucosinolates. Initially many varieties were screened for glucosinolate content and one of them, Bronowski, a Polish spring rape was found to be low in glucosinolates. This variety was introduced into numerous breeding programmes and glucosinolate levels in selected offspring were reduced. However, progress was slow because the inheritance of glucosinolate is complex being controlled by three partially recessive, unlinked genes (Lein, 1972). Currently, levels have been reduced from about $120\,\mu M/g$ defatted flour to less than $30\,\mu M$. In early selections, yields of the crop were disappointingly low, partly because Bronowski was crossed with winter rapes, the principal crop in Western Europe. With further breeding, yields have improved and it is anticipated that low glucosinolate varieties will have comparable yields to commercial types (Thompson, personal communication).

Low glucosinolate rape meal does well in feeding trials although there are still several problems. One is the fibre content of the meal which interferes with the absorption of the protein. Another is the production of foul-smelling eggs by certain strains of brown-egg poultry. This tainting of eggs is less pronounced when poultry are fed low glucosinolate meal but it is still unsatisfactory (Pearson et al., 1980). The chemical causing the smell is triethylamine. It is formed from the choline ester, sinapine, a component of rape meal and choline, present in several other dietary sources (Pearson et al., 1979a). Trimethylamine is normally oxidized by trimethylamine oxidase to its oxide which is then excreted from the body. However goitrin, a breakdown product of glucosinolates, partially inhibits the enzyme. In most strains of poultry triethylamine levels are kept low unless fed very high levels of rape meal. With brown-egged strains the level builds up, even in low glucosinolate varieties, because they synthesize much smaller amounts of triethylamine oxidase.

In relation to plant breeding, tainted eggs are only likely to be eliminated if glucosinolate-free lines can be developed. While this should be attempted perhaps the most satisfactory solution would be to breed commercial flocks which lay brown eggs and synthesize normal levels of triethylamine oxidase (Pearson *et al.*, 1979b). Currently, rapemeal is principally fed to cattle whose alimentary systems are better able to cope with the toxic components.

There is every likelihood that oil-seed rape will be regarded as a plant breeding success in the next few years. In Western Europe, increased production of oil-seed rape using new varieties will reduce the need to import both vegetable oil for human consumption and protein-rich soya-bean meal for animal feeding.

G. LEGUMES WITH SPECIAL REFERENCE TO THE FIELD BEAN

Most of the world's legume protein crop, in the form of soybeans and peanuts (Fig. 2), is grown in the USA and much of it is bought by the Western European countries for incorporation into animal feeds. Western Europe has a wide protein deficit. The UK for instance imports over 70% of its crude protein requirement (other than from grass), 819 kt in 1974 (J.C.O., 1976). Most soybean is imported by West Germany, 3701 kt in 1974, followed by the Netherlands with 1590 kt (Wilson, 1977). A major plant breeding object in Europe is to reduce this protein deficit. Currently, production of the soybean in Europe is largely confined to Romania, Bulgaria and Yugoslavia because the maritime climate of Western Europe is unsuitable for present varieties. A long term solution may be the development of new varieties of soybean which are adapted to the West European climate. It is likely in the short term that any reduction in soybean imports will be achieved by breeding new varieties of legumes that are currently grown in Europe. These are the field bean, pea, lupin, French bean and the lentil.

The recent demand for protein-rich legumes in Europe has arisen from the introduction of intensive rearing of monogastric farm animals. The demand for new crop varieties has hence been sudden after a long period of neglect. In 1973, the major legume crop in the UK, the field bean *Vicia faba*, covered less than 2% of the area grown for wheat and barley (Russell-Eggitt, 1977). Indeed the crop steadily declined in the UK this century until about 1965 when it recovered slightly with its use as a break crop (Wilson, 1977).

It is not therefore surprising that the breeding of improved legume varieties in Europe has received little attention. Most grain legumes have serious shortcomings. The field bean crop is notoriously unreliable owing to its susceptibility to adverse weather conditions, poor seed, set pests and diseases. The average yield of the crop in England and Wales over the seven years 1970–1976 was 2·9 t/ha. The corresponding grain yield for wheat over the same period was 48% higher at 4·3 t/ha (MAFF, 1977). It has been stated that yields of field beans have increased less than 20% in the last century (Russell-Eggitt, 1977) whereas with wheat there has on average been almost a 2% increase in yield per year for the past 30 years (Bingham and Blackman, 1978). In France, the yield of wheat and field beans were similar to each other in 1945 but by 1975 wheat was outyielding field beans by 100% (Picard, 1979).

In other parts of the world, where legumes are principally eaten by man, they are often regarded as secondary crops to cereals. The chickpea is an example. On the Indian subcontinent it is the most important pulse crop although there has been a decrease in its acreage in the past years. During the "green revolution", new high-yielding wheat varieties were introduced and partially replaced the chickpea on more fertile land. The yield of chickpea has barely increased in India in the last 60 years (Van der Maesen, 1972).

As discussed in Section III.B.1, the view of plant breeders is that stability of yield is the most important first goal (Bond, 1977). The yield of legumes must be raised and brought closer to those of cereals. If successful then protein yield will automatically increase also. However, breeding for protein improvement *per se* should continue.

In cereals, an inverse relationship exists between yield and protein content (Fig. 4). In legumes, this relationship is often not apparent. For example, in 15 trials of winter field beans held at PBI, Cambridge over several years, only two showed a significant negative relationship between yield and protein content (Bond, 1977). Unfortunately the genetic relationship between yield and protein content is not known because in legumes there is a large effect of the environment on yield and a small effect of the environment on protein content (Bond, personal communication). Nevertheless, crosses have been made between high yielding varieties and high-protein varieties in the hope of combining both traits in some progeny. Such breeding programmes have met with success in the soyabean (Brim, 1973). With field beans, progeny with combined traits have also been obtained (Bond, 1977). Further selections have unfortunately shown that most of them are late flowering and late maturing (Bond, personal communication), a big disadvantage for use in northern areas of the UK. This breeding programme is continuing.

Breeding for an improved balance of essential amino acids is even more difficult than breeding for protein content. The first limiting amino acid in field beans and other legumes is methionine (+ cysteine) (Table II). In situations where legumes form the major proportion of the diet or where legumes are mixed with root crops or bananas, methionine will be the limiting amino acid. In others where cereals form the bulk of food intake, lysine may well be limiting (Fig. 3). Thus, breeding for protein quality in legumes may mean increasing methionine or lysine, depending on dietary needs.

A major difficulty in breeding for methionine is the lack of a satisfactory screening method using inexpensive apparatus. A total sulphur assay for methionine + cysteine of alcohol-extracted meals has been suggested (Evans and Boulter, 1974) and there are specific methods for cysteine (Herrick *et al.*, 1972) and methionine (Paul, 1977). However, the presence of S-methyl-L-cysteine in some legumes is a complicating factor (Evans and Boulter, 1975). It has been claimed that current, simple techniques show nearly as much error in determining methionine as there is genotypic variation for this amino acid (Evans and Gridley, 1979). So far, no mutants containing elevated lysine or elevated methionine in grain protein have been discovered in legumes, unlike cereals.

Another approach to altering the amino acid balance in legumes is to change the relative proportions of the different protein groups which are laid down in the seed, as has occurred in the Risø barley mutants (Koie and Doll, 1979) and *opaque-2* maize (Mertz *et al.*, 1964). Boulter *et al.* (1973) have shown that in both the field bean and the pea that one of the storage proteins, legumin, is richer in methionine and cysteine than the other, vicilin, though it is lower in lysine, isoleucine and phenylalanine. Since different lines and varieties of both peas (Thompson *et al.*, 1979) and field beans (Martensson, 1980) have different proportions of legumin and vicilin in them, there should be scope for purposely altering the balance of vicilin and legumin by breeding with a view to increasing the limiting essential amino acids. The advantage of this procedure is that both vicilin and legumin are storage proteins so that changing their proportions should not disturb the basic metabolism of the maternal plant (as occurs for instance in the mutant barley Risø 1508) so that yield may not be affected.

Legumes, like oil-seed rape, contain toxic factors. The best known are the trypsin inhibitors and the phytohaemagglutinins (lectins). The great majority of them are destroyed by cooking so they need not be eliminated by plant breeding. On the contrary there may be a case for breeding for increased trypsin inhibitor content as pointed out by Evans and Gridley (1979) for Putztai (1966) has shown that the trypsin inhibitors of *Phaseolus vulgaris* are exceptionally rich in sulphur amino acids. The lectins are discussed in detail by Dr. Putztai in a separate chapter of this book.

The field bean has only small quantities of phytohaemagglutinins and trypsin inhibitors, in the case of the latter about one fiftieth that of the soybean (McNab, 1977). Davidson (1977) recommends a maximum of only 15% bean meal in the diets of laying hens and 20% for growing pigs.

Above these levels egg laying and growth are retarded. Processing beans by heat partially overcomes the anti-nutritive factors but economically this is not feasible. The toxic factors in field beans are greatly lowered if the seed coats are removed (Davidson, 1977), and Martin-Tanguy *et al.* (1977) have presented evidence which indicates that condensed tannins are responsible. Because of this, most breeding work has been associated with zero-tannin types which are invariably white flowered (Picard, 1976) as opposed to white flowers with black patches. These varieties appear to perform better in digestibility tests using rats, though the results (Bond, 1977) also indicated other, unknown factors reducing bean protein digestibility which are superimposed on the main effect of tannin. A white-flowered variety, Triple White has been crossed to high yielding varieties and offspring which yield up to 80–90% of commercial varieties have been selected (Bond, 1977). Further backcrossing to high yielding lines and the creation of more variability within zero-tannin populations is necessary.

The plant breeder is currently restricted in his attempts to further reduce toxicity in the field bean: all the toxic factors have yet to be identified.

IV. Breeding for Bread-making Quality

One of the major crop breeding objectives in Western Europe is the production of new varieties of wheat with improved bread-making quality and high yield. This is because the European Economic Community (EEC) is now in surplus for wheats bred for feeding to animals but it is in deficit for bread-quality wheats. In the UK, the situation is particularly acute for in the past, plant breeders have given priority to yield and other agronomic factors than to bread-making quality (Russell-Eggitt, 1977). The bread-making grists have traditionally included a high proportion of strong flour imported from Canada. In the 1960's a typical British loaf would have been made from 60% Canadian Spring Wheat, 20% poor quality British wheat and 20% of a filler wheat, usually from Australia (Russell-Eggitt, 1977). With Britain joining the EEC the price of Canadian wheat rose sharply because of import tariffs and on top of this, it increased markedly in price on the world markets. Consequently the proportion of Canadian flour in British bread has significantly reduced and that of British flour and quality flours from fellow members of the EEC, particularly France has increased. The percentage of home-grown wheat in the British loaf continues to increase through improvements in technology and in the production of new, more suitable varieties.

To make bread, flour is mixed with water, salt and yeast, worked to form a dough, rested to allow fermentation and aeration of the dough to take place and oven baked. It is universally accepted that gluten, the water and salt-insoluble protein complex of the flour makes a valuable contribution to dough rheology (Wall, 1979). Gluten consists of two proteins, gliadin and glutenin. Gliadin (the prolamin of wheat) consists of numerous different polypeptides. Glutenin (the glutelin of wheat) only partially dissolves in dilute acids and has a very large aggregate molecular weight. Its size is variable, from between 200 000 and 20 000 000 daltons and is built up from about 15 different subunits (Bietz *et al.*, 1975). The two protein types impart different properties to a dough; gliadin is viscous and gives extensibility, allowing the dough to rise during fermentation whereas glutenin gives elasticity, preventing the dough from being over-extended and collapsing either during fermentation or during baking.

One of the difficulties facing the plant breeder in developing new varieties with good bread-making quality is the lack of suitable screening techniques which distinguish protein content from protein quality. Inevitably the breeder in selecting new lines will obtain a combination of the two, resulting in varieties with higher protein contents than biscuit or feed wheats and consquently they are likely to be lower yielding (see Fig. 4).

One of the ways the biochemist can help the breeder is to pull apart this elusive character, protein quality, and to understand in biochemical terms how it can cause the differences in bread-making quality between varieties. The first major breakthrough in understanding protein quality came from the work of Pomeranz (1965) and a little later from Orth and Bushuk (1972). Pomeranz (1965) devised a simple test for bread-making quality: flour was shaken for 2 h at 4° in a solution of 3 M urea buffered to pH 7·0. After centrifugation the absorption of the supernatant at 280 nm was measured. Using the flours of six varieties which varied widely in quality, he demonstrated an inverse relationship between this mesurement and the volume of the loaves baked from the same flours. Evidently the protein component imparting good bread-making quality is insoluble in 3 M urea at low temperature indicating it may be a constituent of glutenin.

In similar though more detailed studies, Orth and Bushuk (1972) fractionated flour proteins into five groups according to their solubility in different solvents: albumin, globulin, gliadin, acid-soluble glutenin and acid-insoluble glutenin (residue protein). The flours were prepared from 26 varieties of widely differing qualities, each grown at four different sites. Although there were marked inter-varietal differences in the proportion of solubility classes, there was no obvious trend between the proportion of albumin, globulin and gliadin and loaf volume. However, loaf volume was inversely related to the proportion of acid-soluble glutenin (Fig. 7) and directly related to the amount of acid-insoluble glutenin (Fig. 8) and these trends were shown to be statistically significant.

A few years later, Huebner and Wall (1976) compared the gel filtration properties of glutenin from a range of wheat varieties which varied in their

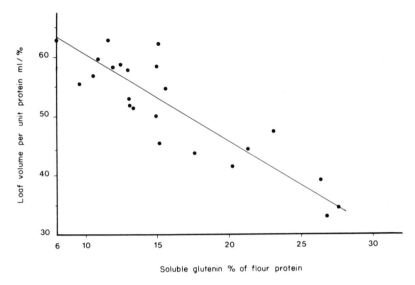

FIG. 7. The relationship between loaf volume per unit protein and % contribution of soluble glutenin to the total protein content of flour. Redrawn from Bushuk (1974) with permission.

bread-making quality. As much glutenin as possible was solubilized using a mixture of dilute acetic acid, 3 M urea and the anionic detergent hexadecyl-trimethyl-ammonium bromide. After centrifugation the proteins in the supernatant were chromatographed in the same solvent mixture through a column of Sepharose 4B. Two glutenin fractions separated, glutenin-I of high molecular weight (HMW) which eluted at the void volume of the column and glutenin-II a smaller protein complex which eluted later. The authors demonstrated that the better the bread-making quality of the flour the higher was the ratio of glutenin-I to glutenin-II.

All the experiments described so far therefore show that protein quality is linked to the glutenin protein complex. The more of it which is of HMW and acid insoluble the better is the bread-making quality of the flour and the smaller and more acid-soluble it is the more inferior is its quality. Now it is known that glutenin is built up from a range of different subunits (Bietz et al., 1975). Therefore, to study this work further, Payne and Corfield (1979) took serial fractions of glutenin as they eluted from a column of Sepharose 4B-CL using the solvent system of Huebner and Wall (1976) and they analysed glutenin subunit composition after reduction of disulphide bonds by SDS-PAGE. They showed a positive correlation between the molecular weight of native glutenin and the proportion of HMW subunits (nominal molecular weight 95 000–140 000) in glutenin. The authors suggested that the HMW subunits interact strongly with other,

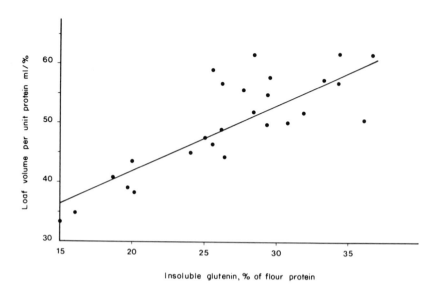

FIG. 8. The relationship between loaf volume per unit protein and the % contribution of insoluble glutenin to the total protein content of flour. Redrawn from Bushuk (1974) with permission.

smaller subunits, and are important in stabilizing the native glutenin structures.

Because of this apparently central role of the HMW subunits in determining glutenin structure they have been the subject of extensive biochemical and genetical studies (Payne et al., 1980; Lawrence and Shepherd, 1980; Payne et al., 1981a) All varieties of bread wheat contain between three and five HMW subunits whose structural genes are located on the long arm of the group one chromosomes; the locus on chromosome 1D codes for two subunits, that on chromosome 1B one or two subunits (according to the variety) and that on 1A either one subunit or none at all. These structural genes display allelic variation and in the 200 varieties analysed, 20 different subunits have been detected in many different combinations.

Can this variation in the HMW subunits amongst varieties be the reason for the intervarietal variation in bread-making quality? To try and answer this question, some 2 to 3 thousand progenies of different crosses have been analysed and further work is in progress. In this research (Payne et al., 1981b) two varieties with contrasting allelic subunits and different bread-making qualities were crossed and the F_1 progeny were grown and self fertilized. The F_3 grains produced from each F_2 plant were removed, milled and tested for bread-making quality by the SDS sedimentation test (Axford et al., 1979) and for HMW subunits by SDS-PAGE. In some crosses, an allelic subunit tended to segregate with the good quality progeny and the alternative allelic subunit derived from the other parent tended to segregate with the poor quality progeny.

We are now in a position to conclude from preliminary findings that allelic variation in HMW subunits does account, in part, for differences in bread-making quality (Payne *et al.*, 1981b). However, calculations reveal that only a proportion of the variation in quality that occurs in segregating populations can be accounted for by these subunits. Other quality factors must be involved. Work is currently in progress to determine the extent of allelic variation in both the low molecular weight glutenin subunits and the gliadin proteins prior to an analysis of their relationship to bread-making quality.

Although all the HMW subunits have not been ranked in relation to bread-making quality, it is known which are the best subunits associated with the 1A and 1D alleles. Consequently at the PBI, Cambridge, these two subunits are being brought together by conventional breeding programmes in four different crosses. One of them involves the cross between Bounty, a bread-making quality wheat from the PBI which has the best 1A subunit but an inferior 1D subunit and Alcedo, a quality wheat from East Germany which only has the best 1D subunit. Those F_2 progeny plants which had good morphology and showed resistance to fungal diseases were marked and the F_3 grain from them was tested for quality by the SDS sedimentation test. The best 10% were then analysed by SDS-PAGE and those which were homozygous for the best 1A and 1D subunits were sown in the field. The five best F_3 plants were selected and allowed to self-fertilize. It is hoped that a line will be developed from this selection which has superior bread-making quality to both of its parents. However, it is feared its yield may not be comparable with modern varieties. Consequently the F_3 plants are also being crossed to very high yielding varieties. The generation after next, progeny will again be selected for bread-making quality and homozygosity for the correct subunits as well as for the more conventional characters. It is hoped that some of these lines will also have inherited the high yielding character of the third parent.

V. Breeding for Biscuit-making Quality

A biscuit dough made from wheat flour must have good extensibility, a property conferred by gliadin, and poor elasticity. Thus a bread-making wheat is quite unsuitable for biscuit-making and vice versa. The rheological properties of the two types of dough, determined using an extensometer, are shown in Fig. 9. The bread-making dough shows a strong resistance to stretching at early stages because of its elastic glutenin.

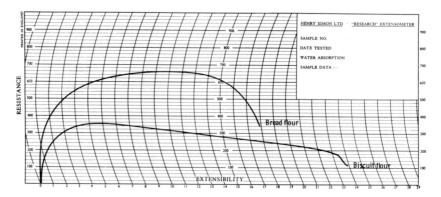

FIG. 9. Extensometer performance of a bread flour and a biscuit flour. Unpublished data of Mr. J. A. Blackman.

After limited extension the dough breaks. In contrast, the biscuit dough is much less resistant to stretching and also extends further before breaking.

In Western Europe, the breeder has little dificulty in breeding a biscuit wheat. To suppress the effect of the glutenin component, the protein content of the flour can be much lower in protein than in bread-quality wheats and they yield correspondingly higher.

VI. Breeding for Malting Quality

Barley used for malting must have a low protein content. The storage protein is released from the protein bodies during sprouting and it then envelopes the starch grains, making the latter less accessible to amylase activity. The net result is reduced or delayed sugar formation and poor fermentation. High protein levels in the grain is also liable to be associated with low yield, another disadvantage. In practice the breeder screens for protein directly by nitrogen determination and indirectly by micro-malting tests.

Another problem associated with protein is haze in beer. Haze is caused by proteins precipitating with polyphenols (Hough et al., 1971). To suppress the formation of haze, polyphenols are partially removed from beer worts by the addition of polyvinylpyrrolidone. An alternative approach is to breed for low or zero polyphenols in barley. However, a survey of varieties for proanthocyanidins and catechins (the principal polyphenols in barley) revealed little quantitative variation (Pollock et al., 1960). More recently, von Wettstein et al. (1977) screened a series of induced mutants of barley which showed suppresssed synthesis of anthocyanin pigments. One mutant, ant-13, was shown to be practically free of proanthocyanidins and catechins in the grains. It was obtained by treating quiescent grains of the variety Forma with the mutagen ethyl methanesulfonate. In a pilot brewing experiment, beer produced from *ant-13* had drastically improved haze stability compared to its parent, Forma. However, field trials showed that *anti-13* yielded 20–25% less than Forma. Currently, *ant-13* is being crossed to high yielding varieties in an attempt to introduce the haze-free trait into a suitable genetic background (von Wettstein *et al.*, 1977).

VII. Conclusions

In the development of a new variety the plant breeder has to select for many different characters. For optimal progress he must first recognize the priorities and rank the characters for improvement accordingly. The number of characters for which selection can be made without detriment to the breeding programme will depend on the size of the plant breeding team.

Probably in all crops, breeding for yield and yield stability will continue to be the major objective. The importance attached to protein content and protein quality will depend on the crop and for what use it is intended. Protein content is often given a higher priority than protein quality because of the simpler and more reliable screening procedures.

The major advances in breeding for protein content and protein quality are likely to come from large breeding organizations which have adequate research facilities.

VIII. Acknowledgements

I thank my colleagues, Drs R. B. Austin, D. A. Bond and K. F. Thompson for helpful discussions.

References

Adams, M. W. (1973). *In* "Nutritional improvement of food legumes by breeding", pp. 143–152. PAG, UN, New York.

Altschul, A. M. (1965). "Proteins: their chemistry and politics". Chapman and Hall, London.

Anon (1980). Genetic improvement of productivity and nutritional quality of wheat. Final Report. University of Nebraska-Lincoln.

Austin, R. B., Ford, M. A., Edrich, J. A. and Blackwell, R. D. (1977). *J. Agric. Sci. (Cambridge)* **88,** 159–167.

Avivi, L. (1978). *In* "Proceedings of the 5th International Wheat Genetics Symp" (S. Ramanujam, ed.), pp. 372–380. Indian Agricultural Research Institute, New Delhi.

Axford, D. W. E., McDermott, E. E. and Redman, D. G. (1979). *Cereal Chem.* **56**, 582–584.

Balaravi, S. P., Bansal, H. C., Eggum, B. O. and Bhaskaran, S. (1976). *J. Sci. Fd. Agric.* **27**, 545–552.

Bietz, J. A., Shepherd, K. W. and Wall, J. S. (1975). *Cereal Chem.* **52**, 513–532.

Bingham, J. and Blackman, J. A. (1978). *Bull. Flour Milling Baking Res. Assoc.* **3**, 125–139.

Blixt, S. (1979). *In* "Seed protein improvement in cereals and grain legumes' Vol. 2, p. 3–21. International Atomic Energy Agency, Vienna.

Bond, D. A. (1977). *In* "Protein quality from leguminous crops" pp. 348–353. Commission of the European Communities, Luxembourg.

Boulter, D. (1976). *In* "Genetic improvement of seed protein" pp. 231–250. National Academy of Science, Washington DC.

Boulter, D., Evans, I. M. and Derbyshire, E. (1973). *Qual. Plant.* **23**, 239–250.

Bressani, R. (1966). *In* "Proc. high lysine corn conference" (E. T. Mertz, ed.), pp. 34–39. Corn Refiners Association, Washington, DC.

Brim, C. A. (1973). *In* "Soybeans: improvement, production and uses", (B. E. Caldwell, ed.), pp. 155–186. Am. Soc. Agron. Inc., Madison, USA.

Bushuk (1974). *Bakers Digest* **48**, 4.

Cameron-Mills, V., Ingversen, J. and Brandt, A. (1978). *Carlsberg Res. Commun.* **43**, 91–102.

Carpenter, K. J. (1975). *In* "Bread, social nutritional and agricultural aspects" (A. Spicer, ed.), pp. 93–114. Applied Science Publishers, London.

Clark, H. E. (1966). *In* "Proc. high lysine corn conference" (E. T. Mertz, ed.), pp. 40–44. Corn Refiners Association, Washington, DC.

Coward, W. A. and Lunn, P. G. (1981). *Br. Med. Bull.* **37**, 19–24.

Davidson, J. (1977). *In* "Protein quality from leguminous crops" pp. 243–251. Commission of the European Communities, Luxembourg.

Derbyshire, E., Wright, D. J. and Boulter, D. (1976). *Phytochem.* **15**, 3–24.

Di Fonzo, N. and Stanca, A. M. (1977). *Genet. Agrar.* **31**, 401–409.

Doll, H. and Kreis, M. (1979). *In* "Crop physiology and cereal breeding" (J. H. J. Spiertz and Th. Kramer, eds.), pp. 173–174. Centre for Agricultural Publishing and Documentation, Wageningen.

El Nockrashy, A. S. (1976). *Fette, Seifen, Anstrichm.* **78**, 311–317.

Evans, I. M. and Boulter, D. (1974). *J. Sci. Fd. Agric.* **25**, 311–322.

Evans, I. M. and Boulter, D. (1975). *Qual. Plant Plant Foods. Hum. Nutr.* **24**, 257–261.

Evans, A. M. and Gridley, H. E. (1979). *Curr. Adv. Plant Sci.* **11**, 32.1–32.17.

FAO (1957). "Protein requirements" FAO Nutritional Studies No. 16. FAO, Rome, Italy.

FAO/WHO (1973). *Wld. Hlth. Org. Tech. Rep. Ser. No. 522.* WHO, Geneva.

Grant, M. N. and McCalla, A. G. (1949). *Can. J. Res.* **27**, 230–240.

Hagberg, A., Persson, G., Ekman, R., Karlsson, K.-E., Tallberg, A. M., Stoy, V., Bertholdsson, N.-O., Mounla, M. and Johansson H. (1979). *In* "Seed protein improvement in cereals and grain legumes" Vol. 2, pp. 303–310. International Atomic Energy Agency, Vienna.

Harlan, J. R. and Starks, K. J. (1980). *In* "Breeding Plants resistant to insects" (F. G. Maxwell and P. R. Jennings, eds) pp. 254–273. John Wiley, New York.

Herrick, H. E., Lawrence, J. M. and Coahran, D. R. (1972). *Anal. Biochem.* **48,** 353–364.
Hoshikawa, K. (1967). *Proc. Crop Sci. Soc.* Japan. **36,** 221–228.
Hough, J. S., Briggs, D. E. and Stevens, R. (1971). Malting and brewing science", pp. 362. Chapman and Hall, London.
Huebner, F. R. and Wall, J. S. (1976). *Cereal Chem.* **52,** 258–269.
J. C. O. (1976). Protein feeds for farm livestock in the U.K. Report no. 2. Joint Consultative Organization for Research and Development in Agriculture and Food, London.
Johnson, V. A., Mattern, P. J., Schmidt, J. W. and Stroike, J. E. (1973). *In* "Proceedings of the 4th International Wheat Genetics Symp." (E. R. Sears and L. M. S. Sears, eds.) pp. 547–556. Columbia University, Missouri.
Johnson, V. A. and Mattern, P. J. (1978). *In* "Nutritional improvement of food and feed proteins" (M. Friedman, ed), pp. 301–316. Plenum Press, New York.
Karlsson, K.-E. (1976). *In* "Barley genetics III" pp. 536–541. Verlag Karl Thiemig, Munich.
Kasarda, D. D., Bernardin, J. E. and Nimmo, C. C. (1976). *In* "Advances in Cereal Science and Technology" (Y. Pomeranz, ed.), *Vol. 1,* pp. 158–236. American Association of Cereal Chemists, St. Paul, Minnesota.
Køie, B. and Doll, H. (1979). *In* "Seed protein improvement in cereals and grain legumes" Vol. 1, pp. 205–214. International Atomic Energy Agency, Vienna.
Laurière, M., Charbonnier, L. and Mossé, J. (1976). *Biochimie* **58,** 1235–1245.
Law, C. N. Young, C. F., Brown, J. W. S., Snape, J. W. and Worland, A. J. (1978). *In* "Seed Protein Improvement by nuclear techniques" pp. 483–502. International Atomic Energy Agency, Vienna.
Lawrence, G. J. and Shepherd, K. W. (1980). *Aust. J. Biol. Sci.* **33,** 221–233.
Lein, L. A. (1972). *Z. Pflzüch* **67,** 243–256.
McCance, R. A. and Widdowson, E. M. (1947). *J. Hyg.* **45,** 59–64.
McClaren, D. S. (1974). *Lancet* **2,** 93–96.
McNab, J. M. (1977). *In* "Protein quality from leguminous crops", pp. 80–86. Commission of the European Communities, Luxembourg.
MAFF (1977). Agricultural statistics, HMSO.
Martensson, P. (1980). *In* "Vicia faba. Feeding value, processing and viruses" (D. A. Bond, ed.), pp. 159–172. Martinus Nijhoff Publishers, The Hague.
Martin-Tanguy, J., Guillaume, J. and Kossa, A. (1977). *In* "Protein quality from leguminous crops", pp. 162–180. Commission of the European Communities, Luxembourg.
Mertz, E. T., Bates, L. S. and Nelson, O. E. (1964). *Science* **145,** 279–280.
Mesdag, J. (1979). *In* "Crop physiology and cereal breeding" (J. H. J. Spiertz and Th. Kramer, eds.), pp. 166–167. Centre for Agricultural Publishing and Documentation, Wageningen.
Munck. L., Karlsson, K.-E. and Hagberg, A. (1969). *J. Swedish Seed Assoc.* **79,** 196–205.
Nelson, O. E., Mertz, E. T. and Bates, L. S. (1965). *Science* **150,** 1469–1470.
Orth, R. A. and Bushuk, W. (1972). *Cereal Chem.* **49,** 268–275.
Paul, C. (1979). *In* "Protein quality from leguminous crops" pp. 370–376. Commission of the European Communities, Luxembourg.
Payne, P. I. and Rhodes, A. P. (1979). *In* "Crop physiology and cereal breeding" (J. H J. Spiertz and Th. Kramer, eds.) pp. 173–174, Centre for Agricultural Publishing and Documentation, Wageningen.
Payne, P. I. and Corfield, K. G. (1979). *Planta* **145,** 83–88.

252   P. I. PAYNE

Payne, P. I., Law, C. N. and Mudd, E. E. (1980). *Theor. Appl. Genet.* **58**, 113–120.
Payne, P. I., Holt, L. M. and Law, C. N. (1981a). *Theor. Appl. Genet.* **60**, 229–236.
Payne, P. I., Corfield, K. G., Holt, L. M. and Blackman, J. A. (1981b). *J. Sci. Fd. Agric.* **32**, 51–60.
Payne, P. R. (1978). *In* "Plant Proteins" (G. Norton, ed.), pp. 247–263, Butterworths, London.
Pearson, A. W., Butler, E. J., Curtis, R. F., Fenwick, G. R. Hobson-Frohock, A. and Land, D. G. (1979a). *J. Sci. Fd. Agric.* **30**, 291–298.
Pearson, A. W., Butler, E. J., Curtis, R. F., Fenwick, G. R., Hobson-Frohock, A. and Land, D. G. (1979b). *J. Sci. Fd Agric.* **30**, 799–804.
Pearson, A. W., Greenwood, N. M., Butler, E. J. and Fenwick, G. R. (1980). *Vet. Rec.* **106**, 560–572.
Pernollet, J. C. (1978). *Phytochemistry* **17**, 1473–1480.
Persson, G. (1975). *In* "Breeding for seed improvement using nuclear techniques", pp. 91–97, International Atomic Energy Agency, Vienna.
Picard, J. (1976). *Ann. Amélior. Plant.* **26**, 101–106.
Picard, J. (1979). *In* "Some current research on *Vicia faba* in Western Europe" pp. 23–37. Commission of the European Communities, Luxembourg.
Pickett, R. A. (1966). *In* "Proc. high lysine corn conference" (E. T. Mertz, ed.), pp. 19–22. Corn Refiners Association, Washington, DC.
Plant Breeding Institute Annual Report 1979, (1980). pp. 30–31. Cambridge.
Pollock, J. R. A., Pool, A. A. and Reynolds, T. (1960). *J. Inst. Brew., London* **66**, 389–394.
Pomeranz, Y. (1965). *J. Sci. Fd. Agric.* **16**, 586–593.
Pushman, F. M. and Bingham, J. (1976). *J. agric. Sci.* (Cambridge) **87**, 281–292.
Putztai, A. (1966). *Biochem. J.* **101**, 379–384.
Rhodes, A. P. and Gill, A. A. (1980). *J. Sci. Fd. Agric.* **31**, 467–473.
Robbelen, G. and Thies, W. (1980). *In* "Brassica crops and wild allies" (S. Tsunoda, K. Minata and C. Gomez-campo, eds), pp. 285–299. Japan Scientific Societies Press, Tokyo.
Russell-Eggitt, P. W. (1977). *Philos. Trans. R. Soc. London, Ser B.* **281**, 93–106.
Sawicki, J. (1952). *Pr. Roln.-lesne* **66**, 1–11.
Soave, C., Righett, P. G., Lorenzoni, C., Gentinetta, E. and Salamini, F. (1976). *Maydica* **21**, 61–75.
Thompson, J. A., Millerd, A. and Schroeder, H. E. (1979). *In* "Seed improvement in cereals and grain legumes" Vol. 1, pp. 231–240. International Atomic Energy Agency, Vienna.
Van der Maesen, L. J. G. (1972). 72–10. Mededelingen Landbouwhogeschool, Wageningen, The Netherlands.
Vasal, S. K., Villegas, E. and Bauer, R. (1979). *In* "Seed protein improvement in cereals and grain legumes" Vol. 2, pp. 127–150. International Atomic Energy Agency, Vienna.
Vogel, K. P., Johnson, V. A. and Mattern, P. J. (1973). *Nebraska Univ. Res. Bull.* **258**, 27 pages.
von Wettstein, D., Jende-strid, B., Ahrenst-Larsen, B. and Sovensen, J. A. (1977). *Carlsberg Res. Commun.* **42**, 341–351.
Wall, J. S. (1979). *In* "Recent advances in the biochemistry of cereals" (D. L. Laidman and R. G. Wyn Jones, eds.), pp. 275–311. Academic Press, London.
Whitehouse, R. N. H. (1973). *In* "The biological efficiency of protein production"

(J. G. W. Jones, ed.), pp. 83–99. Cambridge University Press, Cambridge, U.K.
Wilson, B. J. (1977). *In* "Protein quality from leguminous crops" pp. 183–195, Commission of the European Communities, Luxembourg.
Wolfe, M. J., Cutter, H. C., Zuber, M. S. and Khoo, U. (1972). *Crop Sci.* **12,** 440–442.

CHAPTER 12

Cereal Storage Proteins and Their Effect on Technological Properties

B. J. MIFLIN, J. M. FIELD AND P. R. SHEWRY

Biochemistry Department, Rothamsted Experimental Station, Harpenden, Herts, England

I. INTRODUCTION

Cereal seeds provide the raw material for two of man's oldest technologies—the baking of bread and the fermentation of alcoholic beverages. In this article we wish to examine how these technologies are influenced by the seed proteins. This is made difficult by, on the one hand, the complexity of the technological processes and the difficulty of sub-dividing them into clearly defined constituent parts, and on the other, by the interaction of the different components of the seed. Nevertheless it is generally recognized that the proteins are of positive importance in determining breadmaking quality and have a negative influence on malting and brewing properties.

The proteins of concern in a quantitative sense are the storage proteins and it is these that we shall discuss in this article, although this is not to deny that the activity of certain enzymes (e.g. α-amylase or β-glucanase) may also be important. The definition of the term "storage proteins" is neither clear-cut nor universally agreed; we mean those proteins that are present in protein bodies and which function as a nitrogen store. This latter function is most clearly shown in that they increase disproportionately as the N nutrition of the plant increases. The major storage proteins in the cereals that we are most concerned with in this article (wheat, barley, rye and maize) are alcohol-soluble proteins called prolamins and are deposited in the starchy endosperm; although small amounts of storage globulins are probably present as well. The term prolamin also requires definition—we use it to include those storage proteins whose polypeptides are soluble (but not necessarily completely extractable) in aqueous-alcohol solvents such as 50–60% propan-1-ol or 65–75% ethanol (plus necessary additives such as 2-mercaptoethanol and 1% acetic acid) and whose amino acid composition is rich in glutamine and proline and poor in basic amino acids, particularly lysine. This definition, except for the minor additions in brackets, agrees with that originally proposed by Osborne (1924). Within this fraction we include the specifically named prolamins such as gliadin (of wheat) and hordein (of barley) and most of the glutelin fractions extracted by many workers, particularly the glutenin fraction in wheat (see Miflin and Shewry, 1979a, for a fuller discussion of the reasons for this). The true glutelins of maize and barley are not storage proteins (Wilson *et al.* 1981a,b) and this is probably also true of wheat.

Recent research has increased our knowledge of the components of the prolamins and their synthesis and deposition within the grain. We are now also in a much better position to make comparisons between cereals. In this article we shall try to relate this information to the ability of cereal flours to form a dough and particularly the visco-elastic gluten component of that dough. We will also discuss the possible effects of the storage proteins on the modification of the endosperm during the malting of barley grain.

II. Wheat Gluten and Breadmaking

A. ORIGIN OF GLUTEN

Gluten is generally defined as the visco-elastic mass that remains after thoroughly washing out the starch from a dough and was probably first described by Beccari (1745). Glutens are usually produced from wheat flours and many factors affect the yield (see Bailey, 1944 for detailed discussions); but it is also possible to produce gluten from flours or meals of barley (Shestakova and Vakar, 1979) and rye although the yields are much less (Cunningham and Anderson, 1950). Comparisons of the glutens from the different cereals indicated that whilst wheat gives the most elastic and cohesive gluten, barley gluten is more cohesive and elastic than that of rye (Cunningham et al., 1955). Gluten cannot be washed out of maize although the term "maize gluten" is used to describe the protein left after the starch has been removed by other means. Such gluten does not have the visco-elastic properties referred to above. Correlated with the ability of the different cereal flours to give gluten is their ability to be baked into leavened bread; thus, whilst only wheat flour gives the familiar porous and spongy structure of bread, it is possible to bake bread from rye and barley but not from other cereals.

The amount of gluten and the quality of the bread produced from flour is affected, within a species, by the variety; thus wheat varieties are graded in their usefulness for breadmaking by a number of criteria of which loaf volume is probably the most important. The amount of gluten and the baking performance of a flour is also affected by the protein content of the grain which is in turn affected by the N nutrition of the plant; there is also evidence that the S nutrition, particularly when considered in relation to N supply, can affect the baking properties (Yoshina and McCalla, 1966; Timms et al., 1981). However, for grain produced under broadly similar nutritional conditions with a proper S : N balance the varietal characteristics can be readily identified. These provide a correlative base with which to compare various properties of gluten in an attempt to provide clues as to the determinative properties of the proteins in dough structure. Various properties of gluten will now be considered.

B. SOLUBILITY OF WHEAT SEED PROTEINS

A prerequisite for the detailed study of wheat seed proteins is the need to solubilize them. Since a large proportion (i.e. the gluten) is insoluble in water and dilute salt solutions, solvents which dissociate and/or denature the polypeptide components must be used. A vast number of such solvents have been tried and we will review the results obtained with some of these. Ideally 100% of the seed protein should be brought into solution to ensure that the gluten components are fully represented; this must also be achieved without hydrolysing peptide bonds or uncontrollably modifying amino acid side chains. Evaluation of results obtained is often made difficult because it is not easy to define "in solution". In most instances this means that a visually clear solution is obtained after centrifugation, however, the forces used in centrifugation vary considerably causing varying amounts of colloidal protein, suspended in the solvent, to sediment as a gel at the bottom of the tube.

A further problem is the starting material used for solubility studies (e.g. whole meal or flour and fineness of milling) and the various pretreatments to which it has been subjected. Several studies show that defatting the starting material can affect the subsequent extraction of the proteins. Thus Wren and Nutt (1967) and Charbonnier (1973) reported that defatting affects the recovery of gliadin whereas M. Byers (unpublished results) (see also Miflin *et al.*, 1980a) could find no effect on the gliadin fraction but found a lower recovery of salt-soluble proteins. Comparison is also confounded by whether or not a series of sequential extractions of proteins is carried out and which particular series is used. Thus when investigators are extracting "glutenin" the results are critically dependent on the procedures used to produce it; similarly when alcoholic solvents are used to extract gliadin directly the results will differ (due to the presence of soluble N from the water and salt-soluble fraction) to those from sequential extractions. Similarly, washed gluten preparations can be made in a number of ways (see Bailey, 1944) and these may differ between themselves and also from flour pre-extracted with water and salt solutions. However, even with these difficulties it is possible to make some comparisons between the various techniques which are reviewed below.

1. Alcohol-based Solvents

Sequential extractions have formed the basis for most work. In general about 30% of the N is extracted with water and salt solutions (often 0·5 M NaCl) leaving some 70% in the prolamin, glutelin and residual fractions. The subdivision of these groups from this material has been a source of difficulty for workers since before the turn of the century. Much of the early work has been reviewed by Bailey (1944) and several of these studies were marked by careful accounting of the N in the fractions and attempts to characterize both the extracted and residual fractions (e.g. see Bailey and Blish, 1915; Blish and Sandstedt, 1929). In many cases the workers demonstrated the superiority of alcohols other than ethanol (e.g. propan-1-ol, propan-2-ol) and the advantages to be gained by extracting at temperatures above ambient. Subsequent to Bailey's book much of the work has been based on the use of 70% ethanol either at room temperature or in the cold. Recently Byers has reinvestigated the solubilization of wheat proteins using a range of conditions (M. Byers, unpublished results; Miflin et al., 1980a). She has found that the amount of N extracted in the prolamin fraction can vary from 20%, using 70% ethanol at 4°C, to around 60% of the original seed N, using 50% propanol-1-ol, 1% acetic acid, 1% 2-mercaptoethanol at 60°C; there is a corresponding decrease in the % of N in the "glutelin" plus residue fraction from 50 to 10%. The characteristics of the extracted proteins have been studied by sodium dodecylsulphate polyacrylamide gel electrophoresis (SDS-PAGE) and by amino acid analysis. There is little difference between the procedures in either the qualitative SDS-PAGE patterns or in the quantitative amino acid compositions. The only major difference is the absence of polypeptides of a molecular weight of about 100 kilodaltons (kd) (also termed HMW prolamins) in the extract made with 70% ethanol at 4°C but there is also some variation in the relative proportions of the polypeptides extracted by the other procedures. The results with 70% ethanol at 4°C are not surprising since this is close to the point at which gliadins precipitate out from alcoholic solutions (Dill and Alsberg 1925) and the conclusion to be drawn from the remainder is that the various extraction procedures are all bringing into solution the same class of polypeptides but with varying degrees of efficiency. Thus we consider that these results show that the prolamins represent around 60% of the total seed N and the true glutelin plus residue not more than 15%. Because these results show that 70% ethanol does not extract more than about half of the prolamin fraction it is evident that any remaining glutelin fraction will be heavily contaminated by unextracted prolamins. Such contaminated glutelin fractions will therefore have a predominantly prolamin-like amino acid composition and contain polypeptides and peptide fragments in common with gliadin (e.g. see data quoted in Miflin and Shewry, 1979a; Kasarda et al., 1976a; Ewart, 1966; Bietz and Rothfus, 1970).

2. Acidic Solvents

Several workers have shown that acetic and lactic acids, usually at 0·01–0·05 N, extract a considerable amount of wheat seed proteins (e.g. 60–63%, Danno et al., 1974). These solvents have been widely applied to gluten and a number of workers have used the method of Jones et al. (1959) to produce a fraction termed "glutenin". In this the gluten is dispersed in 0·01 N acetic acid, made 70% with respect to ethanol and then brought to pH 6·5 by the addition of NaOH. The precipitate is considered to be "glutenin". A somewhat similar procedure has been usen by Chen and Bushuk (1970) and Orth and Bushuk (1973) in which, after sequential extraction of protein soluble in water and salt and 70% ethanol at 4°C, the residue is extracted with 0·05 M acetic acid. This acetic-acid-soluble fraction is also termed "glutenin" and represents about 10–30% of the total. In both cases a considerable amount of protein remains insoluble in acetic acid (up to 35% of the total) and there is evidence that the insoluble fraction is of great importance in breadmaking (see below). The "glutenin" has a similar amino acid composition to "gliadin"—i.e. a predominance of glutamine and proline residues with few basic amino acids but also has increased amounts of glycine. When the reduced fraction is separated by SDS-PAGE it is found to contain bands of about 100 kd and 44 kd, the latter co-migrating with "gliadin" bands.

3. Soaps and Detergents

Kobrehel and Bushuk (1977) have shown that when freeze-dried glutenin prepared by the method of Chen and Bushuk (1970) is mixed with various amounts of fatty acids or their sodium salts, water added and the mixture stirred overnight, virtually all of the protein is brought into solution in a "visibly clear liquid" that remains after centrifugation at 17 000 xg for 20 mins. The most effective soaps were sodium stearate and sodium palmitate. Sedimentation velocity ultracentrifugation of the fractions indicated a component polypeptide molecular weight of 40–50 kd (Kobrehel and Bushuk, 1978). From this the authors concluded that the soaps were completely dissociating glutenin into its component parts without disrupting S-S bonds and, because they did not see large molecules, that the glutenin was not an aggregate stabilized by intermolecular disulphide bonds. However further work by Hamauzu *et al.* (1979) has shown that the extract does contain high molecular weight aggregates which after reduction yield polypeptides of 100 kd on SDS-PAGE. They postulate that in the previous experiments the high molecular weight aggregates rapidly sedimented to the bottom of the cell and, therefore, were not seen in the Schlieren pattern. Kobrehel (1980) has also used soaps to dissolve the protein from wheat flour and was again able to achieve solubilization approaching 100%. SDS-PAGE of the reduced protein showed the presence of both the 100 kd series of polypeptides and lower molecular weight material.

Danno *et al.* (1974) used SDS solutions to dissolve flour proteins and with their optimum conditions could extract about 80% of the total protein; some of the extract consisted of high molecular weight aggregates. When 2-mercaptoethanol (1 mM) was included in the solvent virtually 100% of the protein was extracted.

4. Urea Based Solvents

It has been claimed that 2M urea is sufficient to extract all the gluten proteins (Lee, 1968) but this has not been substantiated (Huebner and Rothfus, 1971; Lee and MacRitchie, 1971; Redman, 1973). Urea in conjunction with other denaturing agents has been widely used. Meredith and Wren (1966) reported that AUC (0·1M acetic acid, 3M urea, 0·01M cetyltrimethylammonium bromide) would solubilize about 95% of the gluten protein. Other authors have obtained less successful extraction of wheat flour, Huebner and Wall (1976) extracted between 77 and 91% depending on variety, while Danno et al. (1974) and Payne and Corfield (1979) both extracted 72%. Fractionation of the extracted protein by gel chromatography has shown the presence of high molecular weight aggregates (Huebner and Wall, 1976; Payne and Corfield, 1979).

Recently two groups have modified the AUC solvent and both have reported almost complete solubilization of wheat gluten proteins. Booth et al. (1980) have used 4·5M urea, 40 mM cetyltrimethylammonium bromide, 50 mM sodium citrate pH 5·5 (CUC) and Field and Miflin (1980) and Miflin et al. (1980a) 6M urea, 55 mM cetyltrimethylammonium bromide, 0·01M acetic acid at 40°C. In both cases crude gluten preparations were used as the starting material. The extracted protein consisted of both monomeric polypeptides and high molecular weight aggregates.

5. Miscellaneous

Stanley et al. (1968) claimed that all the gluten proteins could be dissociated in phenol; acetic acid; water and had molecular weights of between 14 000–52 000 in the ultracentrifuge. However, using their techniques these workers determined the molecular weight of bovine serum albumin to be 17 000 whereas the accepted value, authenticated by amino acid sequence studies, is in the order of 69 000. Thus it is possible that this solvent breaks peptide bonds and its use is not recommended. Other workers could not repeat these results (Huebner and Rothfus, 1971).

Several workers have used solutions of $HgCl_2$ as a solvent. Mecham et al. (1972) have shown that it can solubilize those gelatinous proteins ("gel proteins") left after exhaustive extraction of flour with dilute acetic acid. In their studies concentrations of less than 0·1 mM were required and the effect could not be duplicated by N-ethylmaleimide (NEM) or phenylmercuric nitrate. Danno et al. (1974) extended these observations to show that the addition of 1 mM $HgCl_2$ to 0·5% SDS increased the amount of protein extracted from 75 to 98%; again other reagents reacting with sulphydryl groups such as NEM, or p-chloromercuribenzoate were not effective.

Subsequently Danno *et al.* (1975) explained the effect of HgCl$_2$ as being due to its ability to cleave directly disulphide bonds.

C. SEPARATION AND CHARACTERIZATION OF COMPONENTS

The solubilized proteins have been separated by two types of techniques, electrophoresis on gels and column chromatography. In both cases the separations may be done before or after reduction of disulphide bonds.

1. Electrophoretic Separations

Perhaps the most commonly used technique is starch gel electrophoresis (SGE) at low pH (3 to 3·2) as described by Woychik *et al.* (1961). This separates proteins according to their net charge, which at this pH is mainly determined by their content of basic amino acids, although there is also a molecular sieving effect. On the basis of their mobility in this system Woychik *et al.* (1961) classified the gliadins into α, β, γ and ω groups. Similar separations have been obtained using polyacrylamide gel elec-trophoresis at low pH and an example is shown in Fig. 1. However, this system suffers from certain disadvantages. Firstly, like all one-dimensional systems, several proteins may be present in one band and, secondly, not all of the prolamins enter the gel at low pH in the absence of reduction. A more complete analysis is given by SDS-PAGE under reducing conditions and an example of such a separation is given in Fig. 1(B) and 10. In this case the proteins have been alkylated after reduction to stabilize the sulphydryl groups and prevent disulphide bond formation. Under these conditions the α, β, and γ gliadins migrate with apparent molecular weights of around 36 kd and 44 kd. The ω-gliadins are present on the gels as a series of bands with molecular weights between 50 kd and 70 kd. Also present are a series of bands of around 100 kd. The amounts of these bands depends upon the extraction conditions and they are particularly prevalent in so-called "glutenin" preparations. However, their amino acid composition is such that they are clearly prolamins (see Table I and also Miflin and Shewry, 1979a; Miflin *et al.*, 1980a) and we prefer to term them high molecular weight (HMW) prolamins. The final one-dimensional system that has been used is isoelectric focusing (IEF) (Miflin *et al.*, 1980a). This does not readily separate out the different groups referred to above and is really only of great value in two-dimensional systems.

All of the above systems are limited in resolving power and various two-dimensional systems have been developed. The first to be used to any extent was that of Wrigley and Shepherd (1973) in which IEF is used in the first dimension and SGE in the second (see Fig. 2). The more generally

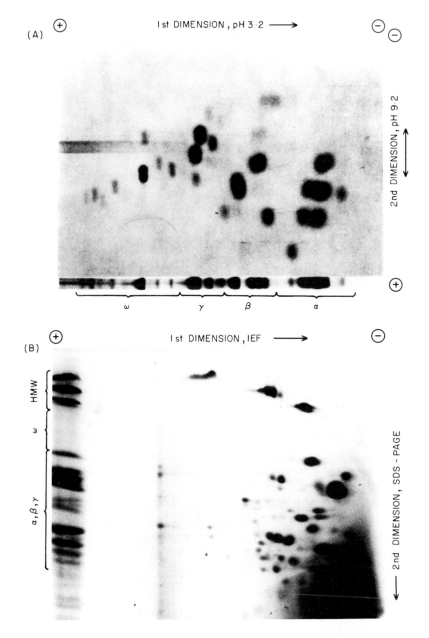

FIG. 1. Two-dimensional analysis of prolamin fractions from wheat. (A) Gliadin fraction from cv. Cheyenne separated by the two pH system of Mecham *et al.* (1978), pH 3·2 in the first dimension and pH 9·2 in the second. Reproduced from Kasarda (1980) with permission of D. D. Kasarda and D. K. Mecham. (B) Total prolamins of cv. Brigand separated by isoelectric focusing in the first dimension and SDS-PAGE in the second.

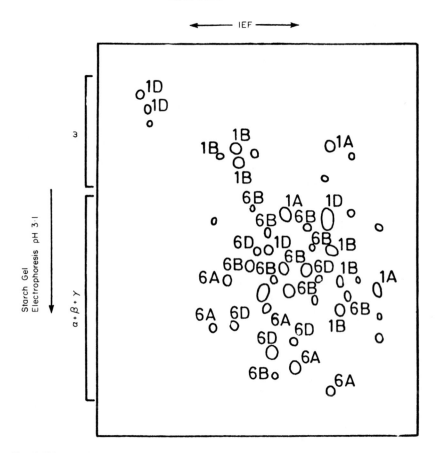

FIG. 2. Diagram of 2-D separation of gliadins from wheat cv. Chinese Spring labelled to show chromosomal control of individual components. Redrawn from Wrigley and Shepherd (1973).

adopted variant of this system is to use SDS-PAGE in the second dimension (e.g. Miflin and Shewry, 1979a and Fig. 1(B)). A third two-dimensional system which has proved successful uses separation at two different pH values (pH 3·2 and pH 9·2) in a single polyacrylamide slab gel (Mecham *et al.*, 1978, Fig. 1(A)).

2. Column Separations

Various column support media have been used to separate solubilized but unreduced wheat proteins into different size classes. All separations show the presence of high molecular weight aggregates—usually termed glutenin—with estimated molecular weights of several millions. Electrophoretic analysis of such material shows it to be rich in the HMW prolamins of ca. 100 kd. The intermediate fractions (termed glutelin-II or included glutelin) have a similar band pattern but with less of the 100 kd polypeptides and greater proportions of lower molecular weight bands. These latter bands are also present in the fraction termed "high molecular weight gliadin" which is that excluded in the void volume when 70% ethanol extracts are chromatographed on Sephadex-G100 (Wright *et al.*, 1964; Beckwith *et al.* 1966). Such a fraction is similar to the low molecular weight glutenin fraction of Neilsen *et al.* (1968). Bietz and Wall (1973, 1980) have shown that their "ethanol-soluble reduced glutenin" also contains many of the same polypeptides and has similar N-terminal amino acid sequences to "high molecular weight gliadin".

3. Amino Acid composition

Individual components of the various fractions have been prepared after reduction (and usually alkylation) by a variety of column chromatography procedures including gel filtration, ion exchange chromatography, preparative IEF and hydrophobic interaction chromatography. This has allowed the purification of many individual fractions and the characteristics of individual bands or groups of bands to be determined. Some of the procedures are described in the references to Table I. A general point should be made that the purity of the fractions isolated must be checked by at least two electrophoretic procedures and even then microheterogeneity might still be present. This is unlikely to be a problem for amino acid analysis but should be borne in mind for sequence studies (Shewry *et al.*, 1980a; Kasarda, 1980). Table I gives detailed amino acid compositions of various purified wheat polypeptides (as well as those from other cereals). From these examples and other data in the literature we can generalize that there are three characteristic groups, all of which have in common a large amount of glutamine (> 20 mol %) and proline (> 10 mol %) and only small amounts of lysine and other basic amino acids. The first group—the HMW prolamins—all contain much glycine, generally greater than 10% with values of up to 20·4% being recorded (Khan and Bushuk, 1979). The second clear-cut group is the ω-gliadins which contain virtually no S-amino acids but have exceptionally large amounts of glutamine (up to 50%) and proline; these latter two residues usually accounting for more than 70% of the total. They also have relatively more phenylalanine (ca. 8 mol %) than other prolamins. The final group of α-, β- and γ-gliadins are more variable in composition but are generally richer in S-amino acids (ca. 3%) and relatively poorer in glutamine and proline. It is probable that many solvents also extract a group of strongly aggregating proteins which are found in "glutelin" but which do not have prolamin-like amino acid compositions (i.e. 20% Glu, 8% Pro, 3·6% Lys) and have carbohydrate associated with them (Khan and Bushuk, 1979; Bushuk *et al.*, 1980). The type of carbohydrates (arabinose, galactose, glucose and xylose) and the fact that they are covalently linked to hydroxyproline in the protein is typical of cell wall glycoproteins (Lamport, 1980) and this is their probable identity. The exact amounts of such proteins in "glutenin" fractions are not known but they are unlikely to be of major importance.

4. Genetics

Some of the first studies on the genetics of wheat storage proteins were done by Shepherd, Wrigley and colleagues who examined the distribution of α, β, γ and ω gliadins in chromosome substitution and addition lines (Boyd *et al.*, 1969; Shepherd, 1968; Wrigley and Shepherd, 1973). These studies are complicated because bread wheat is an allohexaploid with three genomes (A, B and D) each with seven pairs of chromosomes. These genomes are probably derived from related species of *Triticum* and *Aegilops* and in many cases the homoeologous (i.e. homologous but non-pairing) chromosomes carry genes coding for related proteins. A summary of Wrigley and Shepherd's (1973) results is given in Fig. 2 which shows that the genes for these gliadins are on homoeologous chromosomes of groups 1 and 6 with each of the three genomes contributing to the total. Kasarda *et al.* (1976b) confirmed many of these assignments and also, by the use of ditelocentrics, located the genes for the A group of α gliadins on the short (or α) arm of chromosome 6A. Mecham *et al.* (1978) have used 2–D techniques to analyse F_2 seeds of intervarietal crosses and have shown that the amounts of proteins are controlled by gene dosage in the triploid endosperm. Linkage analysis provided evidence for codominant alleles and closely linked genes coding for gliadin proteins that were located on chromosomes of groups 1 and 6. Recently Lawrence and Shepherd (1981) have shown that the genes are on the short arms of the group 1 chromosomes. Similar studies have been carried out by Sosinov and colleagues (see Sosinov and Poperelya, 1980) and by Doekes (1973) which also show that blocks of proteins are inherited as if they were controlled by very tightly linked genes. Some workers have suggested that two genes may be involved in controlling one band (Baker and Bushuk, 1978) but this is almost certainly due to the presence of two or more polypeptides in one position and the improved resolution of a 2-D system allowed Mecham *et al.* (1978) to demonstrate this.

The genetics of the HMW prolamins ("glutelins") has also been studied by many workers. Analysis of SDS-PAGE patterns of chromosome substitution lines and ditelocentric lines suggested that certain of these sub-units are coded for by genes on the long arms of chromosomes 1D and 1B (Orth and Bushuk, 1974: Bietz *et al.*, 1975; Joppa *et al.*, 1975). These results have recently been confirmed by Brown *et al.* (1979) and Payne *et al.* (1980). Payne *et al.* (1980) have also found that two of the HMW prolamin bands present in some, but not all, varieties are coded for by chromosome 1A. Analysis of F_2 seed from crosses between several cultivars has shown no recombination between HMW prolamin genes on the same chromosomes (Payne *et al.*, 1981).

To summarize, we can postulate closely linked groups of genes (complex loci) located on chromosomes 1 and 6 which control all of the storage proteins. The exact number of complex loci on each chromosome is not known since few studies have been done to record recombinational events—however the studies of Mecham *et al.* (1978) would suggest that there is more than one locus on the short arm of chromosome 1B.

D. CORRELATION OF GLUTEN PROPERTIES WITH BREADMAKING QUALITY

The desire to understand the basis of breadmaking quality has led many workers to draw correlations between various parameters and baking quality. Where quality has been compared with extractable groups of proteins it has been shown that generally there is little positive correlation with the amounts of albumins, globulins and gliadins. In contrast many workers have shown strong positive correlations between the amount of residual protein after extraction and various quality factors. Thus the amounts or proportions of protein insoluble in 3M urea (Pomeranz, 1965), in dilute acetic acid (Orth and Bushuk, 1972; Mecham *et al.*, 1972; Orth *et al.*, 1976), in 2-chloroethanol (Jeanjean and Feillet, 1979) or in AUC (Huebner and Wall, 1976) have all been shown to be correlated with breadmaking quality. In a detailed multifactorial analysis Orth *et al.* (1976) showed that the amount of acetic-acid-insoluble residue protein gave the best estimate of protein quality as determined by various rheological properties such as extensigraph resistance, lack of dough breakdown, farinograph development time, Pelshenke time, and extensigraph extensibility. The relationship of "glutenin" to quality is very dependent on the efficiency of extraction and, where only poor extraction is achieved, a negative correlation may be obtained (Orth and Bushuk, 1972). However use of a better solvent (e.g. AUC) gives a positive correlation, particularly when the amount of "glutenin-I" (or high molecular weight aggregated glutenin) is considered (Huebner and Wall, 1976). Whilst the amount of residue protein may have a high predictive value of use in plant breeding it does not aid fundamental studies because it is not possible to determine the properties of the protein present in the flour residue. For this latter purpose it is much more useful if the properties of soluble extracts can be linked to quality and thus the approach of Huebner and Wall (1976), correlating the amount of aggregated "glutenin" with dough strength and mixing times, was of great importance. However, even in these studies, it was obvious that a significant proportion of the total protein (up to 20% for the best quality varieties) was being left behind and this protein was of particular relevance to dough properties. A different approach has been taken by Arakawa and colleagues (Arakawa *et al.*, 1976, 1977) in which the ability of gluten proteins, extracted in acetic acid, to form turbid solutions at pH 5·6 in the presence of 2M NaCl was measured and positively correlated with dough quality. This was found to be due to the "glutelin" fraction, particularly that rich in HMW prolamin bands. The ability to form such precipitates was not dependent on SH groups as it occurred with alkylated proteins. Again, however, these properties were not determined for the total gluten proteins since much protein remained insoluble.

We have made an extract of almost all the flour proteins using our modified AUC solvent (see p. 261) and chromatographed this over a column of controlled-pore-glass. Two fractions are obtained (Fig. 3, Field and Miflin, 1980; Miflin *et al.*, 1980a); the first fraction in the excluded volume of the column. Although the column has a nominal exclusion size of $1·2 \times 10^6$ for molecules in an extended conformation, it is not possible to

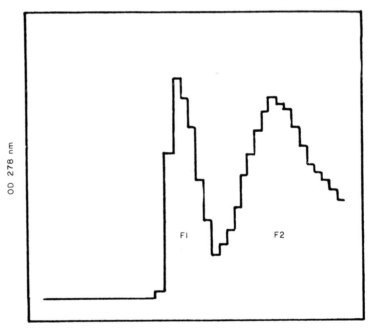

FIG. 3. Elution profile of wheat (cv. Copain) gluten proteins extracted and chromatographed in a modified AUC solvent on controlled pore glass (modified from Miflin *et al.*, 1980a).

state that the aggregates are larger than this because of uncertainties in the exact amount of detergent bound and the effect of the solvent on chromatographic properties. We have, however, established that the aggregate is stabilized by disulphide bonds as it is completely abolished by reduction of the protein with mercaptoethanol. When the amount of aggregate obtained from different varieties is compared to the baking quality of the varieties a strong positive correlation is found. As with previous studies the aggregated fraction contains a predominance of HMW bands (Miflin et al., 1980a). However, since we have extracted almost all of the grain proteins, we can investigate whether the amount of aggregate is related to the amount of HMW prolamin polypeptides. To answer this question we have used three approaches—scanning trichloroacetic acid (TCA) fixed gels, scanning stained gels and column chromatography. Although these three methods do not give the same values for the proportion of HMW polypeptides (they range from 7–19 as a % of the total gluten protein) in each case there was no difference in the amounts between good and bad breadmaking varieties.

Genetical approaches to this problem have also been made (see Konzak, 1977 for review) particularly to see if certain protein bands on SDS-PAGE correlate with quality. Orth and Bushuk (1974) found no evidence for any association between specific HMW prolamin bands ("glutelins") and quality. Similarly Wrigley and colleagues (Wrigley, 1980; Wrigley, 1982) using computer-based inverse pattern analysis of HMW bands from 80 cultivars did not find a correlation between these polypeptides and quality. In contrast Payne et al. (1980, 1981) and Burnouf and Bouriquet (1980) showed that the presence of certain protein bands, particularly the subunit-1 of Payne et al. (1980, 1981) coded for by chromosome 1A, is associated with quality and could be followed through a large number of crosses. Obviously the problem is complex and much may depend upon the background of the genotypes used. Associations between certain of the lower molecular weight prolamins ("gliadins") and quality have also been found. Wrigley (1980), using the same method as above, found associations between gliadins 24, 26, 29 and 31 (the α E group) and hardness and dough strength; the latter was also correlated with the presence of gliadins 2, 4, 14 and 19. Similar associations between gliadins and quality have been made by Autran and colleagues in France (e.g. Damidaux et al., 1978).

Unfortunately it is not yet clear exactly which gliadins are involved. There is also the problem that this type of analysis does not establish cause and effect—it may only indicate linkage between "true quality genes" and the structural genes for certain proteins in much the same way as there is linkage between mildew and rust resistance genes and barley prolamins (Shewry et al., 1979). In this latter case there is no suggestion that the prolamins have anything at all to do with pathogen resistance.

When isolated gluten is added to flour there is an improvement in quality indicating that the total amount of gluten is important (e.g. MacRitchie, 1973). Further, the effect on the dough reflects the characteristics of the flour from which the gluten is obtained (Finney, 1943; Butaki and Dronzek, 1979). Finney and coworkers (summarized in Finney, 1971) have solubilized gluten proteins in lactic acid and fractionated out the "glutenins" and gliadins. They have also isolated the water soluble fraction, lipids and starch. They found that glutenins were responsible for the mixing properties of dough whereas the gliadins were responsible for loaf volume potential. Although the water-soluble fraction played no major part it was proposed that the polar lipids bind to the gliadin by hydrophilic bonds and to the glutenins by hydrophobic bonds and in this way contribute to the gas-retaining complexes in bread. MacRitchie (1973) also tested the effect of residual gluten proteins (left after extraction with 2M urea) and showed that these had a strong positive effect on the strength of the dough.

The results of the fractionation experiments are thus generally consistent with the correlations drawn from extraction experiments in which increases in the amount of residue protein (and/or high molecular weight aggregates) are related to increases in dough strength.

E. THEORIES OF VISCO-ELASTICITY

When a dough is subjected to stress it is deformed. On relaxation of the stress part of the deformation is recovered but part is permanent. This property of "creep" is a standard characteristic of a visco-elastic material. Recent measurements (Hibberd and Parker, 1979) of creep and creep recovery confirm the non-linear, visco-elastic character of dough but do not suggest that there is a yield value (i.e. a stress below which no permanent deformation or strain occurs). Further, within the limits of the experiments, there were no conditions under which the flow became purely viscous. These results therefore suggest that dough normally has both viscous and elastic properties and that its properties are physically complex. Many theories have been put forward to explain these properties. Broadly these fall into two groups—those which emphasize the importance of inter-molecular disulphide bonds and those emphasizing non-covalent interactions between proteins.

1. Role of Inter-molecular Disulphide Bonds

Evidence for the importance of disulphide bonds has accumulated from a number of observations: (a) agents which oxidise or reduce disulphide bonds also affect dough behaviour, (b) cleavage of S-S bonds lowers the apparent molecular weight (Meredith and Wren, 1966) of "glutenin" preparations, (c) the addition of 2-mercaptoethanol markedly increases the solubility of gluten proteins (Danno et al., 1974; Miflin et al., 1980a). One of the first hypotheses put forward to explain the structure and rheology of dough was that of Ewart (1968) which has been modified subsequently (Ewart 1972, 1977a). This theory, the linear glutenin hypothesis, proposes that "glutenin" contains molecules joined by S-S bonds into linear polymers (concatenations) with only a small amount of branching. Additionally, secondary forces are proposed to build up sequentially to produce tension particularly during "work hardening" of dough. Viscous flow is thought to depend primarily on molecular slip but also partly on mechanical scission and S-S interchange. Ewart (1968, 1972, 1977a, 1978) has pointed out that his hypothesis fits many of the observed facts of gluten and dough behaviour, particularly the presence of rheologically active and inactive SS and SH groups. Many studies have tried to define these different categories of sulphydryls and relate them to dough quality (e.g. see Kackowski and Mieleszko, 1980). At least two types of S-S bond can be recognized in gluten—one much more reactive than the other (e.g. see Yoshida et al., 1980). The more reactive bonds are considered to be on the outside of the protein molecules and thus more accessible and more likely to be involved in inter-molecular links. All the evidence is that "glutenin" contains this type of readily reducible S-S bond (Wall, 1979; Ewart, 1979; Yoshida et al., 1980) inferring that it is inter-molecular. The proposed role of dough improvers (e.g. potassium bromate, ascorbic acid) is to shift the level of these S-S bonds to the optimum.

Other workers, subsequent to Ewart, have put forward essentially similar hypotheses. Bloksma (1975) emphasizes the importance of S-S bonds in forming an extensive range of linked (branched) polymers and differs from Ewart in stressing the predominant importance of disulphide interchange in viscous flow. The models of Wall (1979) and Khan and Bushuk (1978, 1979) agree with that of Ewart in stressing the importance of both covalent and non-covalent bonds in gluten structure.

2. Other Hypotheses

An alternative hypothesis for gluten structure is one which suggests that S-S bonds are present only as internal links within the protein and that inter-molecular S-S bonds are of little or no importance. This argument has been put forward most coherently by Kasarda and Bernardin (Kasarda et al., 1967; Kasarda et al., 1976a; Bernardin, 1978). The theory arises out of their observations on the properties of A-gliadin which reversibly aggregates into fibrils under conditions of low ionic strength and low pH. They suggest that, although the inter-molecular S-S theory may be correct, most of the results could be explained by proposing that intra-molecular S-S bonds are required to maintain the 3-D shape of the molecules in a form able to undergo aggregation. Without maintenance of this specific conformation the secondary forces causing aggregation could not occur. In support of this theory Bernardin (1978) has shown that dough improvers such as bromate, iodate and cysteine are equally as effective at pH 5 as KCl in causing A gliadin to aggregate.

Other workers have also suggested that intermolecular S-S bonds are not important. Thus Stanley *et al.* (1968) have suggested that, because virtually all of the proteins in gluten can be solubilized in phenol : acetic acid : water and have an apparent molecular weight of around 50 kd in the ultracentrifuge, only secondary forces are important. Similarly Kobrehel and Bushuk (1977, 1978) put forward a similar idea based on analogous results obtained using soaps to dissolve the protein. In both cases however subsequent work has shown that high molecular weight protein complexes were present in such solutions (see p. 260). Jones and Carnegie (1971) also proposed a hypothesis which did not involve the polymerization of proteins by disulphide linkages. Recently Batey (1980) suggested on the basis of the effect of alkali treatment on gluten, that high molecular weight aggregates remained in the dough even though all of the cystine residues had apparently been destroyed.

3. Summary

As in other fields of wheat protein research, considerable confusion exists. This is due to the complexity of the system and the paucity of facts. In an attempt to clarify our own thinking we will try to present some assessment of the present position.

(a) The basis of elasticity. Rubber and rubber-like materials are distinguished from other substances by a remarkable combination of two characteristics viz. they are capable of sustaining large deformations without rupture and, secondly, possess the capacity to recover spontaneously their initial dimensions upon removal of the stress. Rubber-like bodies resemble liquids in respect to their deformability without rupture; they resemble solids in their capacity to recover. This combination of properties—long-range elasticity—is in no way confined to a restricted group of hydrocarbon polymers. Indeed, virtually any long chain polymer may exhibit typical rubber-like behaviour under appropriate physical conditions.

The unique structural feature common to all rubber-like substances is the presence of long polymer chains which in the unstrained state occur in randomly coiled arrangements, but which are able to rearrange to other configurations, and in particular to more highly extended ones. During elongation, the polymer chains uncoil and tend to become aligned— although incompletely—with the axis of elongation. The restoring force originates in the tendency of the polymer chains to return to their initial configurations. Although long chains are necessary for rubber-like behaviour, they are not the sole requirement. The system must also possess sufficient internal mobility to allow the required rearrangements of chain configuration during deformation and recovery. Additionally a permanence of structure is also required, otherwise plastic flow rather than elastic deformation would be observed. This is usually conferred by the presence of occasional cross-linkages which join the chains into a space network of "infinite" extent. Separate linear molecules could dissipate any orientation induced by deformation by spontaneous rearrangements of their configurations; chains bound at either end to the network cannot do so except through restoration of the initial dimensions of the sample. In other words, the retention of the ability to recover generally depends on the existence of a permanent network structure. However, linear polymers may also exhibit elastic behaviour with good recovery if the molecular weight is very high (10^6 or more), owing to the slow rate of relaxation of very long chains. It is therefore feasible that high molecular weight subunits, linked with each other end-to-end to produce long linear chains, could be elastic. Cross-links are normally considered to be definite covalent bonds (e.g. disulphides). However, non-covalent interactions (hydrophobic, hydrogen bonding) could also provide suitable, although weaker, restraints. Entanglements can also impose configurational restraints and enhance elasticity although such entanglements are only likely to be permanent in the presence of defined cross-linkages. Linear polymers of large molecular weight are normally highly entangled with their neighbours but can with time extricate themselves and so dissipate the restraints imposed by the entanglements (i.e. become permanently deformed).

(b) Nature of cross-linkages. A key question is "are all the secondary non-covalent forces destroyed in solvents such as AUC, CUC or sodium sterate?" If the answer is yes then it is difficult to see how molecules linked other than by S-S bonds could be involved in the high molecular weight aggregates. Unfortunately it is not possible to be absolutely certain. Meredith and Wren's (1966) original studies showed that no difference was obtained by increasing the urea concentration to 8M; had 3M urea been insufficient to fully dissociate the proteins then a doubling of the concentration should have caused a change. Beckwith and Wall (1966) showed that 6M urea was sufficient to ensure that gliadins and glutenins were sufficiently unfolded to expose their S-S bonds to reduction. Other suggestive evidence is that the gluten proteins, like most globular proteins, have the majority of their hydrophobic residues on the inside of the molecule (Batey, 1980). Thus the effect of detergents (e.g. SDS, CTAB) will be to unfold the molecules, at least within the constraints of S-S bonds. This effect would be expected to be of such magnitude as to disrupt the surface structure thus causing changes in potential aggregation sites. Also such sites would not remain hydrophobic since these would bind the detergent. The solvent that we have used with strong concentrations of both urea and detergent and the temperature (45°C) of extraction is almost certain to have disrupted secondary non-covalent bonds; the temperature alone would be sufficient to break down A gliadin fibrils which destablilize above 30°C (Kasarda, 1980).

We therefore conclude that it is unlikely that the high molecular weight aggregates of glutenin are due to non-covalent bonding between molecules stabilized by intra-chain S-S bonds. The theoretical arguments do not favour such a proposal and there is no convincing evidence for the hypothesis. Although various people have claimed to have such evidence subsequent experiment has generally failed to confirm that this is the case. In those cases where non-disulphide linked aggregates have been formed the bonding appears to be relatively weak. Thus we believe that the breakdown in the high molecular weight aggregates in the presence of 2-mercaptoethanol is due to the reduction of intermolecular S-S bonds.

(c) Other relevant properties. Elasticity is enhanced by cross-linking of flexible molecules. The HMW prolamins have an amino acid composition ideally suited to providing such properties. Firstly, they have 12–20% glycine. This amino acid confers flexibility on protein chains since its spectrum of permitted peptide bond angles and thus the conformational space available to the residue is far greater than any other amino acid (see Schulz and Schirmer, 1979). Secondly, the proteins have about six cysteines per molecule which give them the potential for cross-linking. Whether or not such cysteines can participate in cross links depends upon their position and orientation as determined by the amino acid sequence. We have therefore tried to answer this problem by isolating individual HMW-prolamin components, subjecting them to the cysteine cleavage procedure of Jacobson *et al.* (1973) and re-running them on SDS-PAGE. The results show that the largest HMW prolamin subunit has a slightly increased mobility after cleavage of the polypeptide chain at the cysteine residues. This implies that there is only a small change in molecular weight after cleavage and thus the cysteines are at the end of the molecule. This has been partly confirmed by N-terminal sequence analysis of this band in cv. Capain (in collaboration with D. D. Kasarda) which showed the

presence of cysteine residues at positions 10 and 25 (Table III). Thus 2 out of a total of 6 cysteines are in the first 25 amino acids. Further success in predicting the structure and function of these component monomers of the high molecular weight aggregates of gluten will depend on further analysis of their sequence.

(d) Generation of S-S bonds. Besides the structure of the monomers and their potential to form S-S bonds there is also a requirement that such bonds should actually be synthesized. This formation *in vivo* could be spontaneous but probably is catalysed enzymically since the apparent *in vivo* rate is greatly in excess of that obtained *in vitro*. Consideration of the general problem of disulphide bond formation *in vivo* (e.g. see reviews of Ziegler and Poulsen, 1977; Freedman and Hillson, 1980) suggests that very little is known of the process. Poulsen and Ziegler (1977) have suggested cysteamine as a direct oxidant of protein thiols. Cystamine is considered to be generated by a mixed function oxidase using NADPH, O_2 and cystamine as substrates. The final formation of protein disulphide by means of thiol-disulphide interchange with cystamine is thought to be catalysed by the enzyme protein disulphide isomerase. This enzyme has been found in many animal tissues (see Freedman and Hillson, 1980) where it appears to be located in the microsomes. Recently Grynberg *et al.* (1978) have found the enzyme in wheat embryos and Roden (1982) has shown that it is also present in the endosperm during development. When the distribution of the enzyme within the tissue was determined using the methods of Miflin *et al.* (1981) it was found to be clearly associated with the ER (Roden, L. and Miflin, B. J., unpublished results). These results suggest therefore that an enzymic mechanism for disulphide bond synthesis does exist at a site where the storage proteins are synthesized. Whether or not it has any role in controlling the nature of the disulphide linked aggregates, and thus technological quality, is not yet known.

(e) Conclusions. The properties of gluten most closely fit a model in which the elastic component is provided by long linear polymers of the HMW subunits (and possibly some other prolamins) linked end-to-end by di-sulphide links. The extremely high molecular weight of the polymers (10^6 or greater) along with the high degree of flexibility conferred by the large proportions of glycine would be expected to provide an elastic structure. Viscous flow may be explained in part by the rearrangements of entangled polymers but also by movement of the polymers relative to each other due to the presence of hydrophobically associated prolamins interspersed between them.

III. BARLEY PROTEINS, MALTING AND BREWING

The second major technological use of cereals that we wish to discuss is malting. Some 20% of the barley grown in the UK is used for this purpose. Malting is a process in which the grain is germinated under artificial conditions, the main objectives being to encourage the development of flavour componds and to promote the enzymatic hydrolysis, or modification, of the carbohydrate and protein reserves (Palmer and Bathgate, 1976; Palmer, 1980). Although only barley is malted in Western Europe, there do not appear to be any biochemical limitations to the use of other cereals such as wheat, rye, triticale, oats, sorghum and millet (Daiber and Novellie, 1968; Hoeser and Keininger, 1971; Mandl, 1972; Pomeranz and Shands, 1974; Pomeranz *et al.*, 1970; 1975) but the relative efficiency of the different species does not appear to have been assessed. Barley does have the technological advantage that the husks form a bed for filtration of the wort after mashing. A second reason why barley is still most extensively used is possibly cultural. Beer, as drunk in Europe and N. America, probably originated in N. Europe where the climate is more suited to barley than to other cereals. Consequently beer drinkers have grown accustomed to the characteristic flavour imparted by the use of this grain.

A. HORDEIN AND MALTING QUALITY

1. Hordein Electrophoretic Pattern and Malting Quality

It is well established that barley varieties differ in their suitability for malting (malting quality), although the reasons for this are not always obvious. Certainly differences in the ability to produce malt enzymes (Atanda and Miflin, 1970; Gothard, 1974; Morgan *et al.*, 1981) and in endosperm structure and texture (Palmer, 1975, 1980; Palmer and Bathgate, 1976; Palmer and Harvey, 1977) are important in some cases. More recently Baxter and Wainwright (1979) have proposed that malting quality is also affected by the amount and composition of the hordein fraction. They examined hordein fractions from 16 varieties by acid-gel electrophoresis and suggested that varieties with better malting quality tended to have less intensely stained fast-moving "B" hordein polypeptides. We repeated this study (Shewry *et al.*, 1980b) using SDS-PAGE analysis of a larger number of varieties and showed that, although some varieties with similar hordein patterns had similar malting characteristics, this was not always so. This point is illustrated by Fig. 4, which shows the NIAB malting grades of barley varieties with three different hordein patterns. We concluded that, although there was in some cases a loose correlation between malting quality and hordein pattern, this was equally likely to be

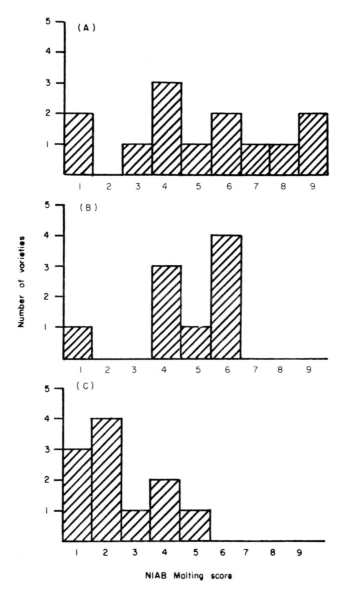

FIG. 4. The NIAB malting scores of European spring barley varieties with three different "B" hordein polypeptide patterns. (A) "B" hordein pattern 1; (B) "B" hordein pattern 7; (C) "B" hordein pattern 15. The data include most of the varieties recommended between 1960 and 1980. For a description of the range of hordein polypeptide patterns in European barley varieties see Shewry *et al.* (1979).

due to common ancestry of varieties with similar patterns as to a direct effect of hordein.

2. Steely and Mealy Grain

The existence of two types of barley grain differing in endosperm texture (called steely or vitreous and mealy or opaque grain) has been known for some time and comparative analyses were reported by Grønlund in 1880. Steely grain has been reported to contain more nitrogen and is generally believed to malt less quickly, than mealy grain (Palmer and Harvey, 1977). Baxter and Wainwright (1979) compared hordein fractions from good (Proctor) and poor (Julia) malting quality varieties. They showed that Proctor steely grain had more intense "B" hordein bands than mealy grain, whereas the Julia steely and mealy grain had similar proportions of B hordein.

We have re-examined steely and mealy grain selected from field-grown samples of these two varieties. In both cases the steely grain had a higher N content (2·09% compared to 1·49% in Proctor; 2·04% compared to 1·57% in Julia) and this was reflected in a higher relative amount of hordein (42 to 35% of the total grain N in Proctor and 48 to 39% in Julia). However visual examination of Coomassie Blue stained SDS-PAGE separations of these fractions showed little difference between the relative amounts of B and C hordein in the mealy and steely grain (Fig. 5).

This was confirmed by scanning of TCA fixed gels (using the method of Kirkman *et al.*, 1982) which showed a slightly higher proportion of C hordein in the steely grain of both varieties (13·6 compared to 11·4% of the total hordein in Julia, 12·3% to 9·1% in Proctor). It is possible that the higher amount of total N and hordein in the steely grain of both varieties is at least partly responsible for the slower rate of modification. It is interesting that in this experiment the steely and mealy grain of Julia, although comparable with those of Proctor in N-content, both had higher proportions of hordein. Little is known of the extent of genetic variation in the amount of hordein, although this could well affect malting quality (as discussed below).

3. Possible Role for Hordein in Malting Quality

"B" hordein polypeptides are comparatively rich in cysteine (2 to 3 mole %, see Table I) and a proportion of them are present in the mature seed as disulphide-linked aggregates (see p. 297). Baxter (1980, 1981) has suggested that this tendency to form disulphide bonds may affect malting quality, possibly because the aggregated protein is less readily digested during malting and forms a matrix around the starch grains which inhibits

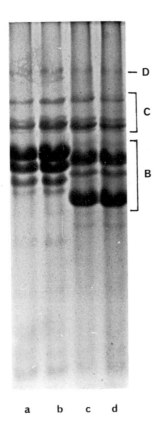

FIG. 5. SDS-PAGE of hordein fractions prepared from steely and mealy grain of barley. (a) mealy grain of cv. Proctor; (b) steely grain of cv. Proctor; (c) mealy grain of cv. Julia; (d) steely grain of cv. Julia.

amylolytic activity during mashing. Differences in the tendency of different 'B'' hordein polypeptides to form aggregates, and in the amounts of these polypeptides in different varieties, would result in the observed loose correlation between hordein pattern and malting quality. In support of this theory Slack *et al.* (1979) showed that large and small starch grains prepared from Proctor contained high levels of firmly bound protein which limited digestion with α-amylase. Electrophoresis of this protein indicated that a high proportion was hordein.

To evaluate the importance of disulphide bonding in malting quality we have compared the relative amounts of hordein extracted in the absence and presence of reducing agent from grain of varieties differing in malting quality and grown under conditions of varying N supply. In these experiments hordein was extracted at 60°C in two sequential fractions (as described by Shewry *et al.*, 1980d) with 50% propan-1-ol to extract hordein I followed by the same solvent + 2% 2-mercaptoethanol to extract hordein II. Although we would expect the presence of disulphide-linked hordein polypeptides in both fractions, the hordein II fractions should contain polypeptides in which such bonds resulted in the polypeptides being tightly bound into the tissue. If such binding is important in malting quality, then this might be reflected in the proportion of hordein present in this fraction.

In the first experiment we determined the amounts of hordein I and II present in grain of 8 barley varieties with a range of NIAB malting scores from 1–6 (see legend to Fig. 6). This seed was obtained from a field experiment set up at Rothamsted in 1979 to compare the effects of N fertilization on the yields of newer barley varieties. Two samples of each variety from separate plots and with N contents between 1·33 and 1·49% were analysed. The total amount of hordein was similar in all the cultivars, but a greater proportion of this was extracted in the hordein II fraction of the poor quality varieties. This difference is emphasized when the results are expressed as the ratio of the amounts of N recovered in the hordein I and hordein II fractions (Fig. 6). SDS-PAGE of the hordein I and hordein

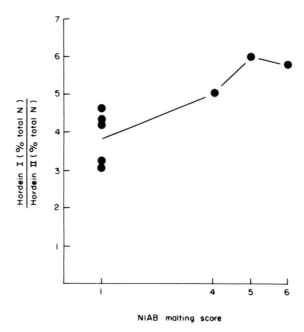

NIAB malting score

FIG. 6. The relationship between NIAB malting score and the ratio of the amounts of nitrogen recovered in the hordein I and hordein II fractions of grain of 8 barley varieties. Each point is the mean of analyses of two samples from separate plots. The varieties, and their NIAB scores in brackets, are Magnum (1), Minak (1), Jupiter (1), Goldmarker (1), Georgie (1), Athos (4), Dram (5), Porthos (6).

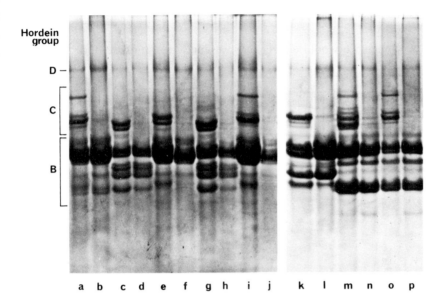

FIG. 7. SDS-PAGE of hordein I and hordein II fractions extracted from seed of 8 barley varieties. (a) Minak hordein I; (b) Minak hordein II; (c) Athos hordein I; (d) Athos hordein II; (e) Jupiter hordein I; (f) Jupiter hordein II; (g) Porthos hordein I; (h) Porthos hordein II; (i) Goldmarker hordein I; (j) Goldmarker hordein II; (k) Dram hordein I; (l) Dram hordein II; (m) Magnum hordein I; (n) Magnum hordein II; (o) Georgie hordein I; (p) Georgie hordein II.

II fractions (Fig. 7) showed that in all cases the hordein II fraction contained predominantly "B" hordein polypeptides.

We then compared the amounts of hordein I and hordein II in the grain of a single cultivar, Julia, grown under a range of N supply resulting in a range of N contents from approximately 1·2 to over 2·0%. These samples have been used previously to study the relationship between seed N, lysine content and the amount and composition of the combined hordein fraction (Kirkman *et al.*, 1982). When separate hordein I and hordein II fractions were extracted it was found that the increased total hordein in the high N seed was due to increased hordein I (Fig. 8). This is perhaps to be expected as it has been shown that, at least under conditions of high N availability but limited availability of S, increased grain N results in an increased proportion of the readily-extracted "C" hordein polypeptides. Although the proportion of seed N in the hordein II fraction does not increase, the total amount of this fraction does and this increase may still be sufficient to contribute to the decreased malting quality of high N grain.

Finally we compared the relative amounts of N in the grain of three different barley varieties grown in the field under a range of N supply.

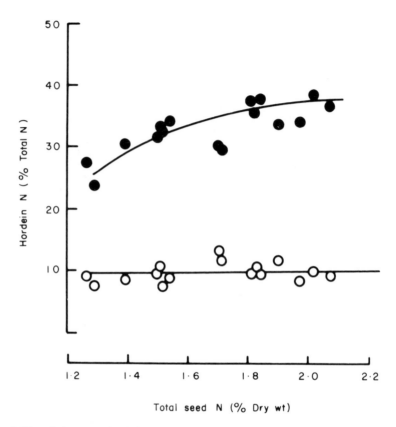

FIG. 8. The relative amounts of nitrogen recovered in the hordein I and hordein II fractions of grain of barley cv. Julia grown under varying conditions of nitrogen availability. ● hordein I; ○ hordein II.

These samples were generously supplied by Dr M. Allison of the Scottish Crop Research Institute. The range of N content in these samples (Fig. 9) was less than in the experiment with Julia (Fig. 8) and as a consequence there was no great effect on the relative amount of hordein in the seed. It did show, however, that whereas the seed of all the varieties had similar amounts of hordein I, the amounts of hordein II were greatest in Vada (NIAB malting score 2), intermediate in Golden Promise (NIAB score 6) and lowest in Maris Mink (NIAB score 5). These results, and those presented in Fig. 6, suggest that the relative amount of hordein II is a specific varietal characteristic which might be related, at least in some cultivars and under certain conditions, to malting quality. Further studies are necessary to clarify this point. Because malting quality is a composite character resulting from the interaction of a number of factors a simple clear correlation would not be expected.

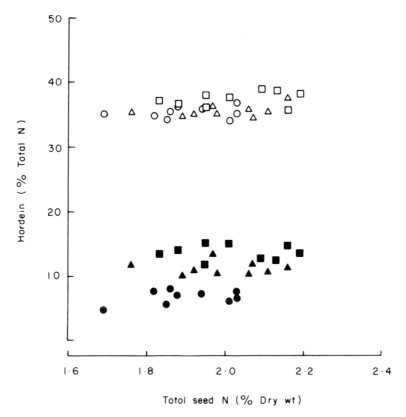

FIG. 9. The relative amounts of nitrogen recovered in the hordein I and hordein II fractions of grain of 3 barley varieties grown in the field under conditions of varying nitrogen supply. □ Vada hordein I; ■ Vada hordein II; △ Golden Promise hordein I; ▲ Golden Promise hordein II; ○ Maris Mink hordein I; ● Maris Mink hordein II.

B. OTHER ASPECTS OF BREWING

It is probable that the amount and composition of the hordein fraction also affects other aspects of brewing. These have been discussed in some detail by Baxter (1980, 1981) and include the filterability of the wort, foam quality and haze formation.

IV. COMPARATIVE STUDIES ON CEREAL PROTEINS

As we have indicated above, the cereals differ in their suitability for baking and the bread made from rye or barley differs markedly from that made from wheat. This variation has stimulated many workers to consider the

similarities and differences between the cereals. Such comparisons are
hindered by the separate development of techniques and nomenclatures
for each of the cereals. However, recent chemical, biological and genetic
studies have shown that the prolamins of wheat, barley and rye have
many similar properties whereas those from maize (another major high-
prolamin cereal) show many differences. In the following sections we
discuss these similarities and differences and, in order to simplify the
comparisons, we propose a unified classification system for the prolamins
of wheat, rye and barley. From this simplified approach we then attempt to
draw correlations with technological properties.

A. EXTRACTION OF CEREAL PROLAMINS

The problems of extraction of wheat proteins have been dealt with above;
similar problems occur with other cereals. Alcoholic solvents on their own
do not extract all of the prolamins, furthermore they preferentially extract
certain groups; to obtain complete extraction requires the addition of a
reducing agent and often acetic acid. When the prolamins of rye are
separated on SDS-PAGE (Fig. 10) four groups of proteins can be
recognized, high molecular weight polypeptides, ω-secalins, 40 kd γ-
secalins and 75 kd γ-secalins (Shewry *et al.*, 1982a,b. Similarly the prolamins of barley have been classified on the basis of their mobility on
SDS-PAGE (see Fig. 10) into A (apparent molecular weight <20 kd), B
(molecular weight 30–46 kd) and C (molecular weight 55–72 kd) hordein
(Koie *et al*, 1976; Miflin and Shewry, 1979b). We have recently separated
and characterized a further fraction called D hordein, which gives a single
band with an apparent molecular weight of around 105 kd. Maize prolamins have much lower molecular weights with major bands around 21 kd
and 23 kd and minor bands at 13·5 kd and 9·6 kd. Because of the different
solubility properties not all of these bands have been extracted by all
workers. The ω-gliadins, ω-secalins and C-hordeins are completely extracted by aqueous alcohol (e.g. 70% ethanol, 50% propanl-ol); similarly
most of the 21 kd and 23 kd zeins are also extracted. In contrast the rest of
the groups are either incompletely (α, β and γ-gliadins, γ-secalins, B
hordeins) or minimally (HMW prolamins of wheat and rye, D hordein,
13·5 kd and 9·6 kd zeins) extracted unless a reducing agent (e.g. 1%
2-mercaptoethanol) and, in some instances, 1% acetic acid is also included
in the solvent (Shewry *et al.*, 1980d; Miflin and Shewry, 1979a; Wilson *et
al.*, 1981a,b).

Because of the many difficulties of obtaining reproducible and complete
extraction of proteins from grain we consider it unsound to use this as a

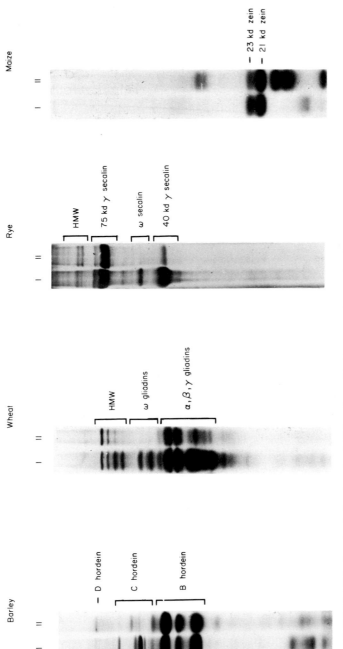

FIG. 10. SDS-PAGE of prolamin I and prolamin II fractions of barley, rye, wheat and maize.

system of detailed classification for comparison of cereal proteins; rather it should be based on chemical and genetical properties as detailed below.

B. CHEMICAL CHARACTERISTICS

The prolamin components of wheat, barley and rye can be classified into 3 groups on the basis of their chemical characteristics. (We have not included the minor low molecular weight prolamin components of wheat and barley recently described by Prado *et al.* (1982) and Aragoncillo *et al.* (1981); these have properties akin to the S-rich prolamins.) The properties of the components in these groups are summarized in Tables I, II and III.

1. HMW Prolamins

The "D" hordein of barley and the HMW prolamins of rye and wheat have the highest subunit molecular weights (in excess of 90 kd by SDS-PAGE and 50 kd by sedimentation equilibrium ultracentrifugation) (Table II) and are least readily extracted from the seed, probably because they are largely present as disulphide-linked aggregates (see p. 296). They have similar amino acid compositions, notably high glycine (over 12%) and relatively low proline.

2. S-poor Prolamins

These include "C" hordein of barley, ω-secalins of rye and ω-gliadins of wheat. They have molecular weights between 40 and 80 kd by SDS-PAGE and are characterized by amino acid compositions rich in glutamine (40–50%), proline (20–35%) and phenylalanine (7–9%) and poor in lysine (0–0·4%), cysteine (0–0·4%) and methionine (0–0·2%) (Table 1). The small cysteine content is probably responsible for the complete extraction of these components by aqueous alcohols in the absence of reducing agents, while the lack of basic amino acids results in very low charge at low pH and consequently slow migration on electrophoresis at pH 3. N-terminal amino acid sequencing of purified components from barley, rye and the diploid wheat *Triticum monococcum* (Table III) confirms the homology. The three components differed at only one position out of the first 10 (Asp, Glu or Ser at position 7), but the *T. monococcum* component contained at least two polypeptides one of which had an additional N-terminal alanine (position-1 in Table III).

3. S-rich Prolamins

α, β and γ gliadins of wheat, "B" hordein of barley and γ-secalin of rye have similar amino acid compositions with 32–42% glutamate plus gluta-mine, 15–24% proline and around 2% cysteine (Table I). This relatively high cysteine content probably results in some intermolecular and intra-molecular disulphide bond formation, hence their increased extraction in the presence of reducing agents. These components contain more basic amino acids than the S-poor prolamins (Table I), resulting in faster electrophoretic migration at low pH (Fig. 1). The α, β and γ gliadins have similar molecular weights of between 36 kd and 44 kd by SDS-PAGE and 30 kd and 37 kd by sedimentation equilibrium ultracentrifugation (Table II). These are in the same range as the "B" hordeins but rye differs in having two disparate groups of γ-secalins, one with molecular weights similar to the wheat and barley components (40 kd by SDS-PAGE) but the other considerably higher (75 kd).

Amino acid sequencing of gliadins shows the presence of at least two different N-terminal sequence types (Table III). One, called the α-type by Autran et al. (1979) is present in several α-gliadins, in β_5 and in γ_1 gliadin while the second, or γ-type, is found in γ_2 and γ_3 gliadin. Bietz et al. (1970) have also shown that the peptide maps resulting from peptic digests of α, β and γ_1 gliadins are closely similar. Similar maps of γ_3 gliadin show some similarities but also a number of differences. This suggests that gliadins with the α-type and γ-type N-terminal sequences have some homologous regions in other parts of the chains. Analysis of mixtures of gliadin-type polypeptides originating from aggregates (commonly called high molecular weight gliadin or ethanol-soluble reduced glutenin) shows that they have a range of molecular weights (36–44 kd by SDS-PAGE) and an amino acid composition (Tables I and II) similar to that of α, β and γ gliadins. N-terminal sequencing of such a mixture shows the presence of a number of amino acids at each position, some of which are the same as those in α and γ-type sequences (Table III) (Bietz and Wall, 1980). It is possible that this fraction contains a mixture of disulphide-linked components, some of which also occur as monomeric α, β and γ gliadins while others are chemically and genetically related but occur solely in the aggregated state. The 40 kd and 75 kd γ-secalins of rye both have N-terminal amino acid sequences which are related to those of the γ_2 and γ_3 gliadins of wheat (Table III). There do not appear to be any rye components with the α-type sequence (Autran et al., 1979).

The only S-rich prolamin fraction of barley is "B" hordein. Although this is similar to the α, β and γ gliadins and the 40 kd γ-secalins in its amino acid composition and molecular weight. (Tables I and II), its homology with these has not been established by N-terminal sequencing as it is blocked to Edman degradation (Shewry et al., 1980a,d).

4. Prolamins of Maize

Amino acid compositions of prolamin components of maize are also presented in Table I. Although the 21 kd and 23 kd subunits differ from the

TABLE I

Amino acid compositions (expressed as mole %) of prolamin fractions from wheat, rye, barley and maize. (a) HMW prolamins and sulphur-poor prolamins

Type of prolamin	HMW Prolamins						Sulphur-poor prolamins				
Species	Barley	Rye	Wheat		Wheat		Barley		Rye	Wheat	Wheat
Cultivar	Sundance	Rheidol	Highbury	Highbury	Manitou		Julia	Julia	Frontier	Cappelle	Triticum monococcum
Component	"D" Hordein	HMW band	HMW band 2	HMW band 12	IEF peak 3	IEF peak 9	Total "C" hordein	"C" hordein band	ω-secalin band	total ω-gliadin	ω-gliadin band
Reference	1	1	1	1	2	2	3	4	5	6	4
Amino Acids											
Asp	1·50	1·30	2·79	1·17	1·5	1·9	1·0	0·83	0·51	0·48	1·18
Thr	8·00	2·72	3·25	2·26	2·8	3·2	1·0	0·99	1·04	1·15	1·08
Ser	9·43	5·01	6·45	5·49	7·4	7·5	4·6	2·61	4·11	3·40	4·89
Glu	29·6	34·0	32·6	36·2	38·8	37·4	41·2	41·1	42·9	50·1	45·2
Pro	11·38	13·72	12·82	13·59	19·9	13·8	30·6	31·9	30·6	22·9	26·2
Gly	13·61	16·46	14·85	14·68	18·1	12·8	0·3	0·41	1·39	1·56	0·89

Ala	3·42	6·43	5·03	3·33	2·2	2·9	0·7	0·69	0·59	0·57	1·57
Cys	1·68	1·42	0·84	1·85	ND	ND	t	t	0	0·36	t
Val	4·81	2·77	3·53	3·25	t	3·0	1·0	1·11	0·97	0·53	0·80
Met	0·60	0·29	0·73	0·63	t	0·3	0·2	0·22	0·10	0·11	0·13
Ile	1·24	1·61	1·72	2·01	0·8	2·1	2·6	3·02	2·43	3·75	2·22
Leu	3·89	4·43	5·27	4·82	3·8	5·8	3·6	4·31	3·92	3·43	4·70
Tyr	3·76	4·12	4·43	4·49	3·5	3·2	2·3	2·29	1·37	1·05	1·17
Phe	1·31	0·92	1·65	1·62	t	1·3	8·8	9·03	7·43	9·10	8·09
His	3·12	2·16	0·86	1·62	t	1·3	1·1	0·60	0·54	0·95	0·56
Lys	0·82	0·25	1·36	0·94	1·4	1·4	0·2	0	0·30	0·41	0·13
Arg	1·78	2·34	1·83	2·06	t	2·0	0·8	0·84	1·79	0·65	1·04

References

1. Unpublished data of the authors. The nomenclature of the Highbury bands follows Payne *et al.*, 1981. 2. Khan and Bushuk., 1979. 3. Shewry *et al.*, 1980d. 4. Shewry *et al.*, 1980a. 5. Shewry *et al.*, unpublished results 6. Charbonnier, 1974.

ND, not determined; t, trace. Trp was not determined in most of the studies and is omitted from the table.

Values reported for aspartate and glutamate include the amides asparagine and glutamine respectively.

It has been shown that between 90 and 100% of the recovered glutamate residues are present in gliadin and hordein as glutamine (Charbonnier, 1974; Shewry *et al.*, 1980d; Ewart, 1981). The same is probably true for the prolamins of other cereals.

TABLE I

Amino acid compositions (expressed as mole %) of prolamin fractions from wheat, rye, barley and maize. (b) Sulphur-rich prolamins

	Barley			Rye		Wheat — Ponca					
Cultivar	Julia	Sundance	Hoppel	Rheidol							
Component	Total "B" hordein	35kd "B" hordein (B-1)	46kd "B" hordein	40kd γ-secalin	75kd γ-secalin	α_8 gliadin	β_5 gliadin	γ_1 gliadin	γ_2 gliadin	γ_3 gliadin	HMW gliadin
Reference	1	2	2	3	3	4	4	4	4	4	5
Amino Acids											
Asp	1·4	0·82	1·48	2·96	1·48	3·0	2·6	2·9	1·8	1·7	2·0
Thr	2·1	1·99	2·27	3·22	1·55	1·6	1·7	1·7	2·0	2·2	2·6
Ser	4·7	4·30	5·10	5·24	5·70	5·2	5·3	5·3	4·9	4·2	7·0
Glu	35·4	32·0	35·1	34·8	40·3	37·2	39·6	41·7	39·1	39·6	37·5
Pro	20·6	20·8	21·6	18·4	23·5	15·5	15·8	15·1	18·7	18·9	15·1
Gly	1·5	2·91	1·81	2·43	1·69	2·5	1·9	2·4	2·7	2·7	4·9
Ala	2·2	2·50	2·65	2·79	2·21	2·9	3·5	3·4	3·0	3·2	2·7
Cys	2·5	2·72	2·24	2·48	1·79	1·9	1·8	1·8	1·9	2·0	1·3
Val	5·6	6·29	5·66	4·73	4·49	4·0	4·0	3·8	3·4	3·7	4·0
Met	0·6	0·94	1·00	1·02	0·62	1·2	1·1	0·9	1·7	1·4	1·3
Ile	4·1	4·13	4·44	4·77	2·83	4·1	3·8	3·9	3·7	3·5	3·4
Leu	7·0	8·22	6·40	7·35	4·92	8·1	7·9	6·9	7·2	6·5	7·4
Tyr	2·5	2·33	2·19	0·72	0·89	3·1	2·5	3·1	0·5	0·4	1·7
Phe	4·8	5·13	4·09	5·28	5·41	3·9	3·7	3·7	5·2	5·6	4·1
His	2·1	1·61	1·38	1·72	1·43	2·5	2·3	1·5	1·4	1·6	1·4
Lys	0·5	0·79	0·25	0·72	0·43	0·5	0·2	0·1	0·7	0·7	1·0
Arg	2·4	2·50	2·32	1·32	0·83	2·4	2·3	1·6	1·5	1·6	2·6

References

1. Shewry et al., 1980d. 2. Unpublished data of the authors. 3. Shewry et al., 1982a. 4. Bietz et al., 1977. 5. Bietz and Wall, 1973.

ND, not determined; t, trace. Trp was not determined in most of the studies and is omitted from the table.

Values reported for aspartate and glutamate include the amides asparagine and glutamine respectively.

It has been shown that between 90 and 100% of the recovered glutamate residues are present in gliadin and hordein as glutamine (Charbonnier, 1974; Shewry et al., 1980d; Ewart, 1981). The same is probably true for the prolamins of other cereals.

TABLE I

Amino acid compositions (expressed as mole %) of prolamin fractions from wheat, rye, barley and maize. (c) Prolamins (zein) of maize inbred line W64A

	Components		
	23 kd + 21 kd	13·5 kd	9·6 kd
Amino Acids			
Asp	5·6	2·5	2·2
Thr	3·2	3·9	4·1
Ser	7·1	6·0	5·8
Glu	17·3	17·9	22·9
Pro	11·3	11·9	11·6
Gly	4·8	9·0	8·6
Ala	11·0	10·4	10·7
Cys	t	t	1·4
Val	3·8	3·4	3·6
Met	t	5·3	3·9
Ile	2·5	1·3	1·1
Leu	19·4	11·6	9·6
Tyr	3·4	4·7	3·4
Phe	4·3	2·8	2·5
His	3·2	1·2	1·2
Lys	0·5	0·3	0·4
Arg	1·9	2·4	2·0

Data of Gianazza *et al.*, 1977

t, trace. Trp was not determined in most of the studies and is omitted from the table.

Values reported for aspartate and glutamate include the amides asparagine and glutamine respectively.

It has been shown that between 90 and 100% of the recovered glutamate residues are present in gliadin and hordein as glutamine (Charbonnier, 1974; Shewry *et al.*, 1980d; Ewart, 1981). The same is probably true for the prolamins of other cereals.

13·5 kd and 9·6 kd subunits in their amino acid compositions, all fractions have very different compositions to the prolamins of barley, rye and wheat, notably lower glutamate + glutamine (17–23%) and proline (11–12%) and much larger amounts of leucine, alanine and serine (sum of 26–38% compared to values of 7–17 mol. % for rye, wheat, or barley). Again, however, it is possible to recognize S-rich (13·5 and 9·6 kd) and S-poor (21 and 23 kd) prolamins.

C. RELATIVE AMOUNTS OF COMPONENTS

Comparison of SDS-PAGE separations of prolamins of barley, wheat and rye (Fig. 10) shows considerable differences in the relative amounts of the HMW, S-poor and S-rich prolamin groups. We have made approximate

TABLE II
Molecular weights of prolamin components from wheat, rye and barley

Species	Cultivar	Component	Molecular weight		Reference
			Ultracentrifugation	SDS-PAGE	
HMW prolamins					
Barley	Sundance	"D" hordein	54 700	105 000	1
Rye	Rheidol	HMW band	67 600	—	1
Wheat					
(*T. aestivum*)	Highbury	HMW bands	63–70 000	—	1
(*T. aestivum*)	Mardler	HMW bands	—	95–136 000	2
S-poor prolamins					
Barley	Julia	Total "C" hordein	52 000	55–72 000	3,4
	Maris Mink	"C" hordein	—	54, 57, 58 000	5
Rye	Frontier	ω-secalin	—	48–53 000	6
Wheat					
T. monococcum	—	ω-gliadin	—	44–74 000	6
T. durum	Edmore				
T. aestivum	Chinese Spring				
	Justin				
T. aestivum	Wichita	ω-gliadin	75 500; 78 700	—	7
T. aestivum	Manitoba No. 2	ω-gliadin	27 000	50, 54, 64 000	8

S-rich prolamins

Barley	Julia	Total "B" hordein	32 000	35–46 000	3,4
	Sundance	B-1 band	32 400	35 000	1,4
	Hoppel	B band	35 200	46 000	1,4
Rye	Rheidol	40k γ-secalin	33 000	40 000	9
		75k γ-secalin	54 000	75 000	9
Wheat *T. aestivum*	4 cultivars including Ponca	A(α) gliadin	—	32 000	10
	Ponca	$\alpha_7, \alpha_8, \alpha_9$ gliadins	30 400, 36 9000	—	11
			30 900		
	Cappelle	α gliadins	—	32–38 000	12,13
	Ponca	β_6 gliadin	33 000	—	11
	Cappelle	β gliadins	—	37 000; 42 000;	14
	Cappelle	$\beta_{13}, \beta_{22}, \beta_{32}$ gliadins	—	34 300; 33 800;	15
				34 900	
	Ponca	γ_1 gliadin	30 300	≃36 500	16,17
	Ponca/Cappelle	γ_2 gliadin	34 600	38 600	11,18
	Ponca/Cappelle	γ_3 gliadin	34 700	42 000	16,18
	Ponca	Subunits of HMW gliadin	—	36–44 000	19

1. Authors' unpublished results. 2. Payne and Corfield, 1979. 3. Shewry *et al.*, 1980d. 4. Faulks *et al.*, 1981. 5. Shewry *et al.*, 1981. 6. Shewry *et al.*, unpublished results. 7. Booth and Ewart, 1969. 8. Hamauzu *et al.*, 1974. 9. Shewry *et al.*, 1982a. 10. Platt *et al.*, 1974. 11. Sexson *et al.*, 1978. 12. Ewart, 1973. 13. Ewart, 1976. 14. Ewart, 1977c,d. 15. Terce-Laforgue *et al.*, 1980. 16. Sexson and Wu, 1972. 17. Bietz and Wall, 1972. 18. Ewart, 1977b. 19. Bietz and Wall, 1973.

TABLE III

N terminal amino acid sequences of prolamin components from wheat, rye and barley

Species	Cultivar	Component	-1	1	2	3	4	5	6	7	8	9	10	Reference
HMW Prolamins														
Wheat	Highbury	Band 2	—	Glu	Gly	Glu	Ala	Ser	Glu	Gln	Leu	Gln	Cys	Authors unpublished results with D. D. Kasarda
Barley	Sundance	"D" hordein	N-terminally blocked											
S-poor Prolamins														
Wheat (T. monococcum)	—	ω-gliadin band	Ala	Arg	Gln	Leu	Asn	Pro	Ser	Asp	Gln	Glu	Leu	Shewry et al., 1980a
Rye	Frontier	ω-secalin band	—	Arg	Gln	Leu	Asn	Pro	Ser	Asp	Gln	Glu	Leu	Shewry et al., unpublished results
		ω-secalin band	—	Arg	Gln	Leu	Asn	Pro	Ser	Glu	Gln	Glu	Leu	
Barley	Julia Maris Mink	4 "C" hordein components	—	Arg	Gln	Leu	Asn	Pro	Ser	Ser	Gln	Glu	Leu	Shewry et al., 1980a; Shewry et al., 1981
S-rich Prolamins														
Barley	Julia	B Hordein	N-terminally blocked											Shewry et al., 1980d
Rye	Rheidol	40kd γ-secalin	—	Asn	Met	Gln	Val	Gly	Pro	Ser	Gly	Gln	Val	Shewry et al., 1982a
		75kd γ-secalin	—	Asn	Met	Gln	Val	Asn	Pro	Ser	Gly	Gln	Val	
Wheat (T. aestivum)	Ponca	α8–α12 gliadins, β5 gliadin	—	Val	Arg	Val	Pro	Val	Pro	Gln	Leu	Gln	Pro	Bietz et al., 1977
		γ1 gliadin	—	Asn	Ile	Gly	Val	Asp	Pro	Trp	Gly	Gln	Val	
		γ2 gliadin	—	Asn	Met	Gly	Val	Asp	Pro	Trp	Gly	Gln	Val	
		γ3 gliadin	—	Val	His	Pro	Glu	Val				Pro	Gln	
	Ponca	Subunits of HMW gliadin	—	Asn	Gln	Ile	Val	Val	Leu	Gln	Leu	Gln	Pro	Bietz and Wall, 1980
				Gln	Met	Val	Gln	Gln	Pro		Gly	Leu	Val	
				Met									Asn	

determinations of the relative amounts of these groups by scanning of TCA-fixed gels. In barley and rye HMW prolamins accounted for only 2–4% of the total fraction but in wheat for 7–9%. The amount of S-poor prolamins is highest in barley where it accounts for approximately 10–30% of the total fraction, depending on the N-nutrition of the plant, with generally between 15 and 20% in field-grown grain. In rye and wheat these components account for less than 5% of the total prolamins of field-grown grain.

D. STRUCTURAL GENES FOR PROLAMINS

In barley and rye all the prolamins are coded for by genes on one chromosome (chromosome 5 of barley and 1R of rye) whereas in wheat the structural genes for prolamins are on at least two chromosomes (1 and 6) of each of the three genomes (see above).

Analysis of aneuploids, addition lines and substitution lines indicates that the high molecular weight components are controlled by genes on the long arms of chromosome 5 of barley, 1R of rye and 1A, 1B and 1D of hexaploid bread wheat (Bietz et al., 1975; Lawrence and Shepherd, 1980; Payne et al., 1980). Other genes on the short arms of the same chromosomes code for other prolamins in the three species (Lawrence and Shepherd, 1981).

In barley the "C" and "B" groups of hordein polypeptides are controlled by single loci, called Hor-1 and Hor-2 respectively, which are located on the short arm of chromosome 5 with Hor-1 closer to the centromere (Shewry et al., 1980c; Jensen et al., 1980). The short arm of chromosome 1R of rye appears to control the γ- and ω-secalins, although the number and exact location of the loci is not known (Shepherd, 1968, 1973; Shepherd and Jennings, 1971; Bernard et al., 1977; Lawrence and Shepherd, 1981). Similarly, in wheat genes on the short arms of chromosomes 1A, 1B and 1D appear to code for ω-gliadins and some β and γ-gliadins (Wrigley and Shepherd, 1973; Kasarda et al., 1976b). Other β and γ-gliadins and probably all the α-gliadins are coded by additional genes on chromosomes 6A, 6B and 6D (Wrigley and Shepherd, 1973; Kasarda et al., 1976b). These loci do not appear to be present in rye and Shepherd (1973) has suggested that they arose in the ancestor of the three genomes of wheat by the duplication of genes on chromsome 1 followed by their translocation to chromosome 6. If so this translocation must have occured after the divergence of the ancestors of wheat and rye. The α-gliadins all appear to have the α-type N-terminal sequence (Kasarda, 1980) and are coded by chromosomes of group 6. It is possible that this sequence type developed

296   B. J. MIFLIN *ET AL.*

after the translocation and therefore occurs only in wheat prolamins coded for by chromosomes of group 6.

E. AGGREGATION PROPERTIES

We have discussed above the aggregation properties of wheat proteins, the question is do the prolamins of other cereals also aggregate? Attempts to resolve this using our modified AUC solvent were not successful because gluten could not be extracted from our barley flour and the solvent is not suitable for direct use with flour. We therefore looked at the aggregation states of prolamins extracted by 50% propan-1-ol (i.e. prolamin I fractions). Comparison of the SDS-PAGE patterns of reduced and unreduced prolamin I fractions shows major differences in banding patterns, indicating that some polypeptides are present in these fractions as disulphide-linked aggregates (Fig. 11). In barley, wheat and rye the S-poor prolamins

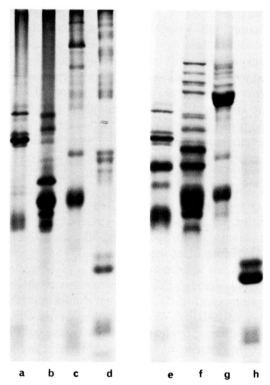

a   b   c   d   e   f   g   h

FIG. 11. SDS-PAGE of unreduced and reduced prolamin-I fractions of barley (cv. Sundance), wheat (cv. Highbury), rye (cv. Rheidol) and maize (cv. Kelvedon Hybrid 59A). (a–d) unreduced; (e–h) reduced; (a, e) barley; (b, f) wheat; (e, g) rye; (d, h) maize. (a–d) and (e–h) were run on separate gels at the same time.

("C" hordein, ω-gliadin, ω-secalin) were clearly separated in both unreduced and reduced samples indicating their presence in a non-aggregated form. The HMW prolamins of wheat and rye were only observed in the reduced samples indicating that these were originally present in an aggregated form. "D" hordein was not present in the prolamin-I fraction of barley, but analysis of total protein body proteins (see p. 306) has shown that this is also aggregated. Certain of the S-rich prolamins are also observed in the separations of reduced but not unreduced samples. In barley the "B" hordein polypeptides are largely absent from the unreduced separations but in wheat the main effect was on the 44 kd subunits, with little apparent effect on those with molecular weights around 36 kd. In rye the 75 kd γ-secalins were apparently absent from the unreduced gels but reduction had little effect on the 40 kd γ-secalins. The unreduced samples of all 3 cereals gave very high background staining on the gels with some material only just entering the 5% stacking gel. This presumably represented a range of disulphide-linked aggregates. These results indicate that the prolamin I fractions contain disulphide linked aggregates formed from "B" hordein in barley, HMW prolamin and mainly 44 000 gliadin subunits in wheat and HMW prolamin and 75 kd γ-secalin subunits in rye.

A similar study of the prolamin I fraction of maize (Fig. 11) also indicated the presence of aggregated material although in this case the aggregates formed more discrete bands on the gel indicating a limited range of molecular weights. However some material again only just entered the stacking gel. These aggregates appeared to be formed from the 21 and 23 kd subunits rather than those of lower molecular weight.

F. SUMMARY

We propose that the three major groups of prolamins— HMW, S-rich and S-poor—are homologous in the three cereals and have probably originated from single genes in the ancestral species (Fig. 12). The differences observed today between the prolamin fractions of wheat, rye and barley have resulted from different degrees of duplication and divergence of these genes. The homology is most clearcut in the case of the HMW prolamins and S-poor prolamins although changes in the relative amounts, presumably resulting from greater duplication of the genes, have occurred. It is these differences in the relative amounts of these three groups of prolamins, and in their properties, which probably account for the different processing properties of the three cereals.

A single S-rich prolamin gene has probably given rise to the B-hordein locus in barley and to genes coding for two distinct groups of S-rich γ-secalins in rye, although it is not known if these are coded by the same or

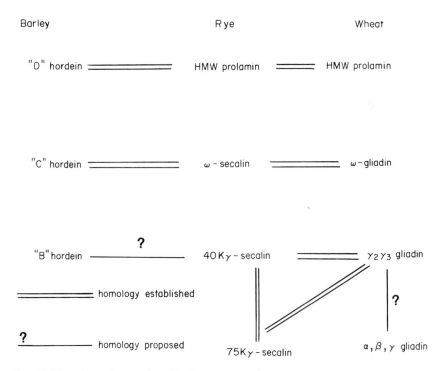

Fɪɢ. 12. Homology of prolamins of barley, rye and wheat.

separate loci. It is of interest that whereas the B hordein polypeptides of barley are partially aggregated, the two groups of γ-secalins differ considerably in their aggregation properties. In wheat the same gene has probably given rise to loci on chromosomes of groups 1 and 6 which code for α, β and γ gliadins as well as genes, which may or may not be at the same loci, which code for similar subunits present predominantly or entirely in aggregated form. The relationships of the latter polypeptides to the α, β and γ gliadins and also to the aggregatable γ-secalins of rye needs to be established.

V. Endosperm Structure and Technological Properties

A. PROPERTIES OF MATURE ENDOSPERMS

The structure of the cereal endosperm is of prime importance in various technological processes. Wheats may be classified as hard or soft, vitreous or opaque. Such parameters are important in quality since they determine milling and baking behaviour. Similarly in barley the grain may be classified into steely (vitreous) or mealy (opaque); the latter is considered better for malting. In general there is a correlation between hardness and vitreosity but these two properties are not due to the same fundamental cause (Simmonds, 1974). More certain is the relationship between high protein content and vitreousness and low protein and mealiness (see p. 278).

The various requirements of the wheat-utilizing industries with respect to hardness and protein content have been classified by Moss (1973). From this type of classification has come the need to develop various tests to measure these properties. Some of those used for wheat (e.g. pearling resistance, particle size index) have been reviewed and discussed by Simmonds (1974). Analogous tests, particularly the milling energy test (Allison et al., 1976), have been developed to measure similar properties of barley as indicators of malting quality. Another series of empirical tests used to estimate the potential quality of both wheat and barley are the sedimentation tests. These tests depend upon uniformly milling the grain, mixing it with a solvent and allowing the suspension to settle under controlled conditions. The size of the sediment, or the speed at which it is formed, is then measured. Whilst the principles underlying these tests are not clearly understood many of them appear to be related to the way in which the endosperm breaks down under milling and to the relationship between the protein, starch and cell wall components of the endosperm. Various attempts have been made to study those relationships in mature grain and these will be discussed below.

Simmonds and colleagues (Barlow et al., 1973; Simmonds et al., 1973; Simmonds, 1974) in a detailed series of studies have shown some of the differences between hard and soft wheats. In soft wheats the cells are readily broken down on milling and the starch granules released. There is little adhesion between the three components (protein, starch and cell walls) and the cell walls tend to form separate sheets of material. When such grains are freeze fractured the fracture lines tend to pass either through or around the starch granules with equal frequency. In contrast hard wheats show strong adhesion between the three components and the cell contents and walls form a coherent whole. Fracture during milling occurs either around the cell wall or through the cell across the starch grain and proteins alike; starch grains are thus rarely released whole. In freeze etching the fracture line passes predominantly through the starch grains (up to 98% of fractures). Correlated with this different behaviour are differences in the amount and type of protein adhering to the starch grains. Hard wheats have much more protein adhering and a considerable proportion of this is storage protein. In contrast, soft wheats have less adhering protein and this is largely buffer-soluble. Various studies, including the use of fluorescent antibodies, revealed that much of the buffer-soluble proteins were concentrated around the starch grains and Simmonds (1974) has suggested that this may act in some way as a cement between the starch grains and the storage protein matrix. Also of importance at the starch protein interface are the membranes and their adherence to the storage protein on one side and to the starch granule on the other. Similar adherance of the storage proteins to the starch grains of barley has been found (Slack *et al.*, 1979)

Grain characters such as hardness and vitreosity are markedly affected by environment during seed development and ripening. However there is also strong evidence for a genetical component and it has proved possible to convert a hard wheat variety into a soft one by back-crossing (Symes, 1965, 1969). The effect seems to be due to a single gene which, on the basis of studies with substitution lines, may be on chromosome 5D (Law *et al.*, 1978). However Symes' studies indicated that hardness was also influenced by several modifying genes and Konzak (1977) summarizes the evidence for genes for "hardness" occurring on chromosomes other than 5D.

Palmer and Harvey (1977) have carried out a scanning electron microscope study of the internal structure of barley grains of good and bad malting varieties and of steely and mealy grains of those varieties. In vitreous kernels of the poor malting variety Julia the starch granules, particularly the small ones, are embedded in a matrix whereas in the high quality variety Proctor the small starch granules and the protein matrix are loosely associated and the endosperm is crumblike in structure. These differences are unrelated to the total protein content since the nitrogen contents of the grains were the same, although there may have been differences in hordein content (see p. 278). Some of the differences between the varieties were present even in mealy grains, those of Julia showing a more compact structure than those of Proctor.

Fractionation of wheat flour into starch and protein granules was achieved by Hess (Hess *et al.* 1952; Hess, 1954). He postulated the existence of two types of proteins: interstitial proteins, which could be separated out, and adherent proteins, which remained attached to the starch grains. The interstitial protein appears equivalent to the storage proteins in that it was largely alcohol-soluble and a cohesive gluten could be produced from it. The adherent protein was present in a layer of about 0·22 μm and was shown by electron microscopy to consist of a series of fibrils (Hess *et al.*, 1955). Adams *et al.* (1976) have also separated out protein particles (bodies) from milled dry grain of several cereals. They showed, by means of electron microscopy, the very different nature of these particles in different cereals. Those from millet (*Eleusine coracana*), babala (*Pennisetum typhoideum*) and maize appear as separated or clumped bodies but with a well defined outline indicating the presence of a clear limiting membrane. In contrast the particles from wheat are irregular and ill defined clumps with much adhering material and containing entrapped starch grains. The protein bodies of barley are less clumped than wheat but like wheat are irregular and not clearly bounded by a membrane.

No clear conclusions can be made from these studies as to the reasons for the differences in properties of the different cereal endosperms and the milled samples obtained from them. However, it is clear that important technological attributes are closely related to the endosperm structure of the mature grain or, in the words of Palmer and Harvey (1977), to "the sub-cellular and physico-chemical features of the grain which determine the relationship between the protein matrix, the starch granules and the cell wall". To obtain an understanding of this relationship requires not only the knowledge of the properties of the final product (the dry seed) but also of the subcellular processes that occur during the formation of that seed. These latter processes will be discussed in the next section.

B. ENDOSPERM DEVELOPMENT

The development of all aspects of the endosperm are relevant to technological use but we will confine our discussion to aspects of protein deposition. A complete review of the morphological and biochemical development of the wheat endosperm has recently been written by Simmonds and O'Brien (1982).

1. Temporal Deposition of Proteins

Several studies have been made of the timing of protein deposition in cereal seeds. Fig. 13 shows the results of a recent study of barley made in our department by S. Rahman. They emphasize the point that the salt-soluble proteins are present from early in development but that they cease accumulating when the endosperm has reached about a half of its final weight. There may, however, be differences in the accumulation patterns of different components of the fraction, for example the globulins probably contain some storage proteins and accumulate later in development than the albumins (Brandt, 1976). In contrast the hordeins only appear as the endosperm enters the phase of rapid growth and they continue to accumulate linearly until maturity. Again consideration of the total fraction conceals the fact that the individual components of the hordein fraction accumulate at different rates with the "C" hordeins forming a relatively greater proportion of the total hordein at early stages and the lower molecular weight B hordeins accumulating relatively later

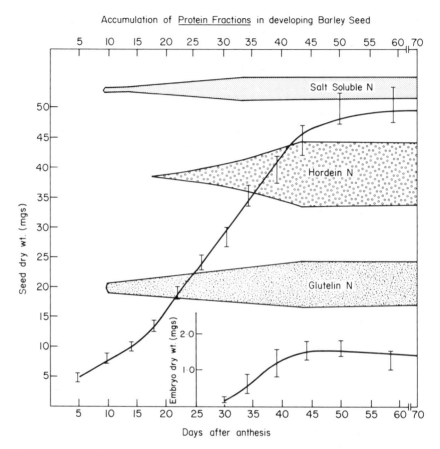

Fɪɢ. 13. The accumulation of protein fractions during the development of the barley endosperm. The heights of the shaded blocks indicate the relative amounts of the various fractions (taken from the results of Rahman *et al.*, 1982).

than the rest. The glutelin fraction, which probably contains mainly the structural proteins of the cell, is formed relatively early and continues to accumulate slowly throughout development. Studies in wheat give a similar picture. Jennings and Morton (1963) showed that wheat storage proteins were not observable until approximately 12 days after flowering and continued to accumulate until day 40. In contrast buffer-soluble cytoplasmic proteins were present at the earliest possible sampling date (8 days) and continued to increase up to day 30 but not beyond. Bushuk and Wrigley (1971) fractionated the proteins soluble in AUC by column chromatography on Sephadex G–150 (not necessarily the most accurate procedure for determining albumin and globulins, see discussion in Miflin and Shewry, 1979a) and obtained similar results. They again showed that the storage proteins (gliadins plus a "glutenin" fraction) formed an increasingly greater proportion of the total from about 14 days after flowering.

An important question in relation to technological properties is whether or not the high molecular weight aggregates are present during seed development. The results of Bushuk and Wrigley (1971) showed that aggregates excluded by Sephadex G–150 were present. However, these may not have been much more than dimers since, although the exclusion limit is 600 kd for globular protein, it is only 200 kd for linear molecules. In contrast Jones and Carnegie (1971) have proposed that inter-protein dilsulphide bonding occurred during the drying out of the grain. We have reinvestigated the problem using a modified AUC solvent (see above) to extract proteins from isolated protein bodies obtained from endosperms approximately 50 to 75% of the way through development. The results obtained (Miflin et al., 1980a) show clearly that the protein bodies do contain very high molecular weight aggregates (exclusion of the column nominally 30×10^6 for globular proteins, $1 \cdot 2 \times 10^6$ for molecules in extended conformation) in the same relative concentration as in the gluten obtained from the mature seed of the same variety. The differences between varieties of different quality are also apparent in the protein bodies of developing seeds. We therefore conclude that the aggregates are present in the developing grain and are not materially affected by the normal maturation and drying that occurs during grain ripening.

We have also used developing protein bodies in comparative studies of the aggregation states of the storage proteins. The use of these overcame the problem of separating out gluten from milled grain (see above) and allowed all of the storage protein to be solubilized without cleavage of disulphide bonds. When the solubilized protein was chromatographed and the proportion present in high molecular weight aggregates determined we found that whereas 40% of the total gluten in wheat (cv. Highbury) was present as aggregates the values for barley (cv. Sundance) and rye (cv. Rheidol) were only 17% and 22% respectively. These results generally confirm those with the partial extracts made in aqueous alcohol (see above, Fig. 11). Interestingly SDS-PAGE of the aggregated barley fraction shows the presence of the HMW prolamin (D hordein).

2. Origin and Development of Protein Bodies

Although it has long been recognized that the storage proteins of seeds are deposited in protein bodies there is no clear consensus of opinion as to how these protein bodies are formed. The most general view, analogous to the formation in animal tissue, is that the proteins are synthesized on the rough endoplasmic reticulum (RER), pass into the lumen and then to the dictysomes where they are packaged into vesicles which move to vacuoles (or larger vesicles) into which the protein is deposited. This view of events is largely accepted for protein bodies in oilseeds (Dieckert and Dieckert, 1976) and legumes (Boulter, 1979) although the role of the dictysomes in legumes has been questioned (Neumann and Weber, 1978). This scheme has also been proposed for the formation of protein bodies in barley (Munck and von Wettstein, 1976; Cameron-Mills and von Wettstein, 1980) and for at least some of the protein bodies in wheat (Barlow *et al.*, 1974; Briarty *et al.*, 1979; Campbell *et al.*, 1981). Furthermore several reviewers imply that this mechanism applies to all seeds (e.g. Matile, 1976; Altschul *et al.*, 1976; Ashton, 1976; Pernollet, 1978). Two other views on the origin of cereal protein bodies exist. Firstly, Morton *et al.* (1964) proposed that protein bodies were independent organelles capable of internally synthesizing protein and containing ribosomes. However, some of the data upon which this conclusion was based has been questioned (Wilson, 1966). The second hypothesis, based upon the work of Khoo and Wolf (1970) with maize, is that the storage proteins are synthesized on, and deposited within, the RER. Evidence in favour of this is that maize protein bodies have polyribosomes attached to the outside of the boundary membrane and that such polysomes, and the mRNA derived from them, are capable of directing the synthesis of zeins in an *in vitro* wheat-germ protein synthesis system (Burr and Burr, 1976; Burr *et al*, 1978; Larkins and Hurkman, 1978; Viotti *et al.*, 1978). Furthermore, the isolated protein bodies are associated with a large proportion of the total cellular NADH-cytochrome-c reductase, a marker enzyme for the endoplasmic reticulum (Larkins and Hurkman, 1978; Miflin *et al.*, 1981). We have proposed that this mode of protein body formation also applies (in a modified form) to barley (Miflin and Shewry, 1979b).

In our studies we have attempted to determine the answers to two questions (1) Does the protein body derive from the vacuole or the endoplasmic reticulum? (2) Is the body totally enclosed in a limiting membrane? We have used both electron microscopy and the separation of protein bodies and other organelles from developing endosperms. The application of electron microscopic techniques to developing endosperms is fraught with difficulties of fixation and sectioning (e.g. see Campbell *et al.*, 1981) and thus subject to artefacts. However we believe that the results we have obtained with our best fixation procedures show a very close relationship between the RER and the developing protein deposits (Fig. 14(A),(B),(C),(D)) in both wheat and barley endosperms. Similar conclusions can be drawn from micrographs published by Jennings *et al.* (1963), Parker (1980) (Plate 3A) and those of Campbell *et al.* (1981), although the authors do not necessarily always interpret them in that way. The micrographs of protein bodies in Fig. 14(E) and (F) could be interpreted as

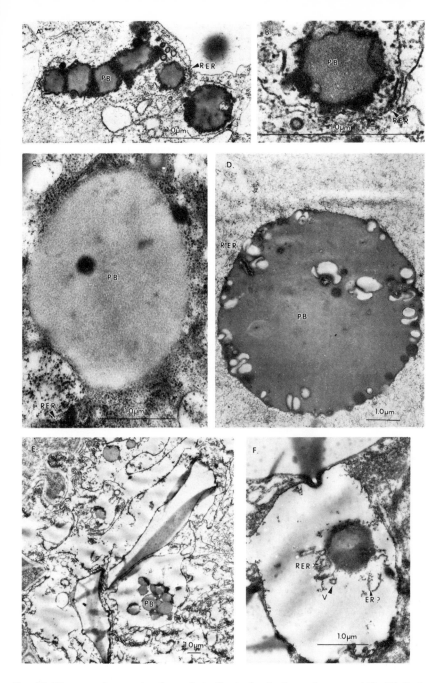

FIG. 14. Electron micrographs of protein bodies in developing endosperms. (A), (B) Barley (fresh weight of caryopsis 30 mg); (C), (D) wheat (fresh wt. respectively 30 and 50 mg); (E), (F) barley (fresh wt. 45 mg). Abbreviations: RER-rough endoplasmic reticulum, PB-protein bodies, V-vesicles. (Previously unpublished work of B. J. Miflin and S. R. Burgess).

showing protein being deposited in a vacuole (e.g. see also Cameron-Mills and von Wettstein, 1980) but we consider that this is due to poor fixation. Close examination of Fig. 14(F) shows that vesicles and fragments of what are apparently RER are present inside the "vacuole" and the general state of preservation within the cell is poor (Fig. 14(E)). However, it is unlikely that the debate can be resolved by electron microscopy alone so we have isolated the protein bodies from several species and shown that the protein deposited in these bodies is predominantly storage prolamin including HMW subunits (Fig. 15). We have also shown that there is an association

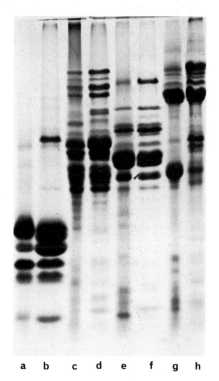

a   b   c   d   e   f   g   h

FIG. 15. SDS-PAGE of reduced and pyridylethylated total prolamins and total protein body proteins. (a), (b) maize (cv. Fronica); (c), (d) wheat (cv. Highbury); (e), (f) barley (cv. Golden Promise); (g), (h) rye (cv. Rheidol). The first track of each pair is total prolamin, the second protein body proteins. Protein bodies were extracted as described by Miflin *et al.* (1981).

between the storage proteins and the main fraction of the RER in wheat and barley. This is to be expected since it is possible to extract the membrane-bound polysomes and show that they direct the synthesis of the prolamins of barley (Brandt and Ingversen, 1976; Fox *et al.*, 1977; Matthews and Miflin, 1980) and wheat (Fig. 16) accordingly.

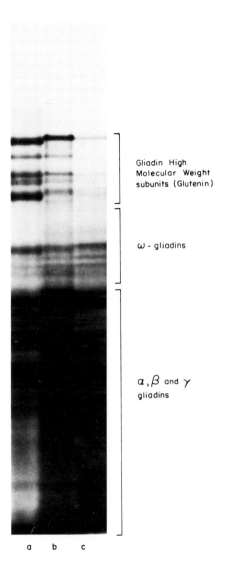

FIG. 16. *In vitro* translation of polysomes derived from developing wheat endosperms. (a) [14C] formaldehyde labelled authentic prolamins; (b) The products of translation in an *in vitro* wheat germ system directed by membrane bound polysomes; (c) products directed by free polysomes (unpublished results of J. Forde and B. J. Miflin).

We have also taken the isolated protein bodies and assayed for the presence of marker enzymes (Miflin *et al*, 1981). Table IV summarizes the results of this work and shows clearly that vacuolar marker enzymes are associated with legume protein bodies (see also van der Wilden *et al.*, 1980)

TABLE IV

Summary of certain properties of protein bodies and endoplasmic reticulum (from Miflin et al., 1981)

	Mg^{2+}	Densities (g cm^{-3}) of endoplasmic reticulum	protein bodies	Proportion (%) of particulate NADH-cytochrome c reductase associated with protein bodies	Proportion (%) of particulate protein associated with protein bodies	Proportion % of the total N-acetylglucosaminidase associated with protein bodies
Barley	+	1·20	1·26	10–15	20–50	≤2
	−	1·11–1·15	1·26			
Wheat	+	1·19	1·27	10–15	70	≤2
	−	1·10–1·17	1·27			
Maize	+	1·15	1·23	40	80	≤2
	−	1·10	1·23			
Peas	+	1·18	1·285	≤5	60	20
	−	1·14	1·285			

but not with those from cereals. Furthermore the protein bodies of the three cereals all co-sediment with NADH-cytochrome-c reductase activity indicating an association of the ER with the protein body. Within the cereals, however, there is a clear distinction in the amount of associated activity between maize and wheat or barley. To test if the membrane completely encloses the protein bodies we have exposed them to protease (Miflin et al., 1980b; Miflin and Burgess, 1982). The results show that the protein bodies of wheat and barley are degraded by this treatment, whereas those of maize and peas are not.

We conclude from these studies that:

(1) The prolamin-containing protein bodies of cereals are not vacuolar in origin.
(2) The prolamins are synthesized on the ER and pass through into the lumen where they aggregate.
(3) In maize the ER remains and completely encloses the aggregate, whereas in wheat and barley the aggregate disrupts the ER and is not completely enclosed by this membrane.

It is important to emphasize that these conclusions are only relevant to the prolamin-containing bodies. All cereals also contain globulin storage proteins (Danielsson, 1949) and these are the major storage form in oats. Rice is also a special case in that the storage protein in the endosperm is chiefly a glutelin, although there is also some prolamin. Tanaka et al. (1980) have recently shown that these two classes of storage proteins are deposited in different protein bodies, which have different morphology (see also Bechtel and Juliano, 1980) and can be physically separated.

C. IMPLICATIONS FOR CEREAL TECHNOLOGY

The results described in the previous two sections clearly show that drying and milling have little effect on the amount and composition of the high molecular weight aggregates that have been implicated in baking quality. Consequently the factors that affect the formation of the aggregates must be those operating during the synthesis and deposition of the storage proteins and a study of these processes is obviously fundamental to an understanding of wheat quality and how it is determined. Thus a clear understanding of the sequence of events between the translation of the mRNA on the polysomes of the RER and the formation of aggregates of deposited proteins is important. Although we have indicated some of the steps there are still many questions unanswered. In particular it is important to understand why the proteins in wheat and barley rupture the ER whilst those of maize remain within it. Our present hypothesis, which is speculative, is that the attractive forces between the lipoprotein membranes of the ER and the storage prolamins are far greater in wheat and barley than in maize. The reason for this difference may lie in the very different amino acid compositions of the proteins (see Table I). These attractive forces would inhibit the free and independent movement of the membranes over the aggregates during cytoplasmic streaming and would also prevent sealing of the membranes around the aggregate during isolation. In both cases the result would be the rupture of the ER and the formation of bodies incompletely surrounded by membranes. However, irrespective of the mechanism, the absence of complete membranes *in vivo* will have certain effects on endosperm structure. Thus as the protein aggregates increase in size they will tend to fuse, form a continuous matrix and entrap small starch grains within them. Protein bodies as such will not exist as recognizable bodies and will tend to force the cytoplasmic contents (buffer-soluble proteins) and the endoplasmic reticulum against the periphery of the amyloplasts. This sequence of events is consistent with many observations, including the lack of discrete protein bodies in mature endosperms of wheat (Parker, 1980) and barley (Bechtel and Pomeranz, 1979) but not maize (Wolf and Khoo, 1970); the inability to recognize S containing protein bodies by elemental analysis in conjunction with scanning electron microscopy of mature endosperms of wheat and barley but not maize, millet, peas or beans (Pernollet and Mossé, 1980; Burgess *et al.*, 1982) and the previously mentioned results of Adams *et al.* (1976) on isolated protein bodies from mature grain. It also provides an explanation for the localization of water-soluble proteins (Simmonds, 1974) and the interstitial protein fibrils (which could well be derived from the ER) of Hess *et al.* (1955) around the starch grains. The adhesion of the storage proteins to the ER would also explain the frequently observed associations between the phospholipids and the storage proteins in flour and the interspersion of membranes within and between aggregates can be clearly seen in many electron micrographs (e.g. Fig. 14(A)). However, although the hypothesis is consistent with the fracture properties of vitreous wheat endosperms (Simmonds, 1974) and with the suggestion that hordein surrounds barley starch grains and prevents modification (Slack *et al.*, 1979) it does not readily suggest an explanation for the differences between steely and opaque grains since this division can also be observed in grains of maize.

A further problem, that may be resolved by the absence of a complete membrane around the protein bodies, is how a gluten mass of almost indefinite size could be formed between proteins that are neatly packaged in separate membrane-bound bodies. Although it could be proposed that this is due to rupture of membranes on drying, the fact that gluten can be prepared from immature grain (Bushuk and Wrigley, 1971), albeit after freeze drying, argues against this objection. However, if there is no membrane the problem disappears, for once the proteins come into contact they are able to interact to form a gluten mass. We have recently been able to reproduce this by isolating protein bodies and then pelleting them at the bottom of a centrifuge tube (unpublished results). When this is done with protein bodies from different cereals those of wheat from a cohesive mass with strong viscoelastic properties while rye and barley bodies also give a cohesive mass but with much less elasticity. In contrast those of maize do not inter-react to form such a cohesive mass and can be resuspended in much the same way as other membrane bound organelles.

In conclusion, therefore, we consider that this view of protein body formation and deposition provides an explanation of many of the technological properties of cereal flours. However, there are still many aspects that are not understood, particularly the relationship between these hydrophobic proteins and the membranes of the ER. One difficult problem is "how do such proteins pass through the ER during processing according to the signal hypothesis of Blobel and Dobberstein (1975)?"—or even "do they always pass completely through?". Current theories of the insertion of membrane proteins into membranes is that they have hydrophobic portions that remain within the membrane as the protein is inserted cotranslationally (Blobel, 1980). If the hydrophobic prolamins pass completely through the membrane, as seems certain in maize, there must be some system that ensures that this occurs.

VI. CONCLUSIONS

The current literature on wheat gluten presents a confused picture with a multitude of different fractions and nomenclatures. We would suggest that there are two origins of this confusion; firstly, the use of techniques that give incomplete extraction resulting in cross-contamination of fractions and, secondly, the absence of a clear classification of components based on the location of their structural loci in the genome and their chemical structure (which is determined by the base sequence of the genes). With the improvement of extraction techniques, the availability of more amino acid and DNA sequences, an increasing knowledge of cereal protein genetics and the recent emphasis on comparative studies between cereals, it is now possible to present a unified picture of the storage proteins of wheat, barley and rye.

1. The major endosperm storage proteins of these cereals are prolamins. The true glutelins are not storage proteins.
2. The storage proteins (of wheat, barley and, possibly, rye) are synthesized on the RER and are deposited in aggregates associated

with, but not completely enclosed by, the ER. The structure of the mature endosperm can be related to these processes of deposition.

3. The prolamins of the three cereals can be classified into the following groups: the HMW (glycine-rich), the S-poor and the S-rich. These three groups differ in their physical and chemical properties, notably in their ability to form aggregates.

4. Each of the above groups consist of a polymorphic series of proteins with a high degree of structural homology but with microheterogeneity in amino acid sequence.

5. The separate groups are coded for by complex multigenic loci that have probably arisen by cycles of gene duplication and divergence. It is likely that single base mutations, insertion and deletions have all contributed to the divergence.

6. The different loci were probably present as ancestral loci in the common progenitor of the three cereals.

7. The differences in the properties of the storage proteins, and thus in the technological properties of the grain dependent on those proteins, is due to evolutionary events that have occurred after the divergence of the ancestors of the three cereals.

The situation in bread wheat is more complex than in the other cereals because it is an allohexaploid with three genomes and storage protein genes are carried on more than one chromosome within a genome. Thus the S-rich prolamins are coded for by loci on chromosome 1 and 6 of each genome. The genes of chromosome 6 have probably been translocated from chromosome 1 after the divergence of the ancestors of modern wheat from those of barley and rye. Further changes in the genes present at these loci could have occurred prior to and since the formation of the polyploid. Similar divergence has also taken place at the different loci on chromosome 1 which code for the HMW prolamins and for the S-poor ω-gliadins. These changes have resulted in the proteins of bread wheat being able to form a visco-elastic gluten. The most important factor in this is the presence of large disulphide-linked aggregates and the key components of these aggregates are the HMW prolamins. We suggest, as have many others, that these molecules from long, flexible, disulphide-linked polymers which provide the elastic network of gluten and that the viscous flow is contributed by the gliadins, some of which may form inter-bonded disulphide-linked aggregates, but most of which are associated due to hydrogen bonding and hydrophobic interactions. Lipids associated with the storage proteins also play a role in determining gluten properties. Genetically determined variations in the length of the aggregates probably accounts for differences in dough strength between varieties. We would suggest that there are two important factors in the formation of the aggregates: (1) the amino acid sequence of the proteins which will determine the position of potential cross-linking residues and also the 3–D conformation of the molecule and (2) those processes that occur during protein deposition which lead to the synthesis of disulphide bonds. We would identify the study of these two factors, of the organization of the prolamin structural genes, and of the relationships of the storage proteins to the ER during synthesis and deposition, as the major areas where further research is required to provide a fundamental understanding of the role of proteins in cereal technology.

ACKNOWLEDGEMENTS

We are grateful to the Agricultural Research Council, the Home Grown Cereals Authority, the EEC and NATO for support of our research. We also acknowledge with thanks the help of many colleagues in our laboratory particularly A. J. Faulks, S. Parmar, S. Burgess and J. Forde and also valuable discussions with D. D. Kasarda, M. Byers, S. Rahman and M. Kirkman. We are also grateful to D. D. Kasarda and D. K. Mecham for permission to use Fig. 1(A) and to M. Allison for supplying barley grain used in malting studies. Finally we wish to thank Susan Wilson for all her efforts in translating this paper into a typescript.

REFERERENCES

Adams, C. A., Novellie, L., and Liebenberg, N. W. (1976). *Cereal Chem.* **53,** 1–12.
Allison, M. J., Cowe, I., and McHale, R. (1976). *J. Inst. Brew., London* **82,** 166–167.
Altschul, A. M., Yatsu, L. Y., Ory, R. L., and Engleman, E. M. (1966) *Annu. Rev. Plant Physiol.* **17,** 113–136.
Arakawa, T., Morishita, H., and Yonezawa, D. (1976). *Agric. Biol. Chem.* **40,** 1217–1220.
Arakawa, T., Yoshida, M., Morishita, H., Honda, J., and Yonezawa, D. (1977). *Agric. Biol. Chem.* **41,** 995–1001.
Aragoncillo, C., Sanchez-Monge, R., and Salcedo, G. (1981). *J. Exp. Bot.* **32,** 1279–1286.
Ashton, F. (1976). *Annu. Rev. Plant Physiol.* **27,** 94–117.
Atanda, O. A., and Miflin, B. J. (1970). *J. Inst. Brew., London* **76,** 51–55.
Autran, J-C., Lew E. J-L., Nimmo, C. C., and Kasarda, D. D. (1979). *Nature* **282,** 527–529.
Bailey, C. H. (1944). *Am. Chem. Soc. Mongr.* **96,** 332.
Bailey, C. H., and Blish, M. J. (1915). *J. Biol. Chem.* **23,** 245.
Baker, R. J., and Bushuk, W. (1978). *Can. J. Plant Sci.* **58,** 325–329.
Barlow, K. K. Buttrose, M. S., Simmonds, D. H., and Vesk, M. (1973). *Cereal Chem.* **50,** 443–454.

314   B. J. MIFLIN *ET AL.*

Barlow, K. K., Lee, J. W., and Vesk, M. (1974). *In* "Mechanisms of the Regulation of Plant Growth" (R. L. Bieleski, A. R. Ferguson, and M. H. Creswell, eds.), pp. 793–797 Bull. 12, R. Soc. NZ., Wellington.
Batey, I. L. (1980). *Ann. Technol. Agric.* **29**, 363–375.
Baxter, E. D. (1980). *Brew. Dig.* Nov. 1980, 45–47.
Baxter, E. D. (1981). *J. Inst. Brew., London* **87**, 173–176.
Baxter, E. D., and Wainwright, T. (1979). *J. Am. Soc. Brew. Chem.* **37**, 8–12.
Beccari (1745). *De Boboniensi Scientiarum et Artium Instituto Atque Academia* **2**, 122–127.
Bechtel, D. B., and Juliano, B. O. (1980). *Ann. Bot. (London)* **45**, 503–509.
Bechtel, D. B., and Pomeranz, Y. (1979). *Cereal Chem.* **56**, 446–452.
Beckwith, A. C., and Wall J. S. (1966). *Biochim. Biophys. Acta* **130**, 155–162.
Beckwith, A. C., Nielsen, H. C., Wall, J. S., and Huebner, F. R. (1966). *Cereal Chem.* **43**, 14–28.
Bernard, M., Autran, J. C., and Joudrier, P. (1977). *Ann. Amelior. Plant.* **27**, 355–362.
Bernardin, J. (1978). *Bakers Dig.* **52**, 20–23.
Bietz, J. A., and Rothfus (1970). *Cereal Chem.* **47**, 381–392.
Bietz, J. A., and Wall, J. S. (1972). *Cereal Chem.* **49**, 416–430.
Bietz, J. A., and Wall, J. S. (1973). *Cereal Chem.* **50**, 537–547.
Bietz, J. A., and Wall, J. S. (1980). *Cereal Chem.* **57**, 415–421.
Bietz, J. A., Huebner, F. R., and Rothfus, J. A. (1970). *Cereal Chem.* **47**, 393–404.
Bietz, J. A., Shepherd, K. W., and Wall, J. S. (1975). *Cereal Chem.* **52**, 513–532.
Bietz, J. A., Huebner, F. R., Sanderson, J. E., and Wall, J. S. (1977). *Cereal Chem.* **54**, 1070–1083.
Blish, M. J., and Sanstedt, R. M. (1929). *J. Biol. Chem.* **85**, 195–206.
Blobel, G. (1980). *Proc. Natl Acad. Sci. USA.* **77**, 1496–1500.
Blobel, G., and Dobberstein, B. (1975). *J. Cell. Biol.* **67**, 852–862.
Bloksma, A. H. (1975). *Cereal Chem.* **52(II)**, 170r–183r.
Booth, M. R., and Ewart, J. A. D. (1969). *Biochim. Biophys. Acta* **181**, 226–233.
Booth, M. R., Bottomly, R. C., Ellis, J. R. S., Malloch, G., Schofield, J. D., and Timms, M. F. (1980). *Ann. Technol. Agric.* **29**, 399–408.
Boulter, D. (1979). *In* "Seed Improvement in Cereals and Grain Legumes" pp. 125–136. Int. Atomic Energy Authority, Vienna.
Boyd, W. J. R., Lee, J. W., and Wrigley, C. W. (1969). *Experientia* **25**, 317–318.
Brandt, A. (1976). *Cereal Chem.* **53**, 890–901.
Brandt, A., and Ingversen, J. (1976). *Carlsberg Res. Commun.* **41**, 312–320.
Briarty, L. G., Hughes, C. E., and Evers, A. D. (1979). *Ann. Bot.* **44**, 641–658.
Brown, J. W., S., Kemble, R. J., Law, C. N., and Flavell, R. B. (1979). *Genetics* **93**, 189–200.
Burgess, S. R., Turner, R. H., Shewry, P. R., and Miflin, B. J. (1982). *J. Exp. Bot.* **33**, 1–11.
Burnouf, T., and Bouriquet, R. (1980). *Theor. Appl. Genet.* **58**, 107–111.
Burr, B., and Burr, F. A. (1976). *Proc. Natl. Acad. Sci. USA.* **73**, 5195–5199.
Burr, B., Burr, F. A., Rubenstein, I., and Simon, M. N. (1978). *Proc. Natl. Acad. Sci. USA.* **73**, 696–700.
Bushuk, W., and Wrigley, C. W. (1971). *Cereal Chem.* **48**, 448–455.
Bushuk, W., Khan, K., and McMaster, G. (1980). *Ann. Technol. Agric.* **29**, 279–294.
Butaki, R. C. and Dronzek, B. (1979). *Cereal Chem.* **56**, 159–161.

Cameron-Mills, V., and von Wettstein, D. (1980). *Carlsberg Res. Commun.* **45,** 577–594.

Campbell, W. P., Lee, J. W., O'Brien, T. P., and Smart, M. G. (1981). *Aust. J. Plant Physiol.* **8,** 5–19.

Charbonnier, L. (1973). *Biochimie* **55,** 1217–1225.

Charbonnier, L. (1974). *Biochim. Biophys. Acta.* **359,** 142–151.

Chen, C. H., and Bushuk, W. (1970). *Can. J. Plant Sci.* **50,** 9–14.

Cunningham, D. K., and Anderson, J. A. (1950). *Cereal Chem.* **27,** 344–355.

Cunningham, D. K., Geddes, D. K., and Anderson, J. A. (1955) *Cereal Chem.* **32,** 91–106.

Daiber, K. H., and Novellie, L. (1968). *J. Sci Fd. Agric.* **19,** 87–90.

Damidaux, R., Autran, J. C., Grignac, P., and Feillet, P. (1978). *C. R. Acad. Sci. Paris Ser. D,* **287,** 701–704.

Danielssson, C. E. (1949). *Biochem. J.* **44,** 387–400.

Danno, G., Kanazawa, K., and Nakate, M. (1974). *Agric. Biol. Chem.* **38,** 1947–1953.

Danno, G., Kanazawa, K., and Nakate, M. (1975). *Agric. Biol. Chem.* **39,** 1379–1384.

Dieckert, J. W., and Dieckert, M. C. (1976). *J. F. Sci.* **41,** 457–482.

Dill, D. B., and Alsberg, C. L. (1925). *J. Biol. Chem.* **65,** 279–304.

Doekes, G. J. (1973). *Euphytica* **22,** 28–33.

Ewart, J. A. D. (1966). *J. Sci. Fd. Agric.* **17,** 30–33.

Ewart, J. A. D. (1968). *J. Sci. Fd. Agric.* **19,** 617–623.

Ewart, J. A. D. (1972). *J. Sci. Fd. Agric.* **23,** 567–579.

Ewart, J. A. D. (1973). *J. Sci. Fd. Agric.* **28,** 849–851.

Ewart, J. A. D. (1976). *J. Sci. Fd. Agric.* **27,** 695–698.

Ewart, J. A. D. (1977a). *J. Sci. Fd. Agric.* **28,** 191–199.

Ewart, J. A. D. (1977b). *J. Sci. Fd. Agric.* **28,** 843–848.

Ewart, J. A. D. (1977c). *J. Sci. Fd. Agric.* **28,** 849–851.

Ewart, J. A. D. (1977d). *J. Sci. Fd. Agric.* **28,** 1080–1083.

Ewart, J. A. D. (1978). *J. Sci. Fd. Agric.* **29,** 551–556.

Ewart, J. A. D. (1979). *J. Sci. Fd. Agric.* **30,** 482–492.

Ewart, J. A. D. (1981). *J. Sci. Fd. Agric.* **32,** 572–578.

Faulks, A. J., Shewry, P. R., and Miflin, B. J. (1981). *Biochem. Genet.* **19,** 841–858.

Field, J. M., and Miflin, B. J. (1980). *Home Grown Cereals Progress Reports on Research and Development for 1978/79,* pp. 26–28.

Finney, K. F. (1943). *Cereal Chem.* **20,** 381.

Finney, K. F. (1971). *Cereal Sci. Today* **16,** 342–356.

Fox, J. E., Pratt, H. M., Shewry, P. R., and Miflin, B. J. (1977). *In* "Nucleic Acids and Protein Synthesis in Plants" pp. 520–524. Centre Nat. Res. Sci., Paris.

Freedman, R. B., and Hillson, D. A. (1980). *In* "The Enzymology of Post-translation Modification of Proteins I." (R. B. Freedman, and H. C. Hawkins, eds) pp. 158–212. Academic Press, London.

Gianazza, E., Viglienghi, V., Righetti, P. G., Salamini, F., and Soave, C. (1977). *Phytochemistry* **16,** 315–317.

Gothard, P. G. (1974). *J. Inst. Brew. London* **80,** 387–390.

Grønlund (1880). *Tidskr. Landokonomi* **1880,** 412–453.

Grynberg, A., Nicholas, J., and Drapon, R (1978). *Biochimie* **60,** 547–551.

Hamauzu, Z., Toyomasu, T., and Yonezawa, D. (1974). *Agric. Biol. Chem.* **38,** 2445–2480.

Hamauzu, Z. Khan, K., and Bushuk, W. (1979). *Cereal Chem.* **56,** 513–516.
Hess, K. (1954). *Kolloid-Z* **136,** 84.
Hess, K., Kiess, G. H., and Hanssen, E. (1952). *Naturwissenschaften,* **30,** 135.
Hess, K., Mahl, H., Gutter, E., and Dodt, E. (1955). *Mickroskopie* **10,** 6.
Hibberd, G. E., and Parker, N. S. (1979). *Cereal Chem.* **56,** 232–236.
Hoeser, K., and Keininger, H. (1971). *Brauwissenschaft 24,* 339–342.
Huebner, F. R., and Rothfus, J. A. (1971). *Cereal Chem.* **48,** 469–478.
Huebner, F. R., and Wall, J. S. (1976). *Cereal Chem.* **53,** 258.
Jacobson, G. R., Schaffer, M. H., Stark, G. R., and Vanaman, T. C. (1973). *J. Biol. Chem.* **248,** 6583–6591.
Jeanjean, M. F., and Feillet, P. (1979). *Getreide Mehl.* **33,** 127–130.
Jennings, A. C., and Morton, R. K. (1963). *Aust. J. Biol. Sci.* **16,** 318–331.
Jennings, A. C., Morton, R. K., and Palk, B. A. (1963). *Aust. J. Biol. Sci.* **16,** 366–374.
Jensen, J., Jorgensen, J. H., Jensen, H. P., Giese, H., and Doll, H. (1980). *Theor. Appl. Genet.* **58,** 27–31.
Jones, I. K., and Carnegie, P. R. (1971), *J. Sci. Fd. Agric.* **22,** 358–364.
Jones, R. W., Taylor N. W., and Senti, F. R. (1959). *Arch. Biochem. Biophys.* **84,** 363–376.
Joppa, L. R., Bietz, J. A., and McDonald, C. (1975), *Can. J. Genet, Cytol.* **17,** 355–365.
Kaczowski, J., and Mieleszko, T. (1980). *Ann. Technol. Agric.* **29,** 377–384.
Kasarda, D. D. (1980). *Ann. Technol. Agric.* **29,** 151–173.
Kasarda, D. D., Bernardin, J. E., and Thomas, R. S. (1967). *Science* **155,** 203–205.
Kasarda, D. D., Bernardin, J. E., and Nimmo, C. C. (1976a). *In* "Advances in Cereal Science and Technology" (Y. Pomeranz, ed.), pp. 158–236. AACC, St. Paul, Minnesota, USA.
Kasarda, D. D., Bernardin, J. E., and Qualset, C. O. (1976b). *Proc. Natl. Acad. Sci. USA.* **73,** 3046–3650.
Khan, K., and Bushuk, W. (1978). *Bakers Dig.* **52,** 14–20.
Khan, K., and Bushuk, W. (1979). *Cereal Chem.* **56,** 505–512.
Khoo, U., and Wolf, M. J. (1970). *Am. J. Bot.* **57,** 1042–1050.
Kirkman, M. A., Shewry, P. R., and Miflin, B. J. (1982). *J. Sci. Fd. Agric.*
Kobrehel, K. (1980). *Ann. Technol. Agric.* **29,** 125–132.
Kobrehel, K., and Bushuk, W. (1977). *Cereal Chem.* **54,** 833–839.
Kobrehel, K., and Bushuk, W. (1978). *Cereal Chem.* **55,** 1060–1064.
Koie, B., Ingversen, J., Anderson, A. J., Doll, H., and Eggum, B. O. (1976). *In* "Evaluation of Seed Protein Alterations by Mutation Breeding." pp. 55–61. IAEA, Vienna.
Konzak, C. F. (1977). *Adv. Genet.* **19,** 407–582.
Lamport, D. T. A. (1980). *In* "The Biochemistry of Plants" (J. Preiss, ed.) Vol. III, pp. 501–543. Academic Press, London and New York.
Larkins, B. A., and Hurkman, W. G. (1978). *Plant Physiol.* **62,** 256–263.
Law, C. N., Young, C. F., Brown, J. W., Snape, J. W., and Worland, A. J. (1978). *In* "Seed Protein Improvement by Nuclear Techniques" pp. 483–502. Int. Atomic Energy Authority, Vienna.
Lawrence, G. J. and Shepherd, K. W. (1980). *Aust. J. Biol. Sci.* **33,** 221–233.
Lawrence, G. J. and Shepherd, K. W. (1981). *Theor. Appl. Genet.* **59,** 23–31.
Lee, J. W. (1968). *J. Sci. Fd. Agric.* **19,** 153–156.
Lee, J. W., and MacRitchie, F. (1971). *Cereal Chem.* **48,** 620.
MacRitchie, F. (1973). *J. Sci. Fd. Agric.* **24,** 1325–1329.

Mandl, B. (1972). *Weihenstephaner* **162**, 170–171.
Matile, P. (1976) *In* "Plant Biochemistry 3rd edition. (Bonner, J. and Varner, J. E. eds.) pp. 189–224. Academic Press, New York.
Matthews, J. A., and Miflin, B. J. (1980). *Planta* **149**, 262–268.
Mecham, D. K., Cole, E. W., and Ng, H. (1972). *Cereal Chem.* **49**, 62–67.
Mecham, D. K., Kasarda, D. D., and Qualset, C. O. (1978). *Biochem. Genet.* **16**, 831–853.
Meredith, O. B., and Wren, J. J. (1966). *Cereal Chem.* **43**, 169–186.
Miflin, B. J., and Shewry, P. R. (1979a). *In* "Seed Improvement in Cereals and Grain Legumes" Vol. I, pp. 137–158. Int Atomic Energy Authority, Vienna.
Miflin, B. J., and Shewry, P. R. (1979b). *In* "Recent Advances in the Biochemistry of Cereals" (D. Laidman and R. G. Wyn Jones, eds.), pp. 239–273. Academic Press, London.
Miflin, B. J., and Burgess, S. R. (1982). *J. Exp. Bot.* **33**, 251–260.
Miflin, B. J., Burgess, S. R., and Shewry, P. R. (1981). *J. Exp. Bot.* **32**, 199–219.
Miflin, B. J., Byers, M., Field, J. M., and Faulks, A. J. (1980a) *Ann. Technol. Agric.* **29**, 133–147.
Miflin, B. J., Matthews, J. A., Burgess, S. R., Faulks, A. J., and Shewry, P. R. (1980b). *In* "Genome Organization and Expression in Plants" (Leaver, C. J. ed.), pp. 233–243. Plenum, New York.
Morgan, A. G., Smith, D. R., and Gill, A. A. (1981). Proceedings of the 4th Int. Barley Genetics Symp. (in press).
Morton, R. K., Palk, B. A., and Raison, J. K. (1964). *Biochem, J.* **91**, 522–528.
Moss, H. J. (1973). *Aust. Inst. Agric. Sci.* **39**, 109–111.
Munck, L., and von Wettstein, D. (1976). *In* "Genetic Improvement of Seed Protein" pp. 71–82. *Natl. Acad. Sci.,* Washington DC.
Neumann, D., and Weber, E. (1978). *Biochem. Physiol. Pflanz.* **173**, 167–180.
Nielsen, H. C., Beckwith, A. C., and Wall, J. S. (1968). *Cereal Chem.* **45**, 37–47.
Orth, R. A., and Bushuk, W. (1972). *Cereal Chem.* **49**, 268–275.
Orth, R. A., and Bushuk, W. (1973). *Cereal Chem.* **50**, 106–114.
Orth, R. A., and Bushuk, W. (1974). *Cereal Chem.* **51**, 118–126.
Orth, R. A., O'Brien, L., and Jardine, R. (1976). *Aust J. Agric. Res.* **27**, 575–582.
Osborne, T. B. (1924). "Vegetable Proteins", pp. 154. Longmans Green, London.
Palmer, G. H. (1975). *J. Inst. Brew., London* **81**, 71–73.
Palmer, G. H. (1980). *In* "Cereals for Food and Beverages" (G. E. Inglett and L. Munck, eds.), pp. 301–338. Academic Press, London.
Palmer, G. H., and Bathgate, G. N. (1976). *In* "Advances in Cereal Science and Technology" (Y. Pomeranz, ed.), Vol. 1, pp. 237–324. American Association of Cereal Chemists, St. Paul, Minnesota, USA.
Palmer, G. H., and Harvey, A. E. (1977). *J. Inst. Brew., London* **83**, 295–299.
Parker, M. L. (1980). *Ann. Bot. (London)* **46**, 29–36.
Payne, P. I., and Corfield, K. G. (1979). *Planta,* **145**, 83–88.
Payne, P. I., Law, C. N., and Mudd, E. L. (1980). *Theor. Appl. Genet.* **58**, 113–120.
Payne, P. I., Corfield, K. G., Holt, L. M., and Blackman, J. A. (1981). *J. Sci. Fd. Agric.* **32**, 51–60.
Pernollet, J–C. (1978). *Phytochemistry* **17**, 1473–1480.
Pernollet, J–C. and Mossé, J. (1980). *C. R. Acad. Sci. (Paris),* **290d**, 267–270.
Platt, S. G., Kasarda, D. D., and Qualset, C. O. (1974). *J. Sci. Fd. Agric.* **25**, 1555–1561.
Pomeranz, Y. (1965). *J. Sci. Fd. Agric.* **16**, 586–593.

Pomeranz, Y., and Shands, H. L. (1974). *J. Fd. Sci.* **39**, 950–952.
Pomeranz, Y., Burkhart, B. A., and Moon, L. C. (1970). *Am. Soc. Brew. Chem. Proc.* **1970**, 40–46.
Pomeranz, Y., Standridge, N. N., Robbins, G. S., Goplin, E. D. (1975). *Cereal Chem.* **52**, 485–492.
Poulsen, L. L., and Ziegler, D. M. (1977). *Arch. Biochem. Biophys.* **183**, 563–570.
Prado, J., Sanchez-Monge, R., Salcedo, G., and Aragoncillo, C. (1981). *Plant Science Lett.* **25**, 281–289.
Redman, D. G. (1973). *Phytochemistry* **12**, 1383–1389.
Roden, L. T. (1982). Ph.D. Thesis, University of Kent.
Schulz, G. E., and Schirmer, R. H. (1979). "Principles of Protein Structure". Springer Verlag, Berlin.
Sexson, K. R., Wu, Y. V., Huebner, F. R., and Wall, J. S. (1978). *Biochim. Biophys. Acta,* **532**, 279–285.
Sexson, K. R., and Wu, Y. V. (1972). *Biochim. Biophys. Acta,* 651–657.
Shepherd, K. W. (1968). *In* "Proc. 3rd Int. Wheat Genet. Symp." pp. 86–96. Australian Acad. Sci., Canberra.
Shepherd, K. W., and Jennings, A. C. (1971). *Experimentia* **27**, 98–99.
Shepherd, K. W. (1973). *In* "Proc. 4th Int. Wheat Genet. Symp." pp. 745–760. Agric. Expt. Station, Univ. Missouri, Columbia.
Shestakova, N. A., and Vakar, A. B. (1979). *Applied Biochem. Microbiol. (Engl. Transl.)* **15**, 136–142.
Shewry, P. R., Pratt, H. M., Faulks, A. J., Parmar, S., and Miflin, B. J. (1979). *J. Natl. Inst. Agric. Bot.* **15**, 34–50.
Shewry, P. R., Autran, J-C., Nimmo, C. C., Lew, E., J-L., and Kasarda, D. D. (1980a). *Nature* **286**, 520–522.
Shewry, P. R., Faulks, A. J., Parmar, S., and Miflin, B. J. (1980b). *J. Inst. Brew.* **86**, 138–141.
Shewry, P. R., Faulks, A. J., Pickering, R. A., Jones, I. T., Finch, R. A., and Miflin, B. J. (1980c). *Heredity,* **44**, 383–389.
Shewry, P. R., Field, J. M., Kirkman, M. A., Faulks, A. J., and Miflin, B. J. (1980d). *J. Exp. Bot.* **31**, 393–407.
Shewry, P. R., Lew, E. J-K., and Kasarda, D. D. (1981). *Planta.* **153**, 246–253.
Shewry, P. R., Field, J. M., Lew, E. J-K., and Kasarda, D. D. (1982a). *J. Exp. Bot.* **33**, 261–268.
Shewry, P. R., Parmar, S., and Miflin, B. J. (1982b). *Cereal Chem.* (in press).
Simmonds, D. H. (1974). *Bakers Digest* **48**, 16–18.
Simmonds, D. H., Barlow, K. K., and Wrigley, C. W. (1973). *Cereal Chem.* **50**, 553–562.
Simmonds, D. H., and O'Brien, T. (1982). *In* "Advances in Cereal Science and Technology" (Y. Pomeranz, ed.), Vol. IV (in press).
Slack, P. T., Baxter, E. D., and Wainwright, T. (1979). *J. Inst. Brew. London* **85**, 112–114.
Sosinov, A. A., and Poperelya, F. A. (1980). *Ann. Technol. Agric.* **29**, 229–245.
Stanley, P. E., Jennings, A. C., and Nicholas, D. J. D. (1968). *Phytochemistry* **7**, 1109–1114.
Symes, K. J. (1965). *Aust J. Agric. Res.* **16**, 113–123.
Symes, K. J. (1969). *Aust. J. Agric. Res.* **20**, 971–979.
Tanaka, K., Sugimoto, T., Ogawa, M., and Kasai, Z. (1980). *Agric. Biol. Chem.* **44**, 1633–1639.

Terce-Laforgue, T., Charbonnier, L., and Mosse, J. (1980). *Biochim. Biophys. Acta* **625,** 118–126.

Timms, M. F., Bottomley, R. C., Ellis, J. R. S., and Schofield, J. D. (1981). *J. Sci. Fd. Agric.* **32,** 684–698.

Viotti, A., Sala, E., Alberi, P., and Soave, C. (1978) *Plant Sci. Lett.* **13,** 365–375.

Wall, J. S. (1979). *In* "Recent Advances in the Biochemistry of Cereals (Laidman, D. L., and Wyn-Jones, R. G. eds.). pp. 275–312. Academic Press, New York.

Wilden, W. v. d., Herman, E. M., and Chrispeels, M. J. (1980). *Proc. Natl. Acad. Sci. USA.* **77,** 428–432.

Wilson, C. M. (1966). *Plant Physiol.* **41,** 325–327.

Wilson, C. M., Shewry, P. R., and Miflin, B. J. (1981a). *Cereal Chem.* **58,** 275–281.

Wilson, C. M., Shewry, P. R., Faulks, A. J., and Miflin, B. J. (1981b). *J. Exp. Bot.* **32,** 1287–1293.

Wolf, M. J., and Khoo, U. (1970). *Stain Technol.* **45,** 277–283.

Woychik, J. H., Boundy, J. A., and Dimler, R. J. (1961). *Arch. Biochem. Biophys.* **94,** 477–482.

Wren, J. J., and Nutt, J. (1967). *J. Sci. Fd. Agric.* **18,** 119–123.

Wright, W. B., Brown, P. J., and Bell, A. V. (1964). *J. Sci. Fd. Agric.* **15,** 56.

Wrigley, C. W. (1980). *Ann. Technol. Agric.* **29,** 213–227.

Wrigley, C. W. (1982). *Qualitas Plantarum.* (in press).

Wrigley, C. W., and Shepherd, K. W. (1973). *Ann. N. Y. Acad. Sci.,* **209,** 154–162.

Yoshida, M. Hamauzu, Z., and Yonezawa, D. (1980). *Agric. Biol. Chem.* **44,** 657–661.

Yoshina, D., and McCalla, A. G. (1966). *Can. J. Biochem. Physiol.* **44,** 339–346.

Ziegler, D. M., and Poulsen, L. L. (1977). *Trends in Biochem. Sci.* **2,** 79–81.

Subject index

322

SUBJECT INDEX

grain desiccation and, 12, 28
isoenzymes, 5–6, 8–9, 10–11, 13, 14, 15, 19, 24–25, 28
maize *see* Maize
oats *see* Oats
pericarp, 2, 4, 7–8, 9, 11–12, 13, 14, 15, 28
premature synthesis of germination type, 9, 12, 13–14, 28
R-enzyme potentiation, 25
resistant glucosidic bonds and activity, 22–23, 28
rice *see* Rice
rye *see* Rye
starch granule degradation, 24–27, 28
synthesis, molecular basis, 16–19
testa, role in synthesis, 18, 28
triticale *see* Triticale
wheat *see* Wheat
β-Amylase
α-amylase potentiation, 25
deficiency, 112–113
localization on starch granules, 119
specific antibody preparation, 123
Animal feeds, 224, 225
cereal breeding strategies and, 231
ant-13, 248, 249
Antibodies, 102–103
Antigen–antibody reaction, 102–103
taxonomy and, 137
Arachin, 159
in vitro reconstitution, 165–166
structure, 163–164
Arachis hypogaea see Peanut
Aralidium, 141
Arylamidase *see* Naphthylamidase
Asteraceae phylogeny, 147, 148
Atlas 66, 234, 235
Avena, 144

B

Babala (*Pennisetum typhoideum*)
protein bodies, 300
Bandeiraea simplicifolia lectins, 108
Barley
alkaline peptidases, 46, 48
α-amylase, 2–6, 18, 22, 23, 24, 25, 26, 105, 119–120
β-amylase specific antibody preparation, 123

breeding programmes, 224, 232, 235–237, 248–249
chemotaxonomy, 144
cultivar identification, 149
endosperm structure, malting quality and, 300
gel immunoprecipitation studies, 106
gluten, 257
haze-free trait, breeding and, 248–249
high lysine strains, 193–194, 201–202, 236–237
high molecular weight aggregates in protein bodies, 303
high molecular weight (HMW) prolamins, 286, 291, 295, 297
immunoelectrophoresis analysis for protein nomenclature, 109
leucine aminopeptidase, 46
leupeptin sensitive proteinases, 39
malate dehydrogenase, antigenic definition, 109
malting quality, 224, 232, 235–237, 248–249, 256, 276–283
mealy grain, 278, 299
mobilization of protein during germination, 47–48
multiple aleurone layers, 233
polyphenols, breeding and, 248–249
protein body development, 199–200, 304, 306, 309, 310
protein body immunoelectrophoretic analysis, 120
proteinase inhibitors, 36, 40
proteinases, 38, 40, 47, 48
rupture of protein body membrane, 174–175, 309, 310
starch granule-storage protein relationship, 174, 300, 310
steely grain, 278, 299
storage protein, 108, 174–175 *see also* Hordein
temporal deposition of proteins, 301–302
tests of potential quality, 299
world production, 224
Berberidaceae taxonomy, 140
Bounty, 247
Brassica
gel electrophoretic chemotaxonomy, 142–143, 145, 146, 147
serotaxonomy, 140

immunohistological techniques,
114–120, 125
immunoprecipitation in gels, 106–113
labelled amino acid incorporation
with immunoelectrophoresis, 113,
125
modified antigen recognition,
112–113, 124, 126–127
physiological studies, 109–113
precipitation of immune complexes
see Immune complex precipitation
precursor protein detection, 112, 126
quantitative evolution of storage
proteins and, 112
separating one protein from acellular
system, 104–105, 126
single antibody precipitation,
104–105, 106
single diffusion gel
immunoprecipitation, 106
urea use and, 128–129
Immunoelectrophoresis, 106, 107
b-32 regulatory protein in opaque 2
endosperm, 215
Brassica serotaxonomy, 140, 142
with labelled amino acid
incorporation, 113
monospecificity of immune serum,
125
sensitivity, 115
serotaxonomy and, 140
taxonomy of *Solanum*, 139
Immunofluorescence, 115, 119–120
sensitivity, 115
storage protein localization, 159
Immunohistological techniques,
114–120, 125
enzyme localization, inactive forms,
119

J

Jack bean
canavalin, α-mannosidase and, 78
lectin receptor protein, 66, 67
lectins, 64, 70
urease content, 185
vicilin *see* Canavalin
Juglans taxonomy, 144, 146
Julia, 278, 281, 300

K

Kola nut allergy, 84
Kwashiorkor, 228, 229

L

Lancer, 234
Lancota, 234
Lasthenia taxonomy, 143
Lathyrus ochrus isolectins, 73
Lectins, 53–78, 157, 171
affinity chromatography purification,
54
animal toxicity, 54
antigenic definition of constituents,
108
biological role, 66–78, 141
carbohydrate reserve transportation
and, 67
cell walls, 66
cytoplasmic, 64
evolutionary conservation of
structure, 73, 78
gel immunoprecipitation studies, 106
glycolytic hydrolase activity and, 73,
78, 108
haemagglutination, 53, 55, 58, 65
host-pathogen interactions and,
67–68, 141
hydrophobicity, 65–66
immunochemical characterization,
59, 69
immunochemical localization, 59, 64,
71
interspecies recognition processes
and, 67–70, 141
lymphocyte mitogenesis and, 54
membrane-associated, 64, 65, 66, 71,
73
mitotic stimulation of callus, 73
protein bodies, 64–65, 66, 71, 73, 92,
159, 185
protein subunit patterns, 55, 58
receptor proteins, 66–67
Rhizobium legume nodulation and,
68–70
in root hairs, 68, 69–70
separation from *in vitro* protein
synthesis system, 127
as storage protein, 71, 73, 171, 185
structural role, 73

Lectins (*contd*)
  sugar binding, 53–54, 66
  synthesis during seed development,
    71, 73
  taxonomy and, 141–142
  transformed cells, selectivity for, 54
  ultrastructural location, 64–66, 71
Legumes
  animal feed, 225, 230, 231
  biochemistry of storage protein
    deposition, 219–220
  breeding programmes, 224, 240–243
  breeding strategies, 231, 232
  dietary compensation for low
    methionine content, 229, 231
  essential amino acids, 228
  food allergy and, 84, 85
  globulin chemotaxonomy by gel
    electrophoresis, 143
  human dietary requirements and, 229
  lectins, 108, 141
  methionine content, breeding and,
    232, 241–242
  protein bodies, 158–160, 304, 307
  protein content, 226
  protein quality, breeding and, 224,
    241–242
  storage proteins, 157–171, 219–222,
    see also Legumin and Vicilin
  taxonomy, 140, 146
  trypsin inhibitor content, breeding
    and, 242
  world protein production and, 224,
    225
  yield, breeding and, 241
Legumin, 108, 109, 156, 157
  absence of cross-reactivity with
    vicilin, 114, 126
  antigenic definition of constituents,
    108
  association–dissociation properties,
    165–167
  cDNA clone studies, 221
  chemotaxonomy and, 143
  conservatism of structure, 167,
    169–170
  cysteine content, 242
  immunohistological localization, 115,
    126
  localization in protein bodies, 159
  methionine content, 242
  peanut *see* Arachin

post-translational modification,
    220–221
  precursory form detection, 126
  quaternary structure, 162–165
  secondary structure predictions,
    169–170
  sequence comparisons, 167–169
  soyabean *see* Glycinin
  specific antibody preparation, 121,
    123, 125
  structure, 162–171
  synthesis, 113, 221
Lens culinaris lectins, 59
Lentil, 240
Leucanthemum taxonomy, 146
Leucine aminopeptidase, 145
Leupeptin, 39
Lupin, 240
  γ-conglutin structure, 162
  protein bodies, 158, 159
  serological investigation, 141
Lupinus luteus, 158
Lys, 95, 236
Lys, 449, 236

M

Magnoliaceae taxonomy, 138–139
Maize
  α-amylase synthesis, 15–16
  breeding for protein quality, 232,
    237–238
  catalase isoenzyme deficiency, 112
  chemotaxonomy, 144
  glutelins, 174
  gluten, 257
  high lysine mutants, 193–194,
    201–202
  mitochondrial protein, 112
  multiple aleurone layers, 233
  protein bodies, 172–173, 199, 300,
    304, 309, 310
  protein content, 226
  storage protein, 171–172, 226, 256,
    see also Zein
  world production, 224
Marasmus, 228
Maris Mink, 282
Maximal packing hypothesis, 157,
    182–183
Medicago taxonomy, 141
Melilotus taxonomy, 141

Pea (*contd*)
crossed immunoelectrophoresis
identification of seed constituents,
109, 112
food allergy, 85
globoids in protein bodies, 159
globulin synthesis, 126–127
lectin receptor protein, 66
lectins, 69, 73, 171, 185
legumin *see* Pea legumin
precursor proteins, 112, 126
protein bodies, 123, 158
Rhizobium attachment, 69
root hair lectins, *Rhizobium
leguminosarum* and, 69
storage protein, 105, 112, 125, 159,
185, 219 *see also* Pea legumin and
Pea vicilin
sulphur amino acid containing
antigen identification, 113
vicilin *see* Pea vicilin
Pea legumin, 219
antibody preparation, 123, 125
chemotaxonomy and, 143
precursor identification, 126
in protein bodies, 159
purification, 123–124
sequence comparison, 167, 169
structure, 164, 171
synthesis initiation, 114, 115
Pea vicilin, 219
antibody preparation, 123
antigenic definition, 108
chemotaxonomy and, 143
in protein bodies, 159
structure, 162
synthesis initiation, 115
Peanut
alkaline peptidases, 46, 48
allergens, 85, 86–87
animal feed, 240
antigens appearing during
germination, 121
conarachin, 159, 161
food allergy, 84, 85
gel immunoprecipitation studies, 106
lectin localization, 63
lectins and *Rhizobium* binding, 69
protein bodies, 158, 159
Rhizobium japonicum nodulation, 68
storage protein, 159, 219
world production, 225

Pennisetin
structure, 173, 183
Peptidases
aleurone layer, 48
germinating seeds, 36, 43–47, 48, 49
Peroxidase, 145, 146
Phaseolin
gene transfer into sunflower crown
gall, 186
specific antibody investigations, 123,
127
structure, 161
Phaseolus
globoids in protein bodies, 59
lectins, 55, 58–59
phaseolin, *see* Phaseolin
Phaseolus aconitifolius, 58
Phaseolus acutifolius latifolius, 58
Phaseolus angularis, 58
Phaseolus aureus
lectins, 58
storage protein, 108
Phaseolus calcaratus, 58
Phaseolus coccineus, 58
Phaseolus darwinius, 58
Phaseolus filiformis, 58
Phaseolus lathyroides, 58
Phaseolus latifolius, 58
Phaseolus leucanthius, 55, 58, 59
Phaseolus lunatus, 55, 58, 59
Phaseolus multiflorus, 58
Phaseolus mungo, 55
Phaseolus ricciardianus, 55
Phaseolus ritensis, 58
Phaseolus semierectus, 58
Phaseolus trilobus, 58
Phaseolus vulgaris, 240
alkaline peptidases, 46, 48
G1 protein, *see* Phaseolin
immobilization of saprophytic
bacteria by lectin binding, 67
lectins, 55, 58, 59, 64, 67, 70, 71, 73,
127, 159, 171, 185
protein bodies, 159
root hair lectins, *Rhizobium*
nodulation and, 68, 69–70
storage protein, 112, 159, 219, *see
also* Vicilin and Legumin
sulphur amino acids in trypsin
inhibitors, 242
vicilin specific antibody preparation,
123, 125